Graduate Texts in Physics

For further volumes:
www.springer.com/series/8431

Graduate Texts in Physics

Graduate Texts in Physics publishes core learning/teaching material for graduate- and advanced-level undergraduate courses on topics of current and emerging fields within physics, both pure and applied. These textbooks serve students at the MS- or PhD-level and their instructors as comprehensive sources of principles, definitions, derivations, experiments and applications (as relevant) for their mastery and teaching, respectively. International in scope and relevance, the textbooks correspond to course syllabi sufficiently to serve as required reading. Their didactic style, comprehensiveness and coverage of fundamental material also make them suitable as introductions or references for scientists entering, or requiring timely knowledge of, a research field.

Series Editors

Professor William T. Rhodes

Department of Computer and Electrical Engineering and Computer Science
Imaging Science and Technology Center
Florida Atlantic University
777 Glades Road SE, Room 456
Boca Raton, FL 33431
USA
wrhodes@fau.edu

Professor H. Eugene Stanley

Center for Polymer Studies Department of Physics
Boston University
590 Commonwealth Avenue, Room 204B
Boston, MA 02215
USA
hes@bu.edu

Professor Richard Needs

Cavendish Laboratory
JJ Thomson Avenue
Cambridge CB3 0HE
UK
rn11@cam.ac.uk

Professor Martin Stutzmann

Walter Schottky Institut
TU München
85748 Garching
Germany
stutz@wsi.tu-muenchen.de

Günter Ludyk

Einstein in Matrix Form

Exact Derivation of the Theory of Special and General Relativity without Tensors

Springer

Professor Dr. Günter Ludyk
Physics and Electrical Engineering
University of Bremen
Bremen, Germany
Guenter.Ludyk@nord-com.net

ISSN 1868-4513 ISSN 1868-4521 (electronic)
Graduate Texts in Physics
ISBN 978-3-642-43906-3 ISBN 978-3-642-35798-5 (eBook)
DOI 10.1007/978-3-642-35798-5
Springer Heidelberg New York Dordrecht London

Softcover reprint of the hardcover 1st edition 2013

Printed on acid-free paper

Springer is part of Springer Science+Business Media (www.springer.com)

To my grandchildren Ann-Sophie
and
Alexander
Hüttermann

Preface

This book is an introduction to the theories of Special and General Relativity. The target audience are physicists, engineers and applied scientists who are looking for an understandable introduction to the topic—without too much new mathematics.

All necessary mathematical tools are provided either directly in the text or in the appendices. Also the appendices contain an introduction to vector or matrices: first, as a refresher of known fundamental algebra, and second, to gain new experiences, e.g. with the Kronecker-product of matrices and differentiation with respect to vectors and matrices.

The fundamental equations of Einstein's theory of Special and General Relativity are derived using matrix calculus without the help of tensors. This feature makes the book special and a valuable tool for scientists and engineers with no experience in the field of tensor calculus. But physicists are also discovering that Einstein's vacuum field equations can be expressed as a system of first-order differential-matrix equations, wherein the unknown quantity is a matrix. These matrix equations are also easy to handle when implementing numerical algorithms using standard software as, e.g. MATHEMATICA or MAPLE.

In Chap. 1, the foundations of Special Relativity are developed. Chapter 2 describes the structure and principles of General Relativity. Chapter 3 explains the Schwarzschild solution of spherical body gravity and examines the "Black Hole" phenomenon. Furthermore, two appendices summarize the basics of the matrix theory and differential geometry.

After completion of the book, I discovered the paper [37], where Einstein's equations of a similar shape are derived.

I would like to thank *Claus Ascheron* (Springer) who has made great effort towards the publication of this book. Finally, I would like to thank my wife *Renate*, without her this book would have never been published!

Bremen Günter Ludyk

Contents

Notation

Important definitions, facts and *theorems* are framed, and important *intermediate results* are double-underlined.

Scalars are written in normal typeface:

$$a, b, c, \alpha, \beta, \gamma, \ldots;$$

Vectors in the 3-dimensional space ($\mathbb{R}^3$) are written in small bold typeface:

$$\boldsymbol{x}, \boldsymbol{v}, \boldsymbol{u}, \boldsymbol{a}, \ldots;$$

Vectors in the 4-dimensional spacetime ($\mathbb{R}^4$) are written in small bold typeface with an arrow:

$$\vec{\boldsymbol{x}}, \vec{\boldsymbol{v}}, \vec{\boldsymbol{u}}, \vec{\boldsymbol{a}}, \ldots;$$

Matrices are written in big bold typeface:

$$\boldsymbol{M}, \boldsymbol{G}, \boldsymbol{R}, \boldsymbol{I}, \ldots.$$

The identity matrix $\boldsymbol{I}_n$ of size $n \times n$ is the matrix in which all the elements on the main diagonal are equal to 1 and all other elements are equal to 0, e.g.,

$$\boldsymbol{I}_4 = \begin{pmatrix} 1 & 0 & 0 & 0 \\ 0 & 1 & 0 & 0 \\ 0 & 0 & 1 & 0 \\ 0 & 0 & 0 & 1 \end{pmatrix}.$$

The *Derivative Operator* ∇ is the 3-dimensional column vector

$$\nabla = \begin{pmatrix} \frac{\partial}{\partial x} \\ \frac{\partial}{\partial y} \\ \frac{\partial}{\partial z} \end{pmatrix},$$

and ∇^{T} is the 3-dimensional row-vector

$$\nabla^{\mathsf{T}} = \left(\frac{\partial}{\partial x} \middle| \frac{\partial}{\partial y} \middle| \frac{\partial}{\partial z} \right).$$

The *Derivative Operator* $\vec{\nabla}$ is the 4-dimensional column vector

$$\vec{\nabla} = \gamma \begin{pmatrix} -\frac{1}{c}\frac{\partial}{\partial t} \\ \frac{\partial}{\partial x} \\ \frac{\partial}{\partial y} \\ \frac{\partial}{\partial z} \end{pmatrix}$$

with

$$\gamma = \frac{1}{\sqrt{1 - \frac{v^2}{c^2}}}.$$

Remark The derivative operators ∇ and $\vec{\nabla}$ are column vector operators and can act both on the right and on the left! Example:

$$\vec{\nabla}^{\mathsf{T}}\vec{a} = \vec{a}^{\mathsf{T}}\vec{\nabla} = \gamma \left(-\frac{1}{c}\frac{\partial a_0}{\partial t} + \frac{\partial a_1}{\partial x} + \frac{\partial a_2}{\partial y} + \frac{\partial a_3}{\partial z} \right).$$

Chapter 1
Special Relativity

This chapter begins with the classical theorems of Galilei and Newton and the Galilei transformation. The special theory of relativity, developed by Einstein in 1905, leads to the four-dimensional spacetime of Minkowski and the Lorentz transformation. After that the relativity of simultaneity of events, the length contraction of moving bodies and the time dilation are discussed. This is followed by the velocity-addition formula and relativistic mechanics. The next topic is the mass–energy equivalence formula $E = mc^2$, where c is the speed of light in a vacuum. Then relativistic electromagnetism is treated and the invariance of special forms of the equations of dynamics and Maxwell's electrodynamic with respect to the Lorentz transformation is shown. The energy–momentum matrix is introduced and discussed.

1.1 Galilei Transformation

1.1.1 Relativity Principle of Galilei

An *event* is anything that can happen in space and time, e.g. the emission of a flash of light in a room corner. Events happen at a single point. We assign to each event a set of four coordinates t, x_1, x_2 and x_3, or with t and the three-dimensional vector

$$\boldsymbol{x} \stackrel{\text{def}}{=} \begin{pmatrix} x_1 \\ x_2 \\ x_3 \end{pmatrix} \in \mathbb{R}^3.$$

The position vector $\boldsymbol{x}$ and the time t form a *reference frame* $\mathcal{X}$. In this frame, Newton's fundamental law of mechanics has the form

$$\frac{\mathrm{d}\boldsymbol{p}}{\mathrm{d}t} = \boldsymbol{f}$$

G. Ludyk, *Einstein in Matrix Form*, Graduate Texts in Physics,
DOI 10.1007/978-3-642-35798-5_1, © Springer-Verlag Berlin Heidelberg 2013

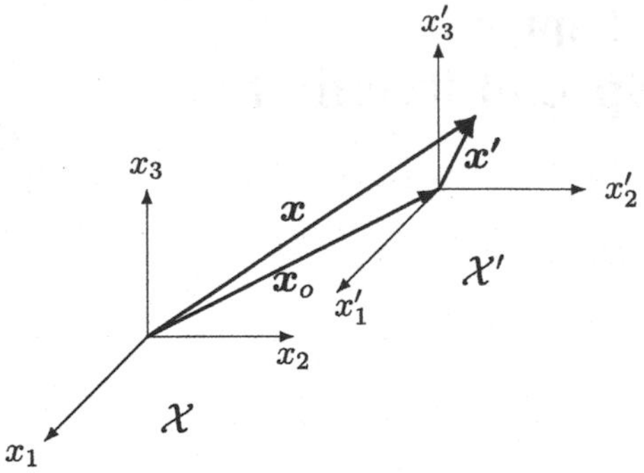

Fig. 1.1 Two against each other shifted reference systems

or, if the mass m in the momentum

$$\boldsymbol{p} = m\frac{\mathrm{d}\boldsymbol{x}}{\mathrm{d}t}$$

is constant,

$$m\frac{\mathrm{d}^2\boldsymbol{x}}{\mathrm{d}t^2} = \boldsymbol{f}. \tag{1.1}$$

An observer may now execute any motion, for example, he makes an experiment in a moving train. We want to find the equation that takes the place of

$$\frac{\mathrm{d}\boldsymbol{p}}{\mathrm{d}t} = \boldsymbol{f}$$

for the moving observer. A coordinate system $\mathcal{X}'$ is connected firmly with the moving observer; it should be axis-parallel to the original coordinate system $\mathcal{X}$. $\boldsymbol{x}_o$ is the location of the origin of $\mathcal{X}'$ measured in $\mathcal{X}$ (Fig. 1.1). Then

$$\boldsymbol{x} = \boldsymbol{x}_o + \boldsymbol{x}',$$

or

$$\boldsymbol{x}' = \boldsymbol{x} - \boldsymbol{x}_o. \tag{1.2}$$

Here $\boldsymbol{x}$ is the position vector of the event measured by an observer at rest in the reference system $\mathcal{X}$, and $\boldsymbol{x}'$ is what an observer measures in the moving reference system $\mathcal{X}'$. Equation (1.2) differentiated with respect to time t,

$$\frac{\mathrm{d}\boldsymbol{x}'}{\mathrm{d}t} = \frac{\mathrm{d}\boldsymbol{x}}{\mathrm{d}t} - \frac{\mathrm{d}\boldsymbol{x}_o}{\mathrm{d}t}, \tag{1.3}$$

results in the speed addition theorem of classical mechanics:

$$\boldsymbol{v}'(t) = \boldsymbol{v}(t) - \boldsymbol{v}_0(t).$$

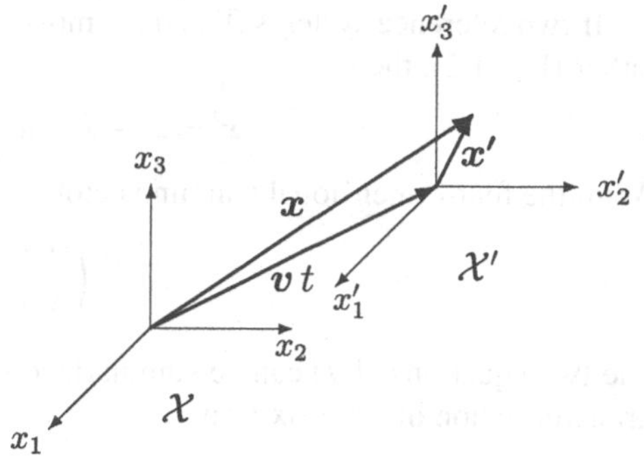

Fig. 1.2 Two reference systems moving against each other

For the acceleration one obtains

$$\frac{d^2\boldsymbol{x}'}{dt^2} = \frac{d^2\boldsymbol{x}}{dt^2} - \frac{d^2\boldsymbol{x}_o}{dt^2}. \tag{1.4}$$

The force $\boldsymbol{f}$ acting on the mass m is independent of the chosen coordinate system, so $\boldsymbol{f}' = \boldsymbol{f}$. This and (1.1) used in (1.4) result in

$$\boldsymbol{f}' - m\frac{d^2\boldsymbol{x}_o}{dt^2} = m\frac{d^2\boldsymbol{x}'}{dt^2}. \tag{1.5}$$

The fundamental law of mechanics has lost its validity! If the moving observer knows the external force $\boldsymbol{f}'$, he can determine with measurements in $\mathcal{X}'$ his acceleration with respect to the rest system $\mathcal{X}$. However, if the motion of $\mathcal{X}'$ with respect to $\mathcal{X}$ is uniform and rectilinear, i.e. $\boldsymbol{x}_o = \boldsymbol{v}_o t$ with a constant $\boldsymbol{v}_o$, then from (1.5)

$$\boldsymbol{f}' = m\frac{d^2\boldsymbol{x}'}{dt^2} \tag{1.6}$$

and the fundamental laws of mechanics have in $\mathcal{X}'$ the same form as in $\mathcal{X}$. Such a uniform and rectilinear moving coordinate system is called an *inertial system*. The moving observer has no possibility to determine his own motion with respect to the coordinate system $\mathcal{X}$ by a mechanical experiment. Any free particle moves in a straight line with constant speed. For example, Galilei considered a uniformly moving ship in a port, whose occupants cannot decide whether the ship moves with respect to the port or whether the port moves with respect to the ship. Today as an example one would take a train in a railway station. This is the *Relativity Principle* of Galilei:

Axiom: All natural laws are the same at every moment in all inertial systems.
All coordinate systems moving uniformly linearly with respect to an inertial system are themselves inertial systems.

If two reference systems $\mathcal{X}$ and $\mathcal{X}'$ move with a constant velocity $\boldsymbol{v}$ against each other (Fig. 1.2), then

$$\boldsymbol{x}' = \boldsymbol{x} - \boldsymbol{v}t \quad \text{and} \quad t' = t. \tag{1.7}$$

With the four-dimensional column vector

$$\vec{\boldsymbol{x}} \stackrel{\text{def}}{=} \begin{pmatrix} t \\ \boldsymbol{x} \end{pmatrix} \in \mathbb{R}^4$$

the two equations (1.7) can be summarized in one equation, and we get the Galilei transformation in a matrix form

$$\underline{\underline{\vec{\boldsymbol{x}}' = \begin{pmatrix} 1 & \boldsymbol{o}^T \\ -\boldsymbol{v} & \boldsymbol{I} \end{pmatrix} \vec{\boldsymbol{x}} = \boldsymbol{T}_{\text{Galilei}} \vec{\boldsymbol{x}}.}} \tag{1.8}$$

Looking at the inverse transformation of $\mathcal{X}'$ to $\mathcal{X}$, one gets

$$t = t' \quad \text{and} \quad \boldsymbol{x} = \boldsymbol{v}t + \boldsymbol{x}',$$

or

$$\vec{\boldsymbol{x}} = \begin{pmatrix} 1 & \boldsymbol{o}^T \\ \boldsymbol{v} & \boldsymbol{I} \end{pmatrix} \vec{\boldsymbol{x}}' = \boldsymbol{T}'_{\text{Galilei}} \vec{\boldsymbol{x}}'.$$

If both transformations are applied in series, the result is

$$\boldsymbol{T}_{\text{Galilei}} \boldsymbol{T}'_{\text{Galilei}} \vec{\boldsymbol{x}}' = \begin{pmatrix} 1 & \boldsymbol{o}^T \\ -\boldsymbol{v} & \boldsymbol{I} \end{pmatrix} \begin{pmatrix} 1 & \boldsymbol{o}^T \\ \boldsymbol{v} & \boldsymbol{I} \end{pmatrix} \vec{\boldsymbol{x}}' = \begin{pmatrix} 1 & \boldsymbol{o}^T \\ \boldsymbol{o} & \boldsymbol{I} \end{pmatrix} \vec{\boldsymbol{x}}' = \vec{\boldsymbol{x}}',$$

i.e. the matrix $\boldsymbol{T}'_{\text{Galilei}}$ is the inverse of the matrix $\boldsymbol{T}_{\text{Galilei}}$.

For the time derivatives of the four-vector $\vec{\boldsymbol{x}}$ we obtain

$$\frac{\mathrm{d}\vec{\boldsymbol{x}}}{\mathrm{d}t} = \begin{pmatrix} 1 \\ \frac{\mathrm{d}\boldsymbol{x}}{\mathrm{d}t} \end{pmatrix} \quad \text{and} \quad \frac{\mathrm{d}^2\vec{\boldsymbol{x}}}{\mathrm{d}t^2} = \begin{pmatrix} 0 \\ \frac{\mathrm{d}^2\boldsymbol{x}}{\mathrm{d}t^2} \end{pmatrix}.$$

With $\vec{\boldsymbol{f}} \stackrel{\text{def}}{=} \binom{0}{\boldsymbol{f}}$ we can write the fundamental equation of mechanics

$$\vec{\boldsymbol{f}} = m \frac{\mathrm{d}^2\vec{\boldsymbol{x}}}{\mathrm{d}t^2}. \tag{1.9}$$

This equation multiplied from the left with the transformation matrix $\boldsymbol{T}_{\text{Galilei}}$, in fact, gives back the same form:

$$\begin{aligned} \underbrace{\boldsymbol{T}_{\text{Galilei}} \vec{\boldsymbol{f}}}_{\stackrel{\text{def}}{=} \vec{\boldsymbol{f}}'} &= m \underbrace{\boldsymbol{T}_{\text{Galilei}} \frac{\mathrm{d}^2\vec{\boldsymbol{x}}}{\mathrm{d}t^2}}_{\stackrel{\text{def}}{=} \frac{\mathrm{d}^2\vec{\boldsymbol{x}}'}{\mathrm{d}t'^2}}, \\ \vec{\boldsymbol{f}}' &= m \frac{\mathrm{d}^2\vec{\boldsymbol{x}}'}{\mathrm{d}t'^2}. \end{aligned} \tag{1.10}$$

The fundamental equation of dynamics is thus *invariant* with respect to this Galilei transformation, i.e. it retains its shape regardless of the reference system.

1.1.2 General Galilei Transformation

Up to this point, the Galilei transformation was considered only under the uniform motion with the speed $\boldsymbol{v}$ of the two inertial systems against each other and a fixed initial time $t_o = 0$, a fixed initial point $\boldsymbol{x}_0 = \boldsymbol{0}$ of the new coordinate system $\mathcal{X}'$ and with no rotation of the coordinate system.

It is now generally assumed that all laws of nature remain constant, therefore, are *invariant* with respect to *time shifting*. If $\boldsymbol{x}(t)$ is a solution of $m\ddot{\boldsymbol{x}} = \boldsymbol{f}$, then for all $t_o \in \mathbb{R}$ is $\boldsymbol{x}(t + t_o)$ also a solution.

Next, it is assumed that the considered spaces are *homogeneous*, thus the same features are available at all points. So, if $\boldsymbol{x}(t)$ is again a solution of $m\ddot{\boldsymbol{x}} = \boldsymbol{f}$, then $\boldsymbol{x}(t) + \boldsymbol{b}$ is also a solution, but now for the starting point $\boldsymbol{x}_0 + \boldsymbol{b}$.

Moreover, it is assumed that the considered spaces are *isotropic*, i.e. there is no directional dependence of properties. So again, if $\boldsymbol{x}(t)$ is a solution of $m\ddot{\boldsymbol{x}} = \boldsymbol{f}$, then also $\boldsymbol{D}\boldsymbol{x}(t)$ is a solution for the initial state $\boldsymbol{D}\boldsymbol{x}_0$. Here, however, also the equality of the distances

$$\rho(\boldsymbol{x}_1, \boldsymbol{x}_2) = \rho(\boldsymbol{D}\boldsymbol{x}_1, \boldsymbol{D}\boldsymbol{x}_2)$$

must be true. For the rotation matrix $\boldsymbol{D}$ this means that it must be *orthogonal* because

$$\begin{aligned}\rho(\boldsymbol{D}\boldsymbol{x}_1, \boldsymbol{D}\boldsymbol{x}_2) &= \sqrt{(\boldsymbol{D}\boldsymbol{x}_2 - \boldsymbol{D}\boldsymbol{x}_1)^\top(\boldsymbol{D}\boldsymbol{x}_2 - \boldsymbol{D}\boldsymbol{x}_1)} \\ &= \sqrt{(\boldsymbol{x}_2 - \boldsymbol{x}_1)^\top \boldsymbol{D}^\top \boldsymbol{D}(\boldsymbol{x}_2 - \boldsymbol{x}_1)} \stackrel{!}{=} \rho(\boldsymbol{x}_1, \boldsymbol{x}_2) \\ &= \sqrt{(\boldsymbol{x}_2 - \boldsymbol{x}_1)^\top(\boldsymbol{x}_2 - \boldsymbol{x}_1)},\end{aligned}$$

so

$$\boldsymbol{D}^\top \boldsymbol{D} \stackrel{!}{=} \boldsymbol{I}.$$

The time invariance, homogeneity and isotropy can be summarized in one transformation, the *general Galilei transformation*, as follows:

If t' is shifted versus t by t_0, i.e. one has

$$t' = t_0 + t, \tag{1.11}$$

furthermore, the new coordinate system relative to the old is moved by $\boldsymbol{x}_0$ and rotated by the rotation matrix $\boldsymbol{D}$, so for $\boldsymbol{v} = \boldsymbol{0}$ one gets

$$\boldsymbol{x}' = \boldsymbol{D}\boldsymbol{x} + \boldsymbol{x}_0, \tag{1.12}$$

and so

$$\begin{pmatrix} t' \\ \boldsymbol{x}' \end{pmatrix} = \begin{pmatrix} 1 & \boldsymbol{o}^\mathsf{T} \\ \boldsymbol{o} & \boldsymbol{D} \end{pmatrix} \begin{pmatrix} t \\ \boldsymbol{x} \end{pmatrix} + \begin{pmatrix} t_o \\ \boldsymbol{x}_o \end{pmatrix}. \tag{1.13}$$

By moving the origin of the new coordinate system with the velocity $\boldsymbol{v}$ as mentioned above, one finally gets the general Galilei transformation:

$$\vec{\boldsymbol{x}}' = \begin{pmatrix} 1 & \boldsymbol{o}^\mathsf{T} \\ -\boldsymbol{v} & \boldsymbol{D} \end{pmatrix} \vec{\boldsymbol{x}} + \vec{\boldsymbol{x}}_o, \tag{1.14}$$

with

$$\vec{\boldsymbol{x}} \stackrel{\text{def}}{=} \begin{pmatrix} t \\ \boldsymbol{x} \end{pmatrix} \quad \text{and} \quad \vec{\boldsymbol{x}}_o \stackrel{\text{def}}{=} \begin{pmatrix} t_o \\ \boldsymbol{x}_o \end{pmatrix}.$$

This is an *affine mapping*, or an *affine transformation*.

One gets a *linear transformation* by introducing the extended vector:

$$\begin{pmatrix} t \\ \boldsymbol{x} \\ 1 \end{pmatrix} \in \mathbb{R}^5, \tag{1.15}$$

namely

$$\begin{pmatrix} t' \\ \boldsymbol{x}' \\ 1 \end{pmatrix} = \begin{pmatrix} 1 & \boldsymbol{o}^\mathsf{T} & t_0 \\ -\boldsymbol{v} & \boldsymbol{D} & \boldsymbol{x}_0 \\ 0 & \boldsymbol{o}^\mathsf{T} & 1 \end{pmatrix} \begin{pmatrix} t \\ \boldsymbol{x} \\ 1 \end{pmatrix}. \tag{1.16}$$

Newton's fundamental laws are invariant also with respect to such a transformation. A general Galilei transformation is determined by 10 parameters: t_o, $\boldsymbol{x}_o \in \mathbb{R}^3$, $\boldsymbol{v} \in \mathbb{R}^3$ and $\boldsymbol{D} \in \mathbb{R}^{3\times 3}$. The rotation matrix $\boldsymbol{D}$, in fact, has only three main parameters since any general rotation consists of successive rotations performed around the x_1-, x_2- and x_3-axis, so the whole rotation is characterized by the three angles φ_1, φ_2 and φ_3, where, for example, the rotation around the x_1-axis is achieved by the matrix

$$\begin{pmatrix} 1 & 0 & 0 \\ 0 & \cos\varphi_1 & \sin\varphi_1 \\ 0 & -\sin\varphi_1 & \cos\varphi_1 \end{pmatrix}.$$

1.1.3 Maxwell's Equations and Galilei Transformation

The situation is completely different with Maxwell's equations of electromagnetic dynamics. They are not invariant with respect to a Galilei transformation! Indeed, in an inertial system $\mathcal{X}$, a static charge q generates only a static electric field; in an inertial frame, moving with the speed $\boldsymbol{v}$, $q\boldsymbol{v}$ is an electric current which generates there a magnetic field!

In the nineteenth century, it was believed that all physical phenomena are mechanical and electromagnetic forces could be traced to the stress states of a world-aether, the Maxwell's tensions. It was assumed that even a vacuum must be filled with aether. This aether then is the carrier of the electromagnetic phenomena.

Suppose an *inertial frame* is a reference system in which Galilei's principle of inertia is valid. Then Einstein in his *general relativity principle* claims:

The laws of nature take on the same form in all inertial systems.

For the fundamental law of mechanics, this is derived above. The principle of relativity applies neither to electrodynamics nor to optics. How should the basic equations of electrodynamics be modified so that the relativity principle is valid? This is the content of Einstein's *Theory of Special Relativity.* He expanded it in the *Theory of General Relativity.* This theory treats how the natural laws must be modified so that they are also valid in *accelerating* or not uniformly against each other moving reference systems.

1.2 Lorentz Transformation

1.2.1 Introduction

At the end of the nineteenth century, experiments were conceived which should determine the velocity of the Earth with respect to the resting cosmic aether. This speed relative to the aether can only be measured by an electromagnetic effect, e.g. the light wave propagation. But in the Michelson–Morley experiment in 1881 and 1887 no drift velocity was found. Einstein concluded:[1]

The speed of light c is always constant.

Independently of the movement of the light source and the observer, light has the same speed value in every inertial frame.

Assume a light pulse is produced at time $t = t' = 0$ in the two axis-parallel reference systems $\mathcal{X}$ and $\mathcal{X}'$ with a common origin. If the light spreads with the speed of light c in the reference system $\mathcal{X}'$, then, for example, it is true that $x_1' = c \cdot t$. From (1.7) it follows for the x_1-direction that if $\boldsymbol{v}$ has the x_1-directional component, then

$$x_1 = x_1' + vt = (c + v)t,$$

yielding, contrary to the Michelson–Morley experiment, a propagation velocity for light of $c + v > c$. Therefore, a transformation must be valid, different from the Galilei transformation. We try a linear transformation:

$$t = ft' + \boldsymbol{e}^{\mathsf{T}}\boldsymbol{x}', \tag{1.17}$$

[1] c from Latin celeritas: speed.

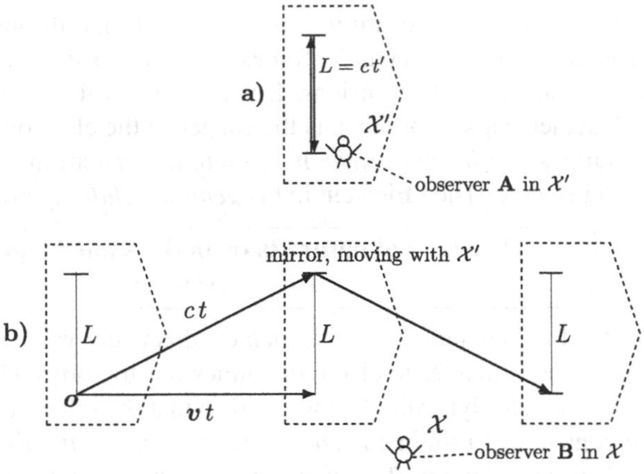

Fig. 1.3 Different ways of light: (**a**) seen by an observer in $\mathcal{X}'$, (**b**) seen by an observer in $\mathcal{X}$

$$\boldsymbol{x} = \boldsymbol{b}t' + \boldsymbol{A}\boldsymbol{x}', \tag{1.18}$$

i.e.

$$\vec{\boldsymbol{x}} = \hat{\boldsymbol{L}}'\vec{\boldsymbol{x}}' \in \mathbb{R}^4,$$

with

$$\hat{\boldsymbol{L}}' \stackrel{\text{def}}{=} \begin{pmatrix} f & \boldsymbol{e}^\top \\ \boldsymbol{b} & \boldsymbol{A} \end{pmatrix} \in \mathbb{R}^{4\times 4}.$$

1.2.2 Determining the Components of the Transformation Matrix

That t' is different from t (according to Galilei this was not the case, he took $t' = t$) is shown by the following reasoning. We consider two observers; observer **A** moves relative to the observer **B** in a spaceship with velocity $\boldsymbol{v}$ (Fig. 1.3). The spaceship with observer **A** has the inertial system $\mathcal{X}'$, and the observer **B** on Earth has the inertial system $\mathcal{X}$. A beam of light moves from the origin $\boldsymbol{x} = \boldsymbol{x}' = \boldsymbol{o}$ of the reference systems $\mathcal{X}$ and $\mathcal{X}'$ at time $t = t' = 0$ perpendicular to the velocity $\boldsymbol{v}$ and reaches after t' seconds for the observer **A** in the moving reference system $\mathcal{X}'$ a mirror moving also with the reference system $\mathcal{X}'$. For the observer **B** in the stationary reference system $\mathcal{X}$, the light beam reaches the mirror, which has been moved in the $\boldsymbol{v}$ direction a distance of $v \cdot t$ after t seconds. Since in all initial systems the speed of light is constant and equal to c, by Pythagorean theorem, one has

$$(ct)^2 = (vt)^2 + (L)^2 = (vt)^2 + \left(ct'\right)^2,$$

or, after solving for t,

$$\underline{\underline{t = \gamma t'}} \tag{1.19}$$

with

$$\boxed{\gamma \stackrel{\text{def}}{=} \frac{1}{\sqrt{1-\frac{v^2}{c^2}}}.} \tag{1.20}$$

A comparison of (1.19) with (1.17) provides for $\boldsymbol{x}' = \boldsymbol{o}$

$$\underline{\underline{f = \gamma.}} \tag{1.21}$$

For $\boldsymbol{x}' = \boldsymbol{o}$ one has $\boldsymbol{x} = \boldsymbol{v}t$, and from (1.18) it follows that $\boldsymbol{x} = \boldsymbol{b}t'$. Thus, $\boldsymbol{b}t' = \boldsymbol{v}t$, i.e. $\boldsymbol{b} = \boldsymbol{v}\frac{t}{t'}$. From (1.19) it follows, on the other hand, that $\frac{t}{t'} = \gamma$, so

$$\underline{\underline{\boldsymbol{b} = \gamma \boldsymbol{v}.}} \tag{1.22}$$

Up to now, the following transformation equations were determined:

$$t = \gamma t' + \boldsymbol{e}^{\mathsf{T}}\boldsymbol{x}', \tag{1.23}$$

$$\boldsymbol{x} = \gamma \boldsymbol{v} t' + \boldsymbol{A}\boldsymbol{x}'. \tag{1.24}$$

Equations (1.23) and (1.24) deliver a transformation of $\mathcal{X}'$ to $\mathcal{X}$. If one wants to invert this transformation, one must replace $\boldsymbol{v}$ by $-\boldsymbol{v}$, $\boldsymbol{x}$ by $\boldsymbol{x}'$, and vice versa, and t by t' and vice versa (as $\boldsymbol{A}$ and $\boldsymbol{e}$ may depend on $\boldsymbol{v}$, in the following we first write $\tilde{\boldsymbol{A}}$ and $\tilde{\boldsymbol{e}}$):

$$t' = \gamma t + \tilde{\boldsymbol{e}}^{\mathsf{T}}\boldsymbol{x},$$

$$\boldsymbol{x}' = -\gamma \boldsymbol{v} t + \tilde{\boldsymbol{A}}\boldsymbol{x},$$

which can be combined to

$$\vec{x}' = \begin{pmatrix} \gamma & \tilde{\boldsymbol{e}}^{\mathsf{T}} \\ -\gamma \boldsymbol{v} & \tilde{\boldsymbol{A}} \end{pmatrix} \vec{x} \stackrel{\text{def}}{=} \hat{\boldsymbol{L}}\vec{x}. \tag{1.25}$$

Both transformations performed one after another must result in the identity matrix:

$$\hat{\boldsymbol{L}}'\hat{\boldsymbol{L}} \stackrel{!}{=} \boldsymbol{I}. \tag{1.26}$$

For the top left (1, 1)-element of the matrix product $\hat{\boldsymbol{L}}'\hat{\boldsymbol{L}}$ one gets:

$$(\gamma, \boldsymbol{e}^{\mathsf{T}}) \begin{pmatrix} \gamma \\ -\gamma \boldsymbol{v} \end{pmatrix} = \gamma^2 - \gamma \boldsymbol{e}^{\mathsf{T}}\boldsymbol{v} \stackrel{!}{=} 1.$$

Hence,

$$\gamma \boldsymbol{e}^{\mathsf{T}}\boldsymbol{v} = \gamma^2 - 1. \tag{1.27}$$

Taking for $\boldsymbol{e}$

$$\boldsymbol{e}^{\mathsf{T}} = \alpha \boldsymbol{v}^{\mathsf{T}} \tag{1.28}$$

and using (1.27) yields

$$\gamma\alpha v^2 = \gamma^2 - 1,$$

and moreover,

$$\gamma\alpha = \frac{1}{v^2}\left(\frac{c^2}{c^2 - v^2} - 1\right) = \frac{1}{c^2 - v^2} = \frac{\gamma^2}{c^2},$$

i.e.

$$\alpha = \frac{\gamma}{c^2}. \tag{1.29}$$

Equation (1.29) used in (1.28) finally yields

$$\underline{\underline{\boldsymbol{e}^{\mathsf{T}} = \frac{\gamma}{c^2}\boldsymbol{v}^{\mathsf{T}}.}} \tag{1.30}$$

Thus, till now we have calculated:

$$\hat{\boldsymbol{L}}' = \begin{pmatrix} \gamma & \frac{\gamma}{c^2}\boldsymbol{v}^{\mathsf{T}} \\ \gamma\boldsymbol{v} & \boldsymbol{A} \end{pmatrix}.$$

Obviously,

$$\tilde{\boldsymbol{e}}^{\mathsf{T}} = -\frac{\gamma}{c^2}\boldsymbol{v}^{\mathsf{T}}.$$

Suppose now that in (1.25) we have $\tilde{\boldsymbol{A}} = \boldsymbol{A}$. Then for the matrix element in the lower right corner of the matrix product $\hat{\boldsymbol{L}}'\hat{\boldsymbol{L}}$ in (1.26) we obtain

$$(\gamma\boldsymbol{v}, \boldsymbol{A})\begin{pmatrix} -\frac{\gamma}{c^2}\boldsymbol{v}^{\mathsf{T}} \\ \boldsymbol{A} \end{pmatrix} = -\frac{\gamma^2}{c^2}\boldsymbol{v}\boldsymbol{v}^{\mathsf{T}} + \boldsymbol{A}^2 \stackrel{!}{=} \boldsymbol{I},$$

i.e.

$$\boldsymbol{A}^2 = \boldsymbol{I} + \frac{\gamma^2}{c^2}\boldsymbol{v}\boldsymbol{v}^{\mathsf{T}}. \tag{1.31}$$

From (1.27), by inserting (1.30), follows

$$\frac{\gamma^2}{c^2}v^2 = \gamma^2 - 1.$$

Plugging this into (1.31) provides

$$\boldsymbol{A}^2 = \boldsymbol{I} + (\gamma^2 - 1)\frac{\boldsymbol{v}\boldsymbol{v}^{\mathsf{T}}}{v^2}. \tag{1.32}$$

Since $(\gamma - 1)^2 = \gamma^2 - 2(\gamma - 1) - 1$, for $\gamma^2 - 1$ one can write

$$\gamma^2 - 1 = (\gamma - 1)^2 + 2(\gamma - 1). \tag{1.33}$$

Equation (1.33) inserted into (1.32) yields

$$A^2 = I + 2(\gamma - 1)\frac{\boldsymbol{v}\boldsymbol{v}^\mathsf{T}}{v^2} + (\gamma - 1)^2\frac{\boldsymbol{v}\boldsymbol{v}^\mathsf{T}}{v^2} = \left(I + (\gamma - 1)\frac{\boldsymbol{v}\boldsymbol{v}^\mathsf{T}}{v^2}\right)^2,$$

where

$$\frac{\boldsymbol{v}\boldsymbol{v}^\mathsf{T}\boldsymbol{v}\boldsymbol{v}^\mathsf{T}}{v^4} = \frac{\boldsymbol{v}(\boldsymbol{v}^\mathsf{T}\boldsymbol{v})\boldsymbol{v}^\mathsf{T}}{v^4} = \frac{\boldsymbol{v}\boldsymbol{v}^\mathsf{T}}{v^2}$$

was used. Therefore,

$$\underline{\underline{A = I + (\gamma - 1)\frac{\boldsymbol{v}\boldsymbol{v}^\mathsf{T}}{v^2}.}} \tag{1.34}$$

It is, in fact, true that $A(-\boldsymbol{v}) = A(\boldsymbol{v})$, i.e. the above assumption that $\tilde{A} = A$ is correct.

Thus, the matrix $\hat{L}$ of the Lorentz transformation is determined completely as

$$\boxed{\hat{L} = \left(\begin{array}{c|c} \gamma & -\frac{\gamma}{c^2}\boldsymbol{v}^\mathsf{T} \\ \hline -\gamma\boldsymbol{v} & I + (\gamma - 1)\frac{\boldsymbol{v}\boldsymbol{v}^\mathsf{T}}{v^2} \end{array}\right).} \tag{1.35}$$

For $c \to \infty$ we get $\gamma = 1$ and the Lorentz transformation turns into the Galilei transformation with T_{Galilei}.

In the often in the textbooks treated special case when the velocity $\boldsymbol{v}$ is towards the x_1-axis, i.e.

$$\boldsymbol{v} = \begin{pmatrix} v \\ 0 \\ 0 \end{pmatrix},$$

one obtains

$$\hat{L} = \left(\begin{array}{c|c} \gamma & -\frac{\gamma}{c^2}(v, 0, 0) \\ \hline -\gamma\begin{pmatrix} v \\ 0 \\ 0 \end{pmatrix} & I + \frac{(\gamma-1)}{v^2}\begin{pmatrix} v^2 & 0 & 0 \\ 0 & 0 & 0 \\ 0 & 0 & 0 \end{pmatrix} \end{array}\right) = \begin{pmatrix} \gamma & -\frac{\gamma v}{c^2} & 0 & 0 \\ -\gamma v & \gamma & 0 & 0 \\ 0 & 0 & 1 & 0 \\ 0 & 0 & 0 & 1 \end{pmatrix},$$

so

$$\boxed{\begin{aligned} t' &= \gamma t - \tfrac{\gamma}{c^2} v x_1, \\ x_1' &= -\gamma v t + \gamma x_1, \\ x_2' &= x_2, \\ x_3' &= x_3. \end{aligned}} \tag{1.36}$$

Introducing as the time component in the vector $\vec{x}$ the time multiplied with the speed of light c, i.e. $x_0 = ct$, we obtain from (1.35) the transformation

$$\vec{x}' \stackrel{\text{def}}{=} \begin{pmatrix} ct' \\ \boldsymbol{x}' \end{pmatrix} = \left(\begin{array}{c|c} \gamma & -\frac{\gamma}{c}\boldsymbol{v}^\mathsf{T} \\ \hline -\frac{\gamma}{c}\boldsymbol{v} & \boldsymbol{I} + (\gamma - 1)\frac{\boldsymbol{v}\boldsymbol{v}^\mathsf{T}}{v^2} \end{array}\right) \begin{pmatrix} ct \\ \boldsymbol{x} \end{pmatrix}, \tag{1.37}$$

i.e. the new transformation matrix

$$\boxed{\boldsymbol{L}(\boldsymbol{v}) \stackrel{\text{def}}{=} \left(\begin{array}{c|c} \gamma & -\frac{\gamma}{c}\boldsymbol{v}^\mathsf{T} \\ \hline -\frac{\gamma}{c}\boldsymbol{v} & \boldsymbol{I} + (\gamma - 1)\frac{\boldsymbol{v}\boldsymbol{v}^\mathsf{T}}{v^2} \end{array}\right)} \tag{1.38}$$

is now a *symmetric matrix*, and so

$$\boxed{\begin{aligned} ct' &= \gamma ct - \frac{\gamma}{c}\boldsymbol{v}^\mathsf{T}\boldsymbol{x}, \\ \boldsymbol{x}' &= \boldsymbol{x} + (\gamma - 1)\frac{\boldsymbol{v}^\mathsf{T}\boldsymbol{x}}{v^2}\boldsymbol{v} - \gamma \boldsymbol{v} t. \end{aligned}} \tag{1.39}$$

1.2.3 Simultaneity at Different Places

It will be seen that events at different places which are simultaneous for an observer in $\mathcal{X}$ are not in general simultaneous for a moving observer in $\mathcal{X}'$. This is caused by the finite speed of light. The reference system $\mathcal{X}'$ may move towards the stationary reference system $\mathcal{X}$ with velocity $\boldsymbol{v}$. If the two events E_1 and E_2 in $\mathcal{X}$ have the coordinates $\vec{x}_1$ and $\vec{x}_2$, then they are *simultaneous* if $t_1 = t_2$. Do these two events then happen also at the same time for an observer in the reference system $\mathcal{X}'$? Due to (1.37),

$$ct_1' = \gamma ct_1 - \frac{\gamma}{c}\boldsymbol{v}^\mathsf{T}\boldsymbol{x}_1$$

and

$$ct_2' = \gamma ct_2 - \frac{\gamma}{c}\boldsymbol{v}^\mathsf{T}\boldsymbol{x}_2.$$

Dividing both equations by c and subtracting one from another yields

$$t_1' - t_2' = \gamma(t_1 - t_2) - \frac{\gamma}{c^2}\boldsymbol{v}^\mathsf{T}(\boldsymbol{x}_1 - \boldsymbol{x}_2),$$

so

$$\underline{\underline{t_2' - t_1' = \frac{\gamma}{c^2}\boldsymbol{v}^\mathsf{T}(\boldsymbol{x}_1 - \boldsymbol{x}_2).}}$$

The events $\vec{x}_1'$ and $\vec{x}_2'$ are also simultaneous only if the velocity $\boldsymbol{v}$ is perpendicular to the local difference $\boldsymbol{x}_1 - \boldsymbol{x}_2$. Conclusion:

Fig. 1.4 Simultaneity

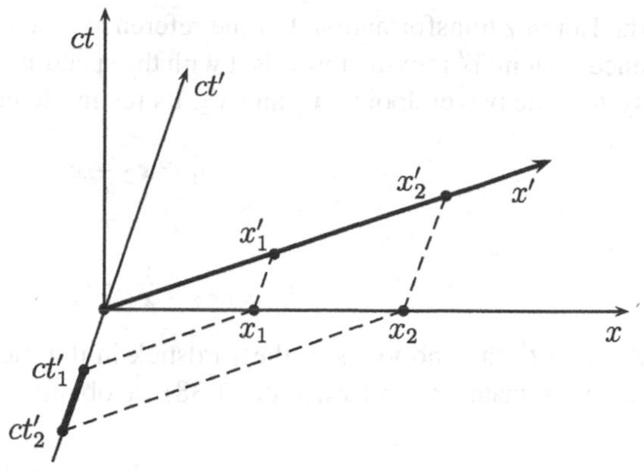

Events at different locations, which in the reference system $\mathcal{X}$ are simultaneous, need *not* be simultaneous when seen from the reference system $\mathcal{X}'$.

Example For the special case $\boldsymbol{v} = [v, 0, 0]^{\mathsf{T}}$ one only needs to consider the ct- and the x-coordinate (Minkowski diagram, Fig. 1.4). The Lorentz transformation does not change the x_2- and x_3-component. It is therefore possible to restrict the consideration to the two-dimensional transformations:

$$\begin{pmatrix} ct' \\ x' \end{pmatrix} = \begin{pmatrix} \gamma & -\frac{\gamma}{c}v \\ -\frac{\gamma}{c}v & \gamma \end{pmatrix} \begin{pmatrix} ct \\ x \end{pmatrix}.$$

The event $\binom{0}{x_1}$ is transformed to

$$\begin{pmatrix} ct'_1 \\ x'_1 \end{pmatrix} = \begin{pmatrix} \gamma & -\frac{\gamma}{c}v \\ -\frac{\gamma}{c}v & \gamma \end{pmatrix} \begin{pmatrix} 0 \\ x_1 \end{pmatrix} = \begin{pmatrix} -\frac{\gamma}{c}vx_1 \\ \gamma x_1 \end{pmatrix}$$

and accordingly the event $\binom{0}{x_2}$, simultaneous in $\mathcal{X}$, is transformed to $\binom{-\frac{\gamma}{c}vx_2}{\gamma x_2}$ in $\mathcal{X}'$. In Fig. 1.4, the difference on the ct'-axis is then

$$ct'_2 - ct'_1 = \frac{\gamma}{c}v(x_1 - x_2) \neq 0,$$

if $v \neq 0$ and $x_1 \neq x_2$, i.e. in the moving reference system $\mathcal{X}'$ the two events E_1 and E_2 are no longer simultaneous.

1.2.4 Length Contraction of Moving Bodies

Einstein (1905) was the first who completely demonstrated that length contraction is an effect due to the change in the notions of space, time and simultaneity brought about by special relativity. Length contraction can simply be derived from the Lorentz transformation. Let the reference system $\mathcal{X}$ be stationary and the reference system $\mathcal{X}'$ moving towards it with the speed $\boldsymbol{v}$. A yardstick has in the stationary system the two endpoints $\boldsymbol{x}_1$ and $\boldsymbol{x}_2$. Its resting length is

$$\boldsymbol{l}_0 = \boldsymbol{x}_2 - \boldsymbol{x}_1, \tag{1.40}$$

i.e.

$$l_0^2 = (\boldsymbol{x}_2 - \boldsymbol{x}_1)^\mathsf{T}(\boldsymbol{x}_2 - \boldsymbol{x}_1). \tag{1.41}$$

At time t', the endpoints of the yardstick in the moving reference system $\mathcal{X}'$ have the coordinates $\boldsymbol{x}_1'$ and $\boldsymbol{x}_2'$. With (1.38) we obtain

$$\boldsymbol{l}_0 = \boldsymbol{x}_2 - \boldsymbol{x}_1 = \boldsymbol{A}(\boldsymbol{x}_2' - \boldsymbol{x}_1') \stackrel{\text{def}}{=} \boldsymbol{A}\boldsymbol{l}'. \tag{1.42}$$

It follows that

$$l_0^2 = \boldsymbol{l}_0^\mathsf{T}\boldsymbol{l}_0 = \boldsymbol{l}'^\mathsf{T}\boldsymbol{A}^2\boldsymbol{l}'. \tag{1.43}$$

With (1.32) one gets

$$l_0^2 = \boldsymbol{l}'^\mathsf{T}\left(\boldsymbol{I} + (\gamma^2 - 1)\frac{\boldsymbol{v}\boldsymbol{v}^\mathsf{T}}{v^2}\right)\boldsymbol{l}' = \boldsymbol{l}'^\mathsf{T}\boldsymbol{l}' + (\gamma^2 - 1)\frac{(\boldsymbol{v}^\mathsf{T}\boldsymbol{l}')^2}{v^2}. \tag{1.44}$$

In the product $\boldsymbol{v}^\mathsf{T}\boldsymbol{l}'$, only the component $\boldsymbol{l}_\parallel'$ of $\boldsymbol{l}'$ parallel to the velocity $\boldsymbol{v}$ comes into effect, i.e. it is true that $\boldsymbol{v}^\mathsf{T}\boldsymbol{l}' = \boldsymbol{v}^\mathsf{T}\boldsymbol{l}_\parallel'$. Thus,

$$l_0^2 = l'^2 + (\gamma^2 - 1)\frac{(\boldsymbol{v}^\mathsf{T}\boldsymbol{l}_\parallel')^2}{v^2} = l'^2 + (\gamma^2 - 1)l_\parallel'^2,$$

so

$$\underline{\underline{l'^2 = l_0^2 - (\gamma^2 - 1)l_\parallel'^2.}} \tag{1.45}$$

Since always $\gamma^2 - 1 \geq 0$, it follows from (1.45) that

$$\underline{\underline{l' \leq l_0.}} \tag{1.46}$$

The result of the yardstick length measurement therefore depends on the reference system in which the length measurement was made.

If the yardstick is parallel to the velocity $\boldsymbol{v}$, then $\boldsymbol{l}' = \boldsymbol{l}_\parallel'$ and (1.45) becomes $\gamma l' = l_0$, or

$$\boxed{l' = \frac{l_0}{\gamma} = l_0\sqrt{1 - \frac{v^2}{c^2}},} \tag{1.47}$$

i.e. as $v \to c$ one gets $l' \to 0$. For example, if the velocity is $v = 0.8c$ (i.e. 80 % of the speed of light), then the length is $l' = 0.6 l_o$.

1.2.5 Time Dilation

Time dilation is an actual difference of elapsed time between two events as measured by observers moving relative to each other. Using (1.19), it is

$$\boxed{\Delta t = \gamma \Delta t',} \tag{1.48}$$

where

$$\gamma = \frac{1}{\sqrt{1-\frac{v^2}{c^2}}} \geq 1.$$

Let t' be the time of a light clock, measured by an observer in the moving inertial frame $\mathcal{X}'$ after covering the distance between the mirrors. Then t is the time required for the light to cover the distance between the two mirrors of the *moving* clock, which is measured by an observer at rest in $\mathcal{X}$. The faster the clock moves, i.e. the greater v is, the longer this time. If, for example, for the moving observer $t' = 1$ second is elapsed, for the stationary observer $t = \gamma \geq 1$ seconds have passed. If v so large that $\gamma = 20$, then, for example, for the stationary observer $t = 20$ years have passed, and for the moving observer only $t' = 1$ year is gone!

Twin Paradox There are two twins **A** and **B** on Earth. **A** starts in a rocket and flies away with high speed, while **B** stays on Earth. During the flight, **A** is aging more slowly than **B**. After some time the rocket is slowed down and returns with **A** at high speed back to Earth. During the flight **A** has aged less than **B**, which remained at rest on Earth, Fig. 1.5. Now comes the paradox: The velocities are relative! One could take also the twin **A** in its entrained coordinate system as stationary and consider **B** as moving with large speeds. That's right; but one key difference is that the twin **A** is not always in the same uniformly moving inertial frame since, at the turning point where the return begins, the inertial system changes! This is not the case for **B**. He always stays in the same inertial frame. Therefore, there is no paradox.

1.3 Invariance of the Quadratic Form

The Michelson–Morley experiment says that in any reference system the light propagates in all directions at the same speed c. If at the origin $\boldsymbol{x} = \boldsymbol{o}$ of $\mathcal{X}$ a flash of light is ignited, it propagates with the speed of light c spherically. After the time t, the light signal reaches all points of the sphere of radius ct. To points on the sphere applies:

$$x_1^2 + x_2^2 + x_3^2 = (ct)^2, \quad \text{i.e.} \quad (ct)^2 - x_1^2 - x_2^2 - x_3^2 = 0. \tag{1.49}$$

If the origins of the two reference systems $\mathcal{X}$ and $\mathcal{X}'$ at the ignition time $t = t' = 0$ of the light flash are at the same space point $\boldsymbol{x}(t=0) = \boldsymbol{X}'(t'=0) = \boldsymbol{o}$, the light in

Fig. 1.5 Twin paradox

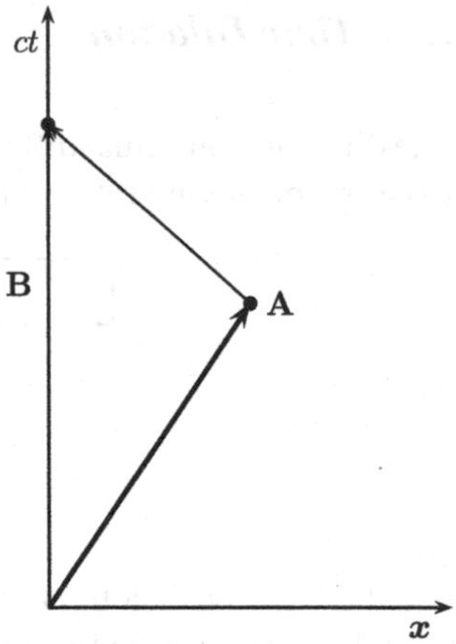

the reference system $\mathcal{X}'$ is also spreading according to the following law:

$$\left(ct'\right)^2 - x_1'^2 - x_2'^2 - x_3'^2 = 0. \tag{1.50}$$

So this quantity is *invariant*.

One can imagine (1.49) generated by the quadratic form

$$\boxed{\vec{x}^\mathsf{T} \boldsymbol{M} \vec{x} = 0,} \tag{1.51}$$

with the Minkowski matrix

$$\boldsymbol{M} \stackrel{\text{def}}{=} \begin{pmatrix} 1 & 0 & 0 & 0 \\ 0 & -1 & 0 & 0 \\ 0 & 0 & -1 & 0 \\ 0 & 0 & 0 & -1 \end{pmatrix}$$

and the *four-dimensional* vector

$$\vec{x} \stackrel{\text{def}}{=} \begin{pmatrix} ct \\ x_1 \\ x_2 \\ x_3 \end{pmatrix}.$$

Minkowski was the first who represented Einstein's Special Theory of Relativity using four-dimensional *spacetime* vectors.

1.3.1 Invariance with Respect to Lorentz Transformation

Now the invariance of the quadratic form (1.51) to a Lorentz transformation is examined. If $\vec{x}' = L\vec{x}$, then

$$\vec{x}'^{\mathsf{T}} M \vec{x}' = \vec{x}^{\mathsf{T}} L^{\mathsf{T}} M L \vec{x}. \tag{1.52}$$

Then for the matrix product $L^{\mathsf{T}} M L$ with the help of (1.31) and (1.34) one obtains

$$L^{\mathsf{T}} M L = \left(\begin{array}{c|c} \gamma & -\frac{\gamma}{c}\boldsymbol{v}^{\mathsf{T}} \\ \hline -\frac{\gamma}{c}\boldsymbol{v} & \boldsymbol{A} \end{array} \right) \left(\begin{array}{c|c} \gamma & -\frac{\gamma}{c}\boldsymbol{v}^{\mathsf{T}} \\ \hline \frac{\gamma}{c}\boldsymbol{v} & -\boldsymbol{A} \end{array} \right)$$

$$= \left(\begin{array}{c|c} \gamma^2 - \frac{\gamma^2}{c^2}\boldsymbol{v}^{\mathsf{T}}\boldsymbol{v} & \frac{\gamma}{c}\boldsymbol{v}^{\mathsf{T}}\boldsymbol{A} - \frac{\gamma^2}{c}\boldsymbol{v}^{\mathsf{T}} \\ \hline \frac{\gamma}{c}\boldsymbol{A}\boldsymbol{v} - \frac{\gamma^2}{c}\boldsymbol{v} & \frac{\gamma^2}{c^2}\boldsymbol{v}\boldsymbol{v}^{\mathsf{T}} - \boldsymbol{A}^2 \end{array} \right) = \begin{pmatrix} 1 & \boldsymbol{o}^{\mathsf{T}} \\ \boldsymbol{o} & -\boldsymbol{I} \end{pmatrix} = M.$$

So for the quadratic forms, in fact, the following is valid:

$$\vec{x}'^{\mathsf{T}} M \vec{x}' = \vec{x}^{\mathsf{T}} M \vec{x},$$

i.e. they are invariant with respect to a Lorentz transformation!

When dealing with the quadratic form, the propagation of light was considered till now. For this the quadratic form $\vec{x}^{\mathsf{T}} M \vec{x}$ is equal to zero. Considering, however, the movement of a particle, the light will spread faster than the particle, i.e. it will always be true that

$$(ct)^2 > \boldsymbol{x}^{\mathsf{T}}\boldsymbol{x}$$

or

$$(ct)^2 - \boldsymbol{x}^{\mathsf{T}}\boldsymbol{x} = \vec{x}^{\mathsf{T}} M \vec{x} > 0.$$

If we denote by $\Delta \boldsymbol{x}$ the travelled way between two events and the elapsed time by Δt, we will obtain the four-dimensional spacetime interval Δs from

$$\Delta s^2 = \Delta \vec{x}^{\mathsf{T}} M \Delta \vec{x}. \tag{1.53}$$

The "distance" Δs between the two events is an *invariant interval* in the four-dimensional spacetime. Since the right-hand side of (1.53) is invariant with respect to a Lorentz transformation, Δs is independent of the chosen inertial frame, it has always the same length. So the theory of relativity does not relativise everything! Δs^2 is negative when the distance $\Delta \vec{x}$ is so big that no light signal can traverse the distance in finite time. This possibility will be examined in the following section.

1.3.2 Light Cone

A flash of light at time $t_0 = 0$ is spreading spherically in the three-dimensional space with the speed of light c. At the time $t_1 > t_0$, the light has reached the surface of a

Fig. 1.6 Light propagation

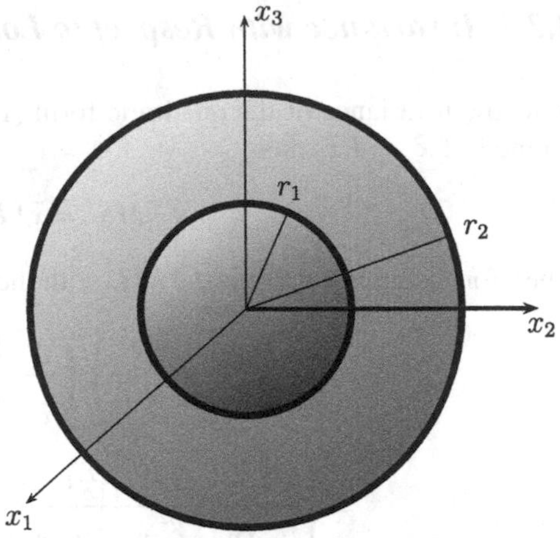

sphere of radius $r_1 = ct_1$. At time $t_2 > t_1$, a spherical surface with a larger radius $r_2 = ct_2$ is reached, and so forth as shown in Fig. 1.6. In this two-dimensional image, the light wave through space is illustrated by a circle that expands with the speed of light. One can implement this movement of light waves in a space–time diagram in which the time coordinate ct is vertical and two of the three spatial coordinates, e.g. x_1 and x_2, are displayed horizontally (Fig. 1.7). Due to the fact that the time coordinate is represented by the time t multiplied by the speed of light c, the photons move in straight lines on this diagram, which are sloped at 45°. For photons the possible paths are in an open top cone whose walls have a slope of 45°. The speed of a moving particle is always less than the speed of light, therefore the path must always be run within the light cone with a slope which is always less than 45° to the time axis.

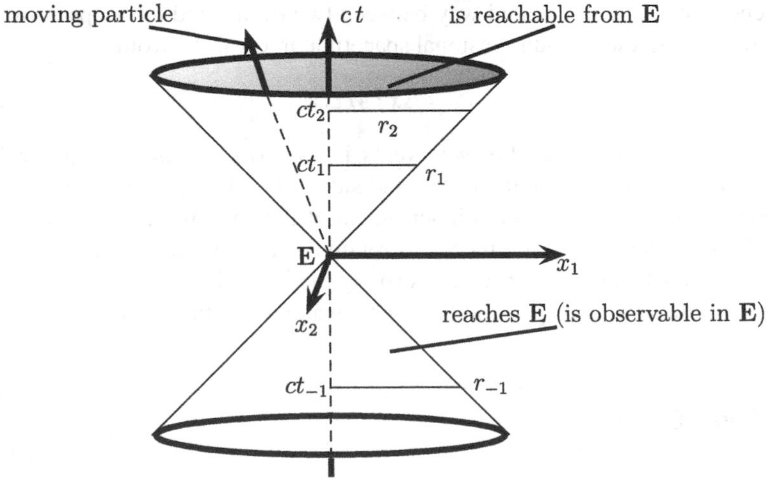

Fig. 1.7 Light cone

The photons which reach the point **E** at a time before $t_0 = 0$, e.g. in the time interval of $t_{-1} < t_0$, all come from a ball of radius $ct_{-1} = r_{-1}$, etc. Altogether we obtain again a cone as in Fig. 1.7, the cone of the events which can reach the event **E**, i.e. these events are observable from **E**.

In the Minkowski spacetime of special relativity, in any event the light cones are aligned in parallel; the central axes of all light cones are parallel to the time axis. In the general theory of relativity, this is no longer always the case due to the curvature of space, i.e. the central axes of the light cones are not always parallel to the time axis.

1.3.3 Proper Time

If one looks at a completely arbitrarily moving clock from any inertial frame, then one can interpret this movement as uniform at any instant of time. If we introduce a coordinate system which is permanently connected to the clock, then this is again an inertial frame. In the infinitesimal time interval $\mathrm{d}t$, measured with the clock of the observer in the inertial system $\mathcal{X}$, the moving clock covers a distance of $(\mathrm{d}x_1^2 + \mathrm{d}x_2^2 + \mathrm{d}x_3^2)^{1/2}$. In the inertial system $\mathcal{X}'$, permanently connected to the clock, the clock does not move, so $\mathrm{d}x_1' = \mathrm{d}x_2' = \mathrm{d}x_3' = 0$, but the elapsed displayed time is $\mathrm{d}t'$. The invariant quadratic form

$$\mathrm{d}s^2 = \mathrm{d}\vec{x}^{\mathsf{T}} \boldsymbol{M}\, \mathrm{d}\vec{x} = \mathrm{d}\vec{x}'^{\mathsf{T}} \boldsymbol{M}\, \mathrm{d}\vec{x}'$$

computed for

$$\mathrm{d}\vec{x}^{\mathsf{T}} = [c\,\mathrm{d}t, \mathrm{d}x_1, \mathrm{d}x_2, \mathrm{d}x_3] \quad \text{and} \quad \mathrm{d}\vec{x}'^{\mathsf{T}} = \left[c\,\mathrm{d}t', 0, 0, 0\right]$$

gives

$$\mathrm{d}s^2 = c^2\,\mathrm{d}t^2 - \mathrm{d}x_1^2 - \mathrm{d}x_2^2 - \mathrm{d}x_3^2 = c^2\,\mathrm{d}t'^2, \tag{1.54}$$

so

$$\begin{aligned}\mathrm{d}t' &= \frac{1}{c}\,\mathrm{d}s = \frac{1}{c}\sqrt{c^2\,\mathrm{d}t^2 - \mathrm{d}x_1^2 - \mathrm{d}x_2^2 - \mathrm{d}x_3^2} \\ &= \mathrm{d}t\sqrt{1 - \frac{\mathrm{d}x_1^2 + \mathrm{d}x_2^2 + \mathrm{d}x_3^2}{c^2\,\mathrm{d}t^2}},\end{aligned}$$

and taking

$$v^2 = \frac{\mathrm{d}x_1^2 + \mathrm{d}x_2^2 + \mathrm{d}x_3^2}{\mathrm{d}t^2},$$

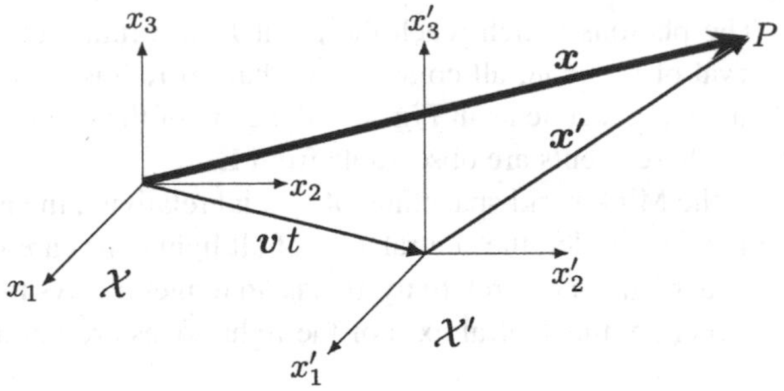

Fig. 1.8 Relativistic velocity addition

where v is the velocity of the moving clock relative to the observer, finally yields the relationship (1.19)

$$dt' = dt\sqrt{1 - \frac{v^2}{c^2}}. \tag{1.55}$$

So, if the stationary clock of the observer shows the time interval $t_2 - t_1$, then the moving clock shows the interval $t'_2 - t'_1$ of the *proper time*,

$$t'_2 - t'_1 = \int_{t_1}^{t_2} \sqrt{1 - \frac{v(t)^2}{c^2}}\, dt. \tag{1.56}$$

The proper time interval of a moving mass is due to (1.55) and (1.56) always smaller than the time interval in the stationary system. In general, the proper time of a moving mass is called τ instead of t'. Due to (1.54), the proper time is thus

$$\underline{\underline{d\tau = ds/c.}} \tag{1.57}$$

1.4 Relativistic Velocity Addition

1.4.1 Galilean Addition of Velocities

Galilei observed that if a ship is moving relative to the shore at velocity v, and a sailor is moving with velocity u measured on the ship, calculating the velocity of the sailor measured on the shore is what is meant by the addition of the velocities v and u. When both the sailor and the ship are moving slowly compared to light, it is accurate enough to use the vector sum

$$\boldsymbol{w} = \boldsymbol{v} + \boldsymbol{u}$$

where $\boldsymbol{w}$ is the velocity of the sailor relative to the shore.

Consider the two inertial systems $\mathcal{X}$ and $\mathcal{X}'$ in Fig. 1.8. The reference system $\mathcal{X}'$ is moving with respect to the other reference system $\mathcal{X}$ with the velocity $\boldsymbol{v}$. Suppose that the vector $\boldsymbol{x}'$ describes the motion of a point P in the inertial system $\mathcal{X}'$, and this point is moving with respect to the reference system $\mathcal{X}'$ with the velocity $\boldsymbol{u} = \frac{\mathrm{d}\boldsymbol{x}'}{\mathrm{d}t'}$. What is the speed $\boldsymbol{w} \stackrel{\text{def}}{=} \frac{\mathrm{d}\boldsymbol{x}}{\mathrm{d}t}$ of the point P relative to the reference system $\mathcal{X}$?

It is

$$\boldsymbol{w} = \frac{\mathrm{d}\boldsymbol{x}}{\mathrm{d}t} = \frac{\mathrm{d}\boldsymbol{x}}{\mathrm{d}t'}\left(\frac{\mathrm{d}t}{\mathrm{d}t'}\right)^{-1}. \tag{1.58}$$

Between $\vec{\boldsymbol{x}}$ and $\vec{\boldsymbol{x}}'$ the relationship $\vec{\boldsymbol{x}}' = \boldsymbol{L}(v)\vec{\boldsymbol{x}}$ holds, or, solving for $\vec{\boldsymbol{x}}$,

$$\vec{\boldsymbol{x}} = \boldsymbol{L}^{-1}(v)\vec{\boldsymbol{x}}'. \tag{1.59}$$

In detail,

$$t = \gamma_v t' + \frac{\gamma_v}{c^2}\boldsymbol{v}^\mathsf{T}\boldsymbol{x}', \tag{1.60}$$

$$\boldsymbol{x} = \boldsymbol{x}' + (\gamma_v - 1)\frac{\boldsymbol{v}^\mathsf{T}\boldsymbol{x}'}{v^2}\boldsymbol{v} + \gamma_v \boldsymbol{v}t'. \tag{1.61}$$

From (1.60) follows

$$\frac{\mathrm{d}t}{\mathrm{d}t'} = \gamma_v + \frac{\gamma_v}{c^2}\boldsymbol{v}^\mathsf{T}\frac{\mathrm{d}\boldsymbol{x}'}{\mathrm{d}t'} = \gamma_v\left(1 + \frac{\boldsymbol{v}^\mathsf{T}\boldsymbol{u}}{c^2}\right), \tag{1.62}$$

and from (1.61)

$$\begin{aligned}\frac{\mathrm{d}\boldsymbol{x}}{\mathrm{d}t'} &= \frac{\mathrm{d}\boldsymbol{x}'}{\mathrm{d}t'} + (\gamma_v - 1)\frac{\boldsymbol{v}\boldsymbol{v}^\mathsf{T}}{v^2}\frac{\mathrm{d}\boldsymbol{x}'}{\mathrm{d}t'} + \gamma_v\boldsymbol{v} \\ &= \boldsymbol{u} + (\gamma_v - 1)\frac{\boldsymbol{v}^\mathsf{T}\boldsymbol{u}}{v^2}\boldsymbol{v} + \gamma_v\boldsymbol{v}.\end{aligned} \tag{1.63}$$

Equations (1.62) and (1.63) used in (1.58) provide

$$\boldsymbol{w} = \frac{\boldsymbol{v} + \frac{1}{\gamma_v}\boldsymbol{u} + (1 - \frac{1}{\gamma_v})\frac{\boldsymbol{v}^\mathsf{T}\boldsymbol{u}}{v^2}\boldsymbol{v}}{1 + \frac{\boldsymbol{v}^\mathsf{T}\boldsymbol{u}}{c^2}}, \tag{1.64}$$

or, after addition of $\boldsymbol{u} - \boldsymbol{u} = \boldsymbol{o}$ in the numerator,

$$\boxed{\boldsymbol{w} = \frac{\boldsymbol{v} + \boldsymbol{u} + (\frac{1}{\gamma_v} - 1)(\boldsymbol{u} - \frac{\boldsymbol{v}^\mathsf{T}\boldsymbol{u}}{v^2}\boldsymbol{v})}{1 + \frac{\boldsymbol{v}^\mathsf{T}\boldsymbol{u}}{c^2}}.} \tag{1.65}$$

So this is the speed of the point P with respect to the reference system $\mathcal{X}$. For a double vectorial product one has

$$\boldsymbol{a} \times (\boldsymbol{b} \times \boldsymbol{c}) = (\boldsymbol{a}^\mathsf{T}\boldsymbol{c})\boldsymbol{b} - (\boldsymbol{a}^\mathsf{T}\boldsymbol{b})\boldsymbol{c}. \tag{1.66}$$

This allows one to transform the last parenthesis in the numerator of (1.65) as follows:

$$\boldsymbol{u} - \frac{\boldsymbol{v}^{\mathsf{T}}\boldsymbol{u}}{v^2}\boldsymbol{v} = \frac{\boldsymbol{v}^{\mathsf{T}}\boldsymbol{v}}{v^2}\boldsymbol{u} - \frac{\boldsymbol{v}^{\mathsf{T}}\boldsymbol{u}}{v^2}\boldsymbol{v} = \frac{1}{v^2}\big(\boldsymbol{v}\times(\boldsymbol{u}\times\boldsymbol{v})\big),$$

i.e. instead of (1.65) one can also write

$$\boxed{\boldsymbol{w} = \frac{\boldsymbol{v}+\boldsymbol{u}+\frac{1}{v^2}(\frac{1}{\gamma_v}-1)(\boldsymbol{v}\times(\boldsymbol{u}\times\boldsymbol{v}))}{1+\frac{\boldsymbol{v}^{\mathsf{T}}\boldsymbol{u}}{c^2}}.} \tag{1.67}$$

If the two velocities $\boldsymbol{v}$ and $\boldsymbol{u}$ are parallel, then $\boldsymbol{v}\times\boldsymbol{u}=\boldsymbol{o}$ and from (1.67) it is clear that the sum of the two velocities is

$$\boldsymbol{w} = \frac{\boldsymbol{v}+\boldsymbol{u}}{1+\frac{\boldsymbol{v}^{\mathsf{T}}\boldsymbol{u}}{c^2}}. \tag{1.68}$$

If, in contrast, the two velocities $\boldsymbol{v}$ and $\boldsymbol{u}$ are perpendicular to each other, then $\boldsymbol{v}^{\mathsf{T}}\boldsymbol{u}=\boldsymbol{o}$ and from (1.65) it follows that the sum of the two velocities is

$$\boldsymbol{w} = \boldsymbol{v} + \frac{1}{\gamma_v}\boldsymbol{u}. \tag{1.69}$$

In a further special case which is almost exclusively treated in textbooks and in which both the vector $\boldsymbol{u}$ and the vector $\boldsymbol{v}$ have only one component in the x_1-direction, i.e.

$$\boldsymbol{u} = \begin{pmatrix} u_1 \\ 0 \\ 0 \end{pmatrix} \quad \text{and} \quad \boldsymbol{v} = \begin{pmatrix} v_1 \\ 0 \\ 0 \end{pmatrix},$$

one has $w_2 = w_3 = 0$ and

$$w_1 = \frac{v_1+u_1}{1+\frac{u_1 v_1}{c^2}}. \tag{1.70}$$

1.5 Lorentz Transformation of the Velocity

How does one write the basic laws of mechanics so that they remain invariant under a Lorentz transformation? We start from the transformation equation

$$\vec{x}' = \boldsymbol{L}\vec{x} \tag{1.71}$$

with the four-dimensional spacetime vector $\vec{x} \stackrel{\text{def}}{=} \binom{ct}{\boldsymbol{x}}$ and the transformation matrix (1.38)

$$\boldsymbol{L} = \left(\begin{array}{c|c} \gamma & -\frac{\gamma}{c}\boldsymbol{v} \\ \hline -\frac{\gamma}{c}\boldsymbol{v}^{\mathsf{T}} & \boldsymbol{I}+(\gamma-1)\frac{\boldsymbol{v}\boldsymbol{v}^{\mathsf{T}}}{v^2} \end{array}\right).$$

Differentiating (1.71) with respect to time t', we obtain

$$\frac{\mathrm{d}\vec{x}'}{\mathrm{d}t'} = \begin{pmatrix} c \\ \frac{\mathrm{d}\boldsymbol{x}'}{\mathrm{d}t'} \end{pmatrix} = \boldsymbol{L}\frac{\mathrm{d}\vec{x}}{\mathrm{d}t}\frac{\mathrm{d}t}{\mathrm{d}t'}.$$

Using (1.35) with $\boldsymbol{u} \stackrel{\text{def}}{=} \frac{\mathrm{d}\boldsymbol{x}}{\mathrm{d}t}$ (which now has a different $\boldsymbol{u}$ than in the previous section),

$$\frac{\mathrm{d}t'}{\mathrm{d}t} = \frac{\mathrm{d}}{\mathrm{d}t}\left(-\frac{\gamma}{c^2}\boldsymbol{v}^\top\boldsymbol{x} + \gamma t\right) = -\frac{\gamma}{c^2}\boldsymbol{v}^\top\boldsymbol{u} + \gamma. \tag{1.72}$$

Thus we obtain

$$\begin{pmatrix} c \\ \boldsymbol{u}' \end{pmatrix} = \boldsymbol{L}\begin{pmatrix} c \\ \boldsymbol{u} \end{pmatrix}\frac{1}{\gamma}\frac{1}{1-\frac{\boldsymbol{v}^\top\boldsymbol{u}}{c^2}}. \tag{1.73}$$

In (1.73), we see that the velocity vector $\binom{c}{\boldsymbol{u}}$ is *not* transformed into the velocity vector $\binom{c}{\boldsymbol{u}'}$ using a Lorentz transformation matrix! For this to be the case, the definition of the velocity has to be modified. For this purpose, a short interim statement:

The second block row of (1.73) provides

$$\boldsymbol{u}' = \frac{1}{\gamma(1-\frac{\boldsymbol{v}^\top\boldsymbol{u}}{c^2})}(\boldsymbol{A}\boldsymbol{u} - \gamma\boldsymbol{v}).$$

Taking the scalar product of this vector with itself, we obtain

$$u'^2 \stackrel{\text{def}}{=} \boldsymbol{u}'^\top\boldsymbol{u}' = \left(\frac{1}{\gamma(1-\frac{\boldsymbol{v}^\top\boldsymbol{u}}{c^2})}\right)^2(\boldsymbol{u}^\top\boldsymbol{A}^\top - \gamma\boldsymbol{v}^\top)(\boldsymbol{A}\boldsymbol{u} - \gamma\boldsymbol{v}).$$

Since $\boldsymbol{A} = \boldsymbol{A}^\top$ and $\boldsymbol{A}^2 = \boldsymbol{I} + \frac{\gamma^2}{c^2}\boldsymbol{v}\boldsymbol{v}^\top$, the above is

$$\begin{aligned} u'^2 &= \frac{1}{\gamma^2(1-\frac{\boldsymbol{v}^\top\boldsymbol{u}}{c^2})^2}\left(u^2 - \frac{\gamma^2}{c^2}(\boldsymbol{u}^\top\boldsymbol{v})^2 - 2\gamma^2(\boldsymbol{u}^\top\boldsymbol{v})^2 + \gamma^2 v^2\right) \\ &= \frac{1}{(1-\frac{\boldsymbol{v}^\top\boldsymbol{u}}{c^2})^2}\left(\frac{1}{\gamma^2}u^2 - \frac{1}{c^2}(\boldsymbol{u}^\top\boldsymbol{v})^2 - 2(\boldsymbol{u}^\top\boldsymbol{v})^2 + v^2\right). \end{aligned} \tag{1.74}$$

With

$$\frac{u^2}{\gamma^2} = \left(1 - \frac{v^2}{c^2}\right)u^2 = u^2 - \frac{v^2u^2}{c^2}$$

from (1.74) (without γ) one gets

$$u'^2 = \frac{1}{(1-\frac{\boldsymbol{v}^\top\boldsymbol{u}}{c^2})^2}\left(u^2 + \frac{-(\boldsymbol{u}^\top\boldsymbol{v})^2 - v^2u^2}{c^2} - 2(\boldsymbol{u}^\top\boldsymbol{v})^2 + v^2\right). \tag{1.75}$$

With the help of (1.75) one obtains

$$1-\frac{u'^2}{c^2}=\frac{1}{(1-\frac{\boldsymbol{v}^\top\boldsymbol{u}}{c^2})^2}\left(1-\frac{u^2}{c^2}-\frac{v^2}{c^2}+\frac{v^2u^2}{c^4}\right)$$
$$=\frac{1}{(1-\frac{\boldsymbol{v}^\top\boldsymbol{u}}{c^2})^2}\left(1-\frac{u^2}{c^2}\right)\left(1-\frac{v^2}{c^2}\right),$$

and this implies

$$\sqrt{1-\frac{u'^2}{c^2}}=\sqrt{1-\frac{u^2}{c^2}}\sqrt{1-\frac{v^2}{c^2}}\Big/\left(1-\frac{\boldsymbol{v}^\top\boldsymbol{u}}{c^2}\right),$$

or with

$$\gamma_u\stackrel{\text{def}}{=}\frac{1}{\sqrt{1-\frac{u^2}{c^2}}}$$

and

$$\gamma_{u'}\stackrel{\text{def}}{=}\frac{1}{\sqrt{1-\frac{u'^2}{c^2}}},$$

we finally get

$$\underline{\underline{\gamma\left(1-\frac{\boldsymbol{v}^\top\boldsymbol{u}}{c^2}\right)=\frac{\gamma_{u'}}{\gamma_u}}}. \tag{1.76}$$

Inserting (1.76) into (1.73), we obtain

$$\begin{pmatrix}c\\ \boldsymbol{u}'\end{pmatrix}=\boldsymbol{L}\begin{pmatrix}c\\ \boldsymbol{u}\end{pmatrix}\frac{\gamma_u}{\gamma_{u'}}, \tag{1.77}$$

or

$$\boxed{\begin{pmatrix}\gamma_{u'}c\\ \gamma_{u'}\boldsymbol{u}'\end{pmatrix}=\boldsymbol{L}\begin{pmatrix}\gamma_u c\\ \gamma_u\boldsymbol{u}\end{pmatrix}.} \tag{1.78}$$

The thus modified new velocity vector

$$\vec{\boldsymbol{u}}\stackrel{\text{def}}{=}\gamma_u\begin{pmatrix}c\\ \boldsymbol{u}\end{pmatrix} \tag{1.79}$$

now is transformed by a Lorentz transformation $\boldsymbol{L}$ into the velocity vector $\vec{\boldsymbol{u}}'$:

$$\boxed{\vec{\boldsymbol{u}}'=\boldsymbol{L}\vec{\boldsymbol{u}}.} \tag{1.80}$$

The so-defined velocity $\vec{\boldsymbol{u}}$ is much better suited for the formulation of physics laws, as they have the same shape in every inertial frame. With $\boldsymbol{u}$ this would not be the case. For clarification we re-emphasize that a particle or a focal point moves with the velocity $\boldsymbol{u}$ in an inertial system $\mathcal{X}'$, which itself moves or can move at the speed $\boldsymbol{v}$ with respect to another reference system. That is the difference between $\boldsymbol{u}$ and $\boldsymbol{v}$!

By the way, the quadratic form for the velocity

$$\vec{\boldsymbol{u}}^\mathsf{T} \boldsymbol{M} \vec{\boldsymbol{u}} = \gamma_u^2 c^2 - \gamma_u^2 \boldsymbol{u}^\mathsf{T} \boldsymbol{u} = \frac{c^4}{c^2 - u^2} - \frac{c^2 u^2}{c^2 - u^2} = c^2 \tag{1.81}$$

is, of course, invariant with respect to a Lorentz transformation because—as one can easily show—even $\vec{\boldsymbol{u}}'^\mathsf{T} \boldsymbol{M} \vec{\boldsymbol{u}}' = c^2$ holds.

In (1.57), the proper time $\mathrm{d}\tau = \frac{1}{c}\,\mathrm{d}s$ was introduced. We have

$$\mathrm{d}s^2 = \vec{\boldsymbol{x}}^\mathsf{T} \boldsymbol{M} \vec{\boldsymbol{x}} = c^2\,\mathrm{d}t^2 - \boldsymbol{x}^\mathsf{T} \boldsymbol{x} = c^2\,\mathrm{d}t^2 \left(1 - \frac{1}{c^2} \frac{\mathrm{d}\boldsymbol{x}^\mathsf{T}}{\mathrm{d}t} \cdot \frac{\mathrm{d}\boldsymbol{x}}{\mathrm{d}t}\right),$$

so with $\frac{\mathrm{d}\boldsymbol{x}}{\mathrm{d}t} = \boldsymbol{u}$

$$\mathrm{d}\tau = \mathrm{d}t \left(1 - \frac{1}{c^2} \boldsymbol{u}^\mathsf{T} \boldsymbol{u}\right)^{\frac{1}{2}} = \mathrm{d}t \left(1 - \frac{u^2}{c^2}\right)^{\frac{1}{2}},$$

or with $\gamma_u = (1 - \frac{u^2}{c^2})^{-\frac{1}{2}}$

$$\underline{\underline{\mathrm{d}t = \gamma_u\,\mathrm{d}\tau}}. \tag{1.82}$$

This is the same relationship as in the time dilation in (1.73). τ is therefore the time that a comoving clock displays, while t is the time that an observer at rest measures. However, the moving clock must no longer move rectilinearly and uniformly!

In (1.79), one has $\boldsymbol{u} = \frac{\mathrm{d}\boldsymbol{x}}{\mathrm{d}t}$. Replacing $\mathrm{d}t$ by (1.82) in it gives $\boldsymbol{u} = \frac{1}{\gamma_u} \frac{\mathrm{d}\boldsymbol{x}}{\mathrm{d}\tau}$, or equivalently, $\gamma_u \boldsymbol{u} = \frac{\mathrm{d}\boldsymbol{x}}{\mathrm{d}\tau}$. Furthermore,

$$\mathrm{d}\vec{\boldsymbol{x}} = \begin{pmatrix} c\,\mathrm{d}t \\ \mathrm{d}\boldsymbol{x} \end{pmatrix} = \begin{pmatrix} c\gamma_u\,\mathrm{d}\tau \\ \mathrm{d}\boldsymbol{x} \end{pmatrix},$$

so

$$\underline{\underline{\frac{\mathrm{d}\vec{\boldsymbol{x}}}{\mathrm{d}\tau}}} = \begin{pmatrix} \gamma_u c \\ \gamma_u \frac{\mathrm{d}\boldsymbol{x}}{\mathrm{d}\tau} \end{pmatrix} = \begin{pmatrix} \gamma_u c \\ \gamma_u \boldsymbol{u} \end{pmatrix} = \underline{\underline{\vec{\boldsymbol{u}}}}. \tag{1.83}$$

When the trajectory in spacetime is parameterized by the proper time τ, $\vec{\boldsymbol{x}} = \vec{\boldsymbol{x}}(\tau)$, then $\vec{\boldsymbol{u}} = \frac{\mathrm{d}\vec{\boldsymbol{x}}}{\mathrm{d}\tau}$ is the four-velocity along the trajectory.

1.6 Momentum and Its Lorentz Transformation

Multiplying the equation $\vec{\boldsymbol{u}}' = \boldsymbol{L}\vec{\boldsymbol{u}}$ with the rest mass m_0, we obtain

$$\begin{pmatrix} m_0 \gamma_{u'} c \\ m_0 \gamma_{u'} \boldsymbol{u}' \end{pmatrix} = \boldsymbol{L} \begin{pmatrix} m_0 \gamma_u c \\ m_0 \gamma_u \boldsymbol{u} \end{pmatrix}. \tag{1.84}$$

Herein we define as usual the *momentum*

$$\boldsymbol{p} \stackrel{\text{def}}{=} m_0 \gamma_u \boldsymbol{u} = m_u \boldsymbol{u} = m_u \frac{\mathrm{d}\boldsymbol{x}}{\mathrm{d}t} \tag{1.85}$$

by

$$m_u \stackrel{\text{def}}{=} m_0 \gamma_u = \frac{m_0}{\sqrt{1 - \frac{u^2}{c^2}}}.$$

The momentum vector

$$\vec{\boldsymbol{p}} \stackrel{\text{def}}{=} \begin{pmatrix} m_u c \\ \boldsymbol{p} \end{pmatrix} = m_0 \vec{\boldsymbol{u}} = m_0 \gamma_u \begin{pmatrix} c \\ \boldsymbol{u} \end{pmatrix} \tag{1.86}$$

is due to (1.84) transformed as

$$\boxed{\vec{\boldsymbol{p}}' = \boldsymbol{L} \vec{\boldsymbol{p}}.} \tag{1.87}$$

Also, the quadratic form associated with the momentum vector

$$\vec{\boldsymbol{p}}^{\mathsf{T}} \boldsymbol{M} \vec{\boldsymbol{p}} = m_0^2 \vec{\boldsymbol{u}}^{\mathsf{T}} \boldsymbol{M} \vec{\boldsymbol{u}} = m_0^2 c^2 \tag{1.88}$$

is invariant with respect to a Lorentz transformation because also $\vec{\boldsymbol{p}}'^{\mathsf{T}} \boldsymbol{M} \vec{\boldsymbol{p}}' = m_0^2 c^2$.

1.7 Acceleration and Force

1.7.1 Acceleration

The acceleration is generally defined as the time derivative of speed. Differentiating the modified velocity vector $\vec{\boldsymbol{u}} = \binom{\gamma_u c}{\gamma_u \boldsymbol{u}} \in \mathbb{R}^4$ with respect to time t, one receives as the derivative of the second component in this vector

$$\frac{\mathrm{d}}{\mathrm{d}t}(\gamma_u \boldsymbol{u}) = \frac{\mathrm{d}\gamma_u}{\mathrm{d}t} \boldsymbol{u} + \gamma_u \frac{\mathrm{d}\boldsymbol{u}}{\mathrm{d}t}. \tag{1.89}$$

In particular, for $\frac{\mathrm{d}\gamma_u}{\mathrm{d}t}$ we obtain with

$$\boldsymbol{a} \stackrel{\text{def}}{=} \frac{\mathrm{d}\boldsymbol{u}}{\mathrm{d}t} \in \mathbb{R}^3$$

and

$$\frac{\mathrm{d}u^2}{\mathrm{d}t} = \frac{\mathrm{d}\boldsymbol{u}^{\mathsf{T}} \boldsymbol{u}}{\mathrm{d}t} = \frac{\mathrm{d}\boldsymbol{u}^{\mathsf{T}}}{\mathrm{d}t} \boldsymbol{u} + \boldsymbol{u}^{\mathsf{T}} \frac{\mathrm{d}\boldsymbol{u}}{\mathrm{d}t} = 2\boldsymbol{u}^{\mathsf{T}} \boldsymbol{a}$$

the result

$$\frac{\mathrm{d}\gamma_u}{\mathrm{d}t} = \frac{\mathrm{d}}{\mathrm{d}t}\left(1 - \frac{u^2}{c^2}\right)^{-1/2} = -\frac{1}{2}\left(1 - \frac{u^2}{c^2}\right)^{-3/2} \cdot \frac{\mathrm{d}}{\mathrm{d}t}\left(1 - \frac{u^2}{c^2}\right)$$
$$= -\frac{1}{2}\gamma_u^3 \cdot \frac{-2\boldsymbol{u}^\mathsf{T}}{c^2} \cdot \boldsymbol{a}, \tag{1.90}$$

i.e.

$$\frac{\mathrm{d}\gamma_u}{\mathrm{d}t} = \frac{\gamma_u^3}{c^2} \cdot \boldsymbol{u}^\mathsf{T}\boldsymbol{a}. \tag{1.91}$$

With (1.91) one obtains for (1.89)

$$\frac{\mathrm{d}}{\mathrm{d}t}(\gamma_u \boldsymbol{u}) = \frac{\gamma_u^3}{c^2} \cdot \boldsymbol{u}^\mathsf{T}\boldsymbol{a} \cdot \boldsymbol{u} + \gamma_u \cdot \boldsymbol{a}. \tag{1.92}$$

Differentiating the velocity transformation equation (1.78),

$$\begin{pmatrix} \gamma_{u'} c \\ \gamma_{u'} \boldsymbol{u}' \end{pmatrix} = \boldsymbol{L} \begin{pmatrix} \gamma_u c \\ \gamma_u \boldsymbol{u} \end{pmatrix},$$

with respect to time t', we obtain

$$\frac{\mathrm{d}}{\mathrm{d}t'}\begin{pmatrix} \gamma_{u'} c \\ \gamma_{u'} \boldsymbol{u}' \end{pmatrix} = \boldsymbol{L} \cdot \frac{\mathrm{d}}{\mathrm{d}t}\begin{pmatrix} \gamma_u c \\ \gamma_u \boldsymbol{u} \end{pmatrix} \cdot \frac{\mathrm{d}t}{\mathrm{d}t'}. \tag{1.93}$$

From (1.72) and (1.76) follows

$$\frac{\mathrm{d}t}{\mathrm{d}t'} = \frac{\gamma_u}{\gamma_{u'}}. \tag{1.94}$$

This, used in (1.93), results for the newly defined four-dimensional acceleration vector

$$\boxed{\vec{\boldsymbol{a}} \stackrel{\text{def}}{=} \gamma_u \cdot \frac{\mathrm{d}}{\mathrm{d}t}\vec{\boldsymbol{u}} \in \mathbb{R}^4} \tag{1.95}$$

in the Lorentz transformation of the acceleration vector $\vec{\boldsymbol{a}}$

$$\boxed{\vec{\boldsymbol{a}}' = \boldsymbol{L}\vec{\boldsymbol{a}}.} \tag{1.96}$$

The acceleration vector defined in (1.95),

$$\vec{\boldsymbol{a}} = \gamma_u \cdot \frac{\mathrm{d}}{\mathrm{d}t}\begin{pmatrix} \gamma_u c \\ \gamma_u \boldsymbol{u} \end{pmatrix}, \tag{1.97}$$

is therefore suitable to formulate physical laws in relativistic form! With $\mathrm{d}t = \gamma_u \, \mathrm{d}\tau$ and $\vec{\boldsymbol{u}} = \mathrm{d}\vec{\boldsymbol{x}}/\mathrm{d}\tau$ one can write for (1.97)

$$\boxed{\vec{\boldsymbol{a}} = \frac{\mathrm{d}\vec{\boldsymbol{u}}}{\mathrm{d}\tau} = \frac{\mathrm{d}^2\vec{\boldsymbol{x}}}{\mathrm{d}\tau^2}.} \tag{1.98}$$

The vector $\vec{\boldsymbol{a}}$ is also obtained using (1.91) and (1.92) as

$$\boxed{\vec{\boldsymbol{a}} = \begin{pmatrix} \frac{\gamma_u^4}{c} \cdot \boldsymbol{u}^\top \boldsymbol{a} \\ \frac{\gamma_u^4}{c^2} \cdot \boldsymbol{u}^\top \boldsymbol{a} \cdot \boldsymbol{u} + \gamma_u^2 \cdot \boldsymbol{a} \end{pmatrix}.} \tag{1.99}$$

If $\vec{\boldsymbol{u}}^\top \boldsymbol{M} \vec{\boldsymbol{u}} = c^2$ is differentiated with respect to the proper time τ, we obtain

$$\frac{\mathrm{d}\vec{\boldsymbol{u}}^\top}{\mathrm{d}\tau} \boldsymbol{M} \vec{\boldsymbol{u}} + \vec{\boldsymbol{u}}^\top \boldsymbol{M} \frac{\mathrm{d}\vec{\boldsymbol{u}}}{\mathrm{d}\tau} = 2\vec{\boldsymbol{u}}^\top \boldsymbol{M} \frac{\mathrm{d}\vec{\boldsymbol{u}}}{\mathrm{d}\tau} = 2\vec{\boldsymbol{u}}^\top \boldsymbol{M} \vec{\boldsymbol{a}} = 0,$$

i.e. the two four-dimensional vectors $\vec{\boldsymbol{u}}$ and $\boldsymbol{M}\vec{\boldsymbol{a}}$ are orthogonal in $\mathbb{R}^4$!

For each point of time of an arbitrarily accelerated motion, one can always specify a reference system $\mathcal{X}'$ which is an inertial system, named "local inertial system". We obtain the corresponding Lorentz transformation by selecting $\boldsymbol{L}(\boldsymbol{u})$ as the transformation matrix. Then, with

$$\boldsymbol{A}(\boldsymbol{u}) \stackrel{\text{def}}{=} \boldsymbol{I} + (\gamma_u - 1)\frac{\boldsymbol{u}\boldsymbol{u}^\top}{u^2},$$

it is indeed true that

$$\vec{\boldsymbol{a}}' = \boldsymbol{L}(\boldsymbol{u})\vec{\boldsymbol{a}} = \left(\begin{array}{c|c} \gamma_u & -\frac{\gamma_u}{c}\boldsymbol{u}^\top \\ \hline -\frac{\gamma_u}{c}\boldsymbol{u} & \boldsymbol{A}(\boldsymbol{u}) \end{array}\right) \begin{pmatrix} \frac{\gamma_u^4}{c} \cdot \boldsymbol{u}^\top \boldsymbol{a} \\ \gamma_u^2 \cdot \boldsymbol{a} + \frac{\gamma_u^4}{c^2} \cdot \boldsymbol{u}^\top \boldsymbol{a} \cdot \boldsymbol{u} \end{pmatrix}$$

$$= \begin{pmatrix} -\gamma_u^3 \frac{\boldsymbol{u}^\top \boldsymbol{a}}{c} - \gamma_u^5 \frac{(\boldsymbol{u}^\top \boldsymbol{a})(\boldsymbol{u}^\top \boldsymbol{u})}{c^3} + \gamma_u^5 \frac{\boldsymbol{u}^\top \boldsymbol{a}}{c} \\ \gamma_u^2 \boldsymbol{A}(\boldsymbol{u})\boldsymbol{a} + \gamma_u^4 \frac{\boldsymbol{u}(\boldsymbol{u}^\top \boldsymbol{a})}{c^2} + (\gamma_u - 1)\gamma_u^4 \frac{\boldsymbol{u}(\boldsymbol{u}^\top \boldsymbol{u})(\boldsymbol{u}^\top \boldsymbol{a})}{c^2 u^2} - \gamma_u^5 \frac{(\boldsymbol{u}^\top \boldsymbol{a})\boldsymbol{u}}{c^2} \end{pmatrix}$$

$$= \begin{pmatrix} 0 \\ \gamma_u^2 \boldsymbol{A}(\boldsymbol{u})\boldsymbol{a} \end{pmatrix} = \begin{pmatrix} 0 \\ \gamma_u^2 \boldsymbol{a} + \gamma_u^2(\gamma_u - 1)\frac{\boldsymbol{u}^\top \boldsymbol{a}}{u^2}\boldsymbol{u} \end{pmatrix} = \begin{pmatrix} 0 \\ \boldsymbol{a}' \end{pmatrix}.$$

1.7.2 Equation of Motion and Force

The relativistic equation of motion for a particle has to be Lorentz-invariant, and, in the inertial system of the considered particle, Newton's equation of motion has to be true:

$$m_0 \frac{\mathrm{d}\boldsymbol{u}}{\mathrm{d}t} = \boldsymbol{f} \in \mathbb{R}^3. \tag{1.100}$$

Let the accompanying inertial system be $\mathcal{X}$. Furthermore, suppose $\mathcal{X}'$ is the inertial system which moves relative to $\mathcal{X}$ with the constant speed $\boldsymbol{u}(t_0)$. The particle rests momentarily at time $t = t_0$ in $\mathcal{X}'$. The equation of motion (1.100) refers to a point of time and its neighbourhood. For this neighbourhood, $t = t_0 \pm \mathrm{d}t$ is the desired arbitrary small speed in $\mathcal{X}'$. For speeds $v \ll c$ we have (1.100). Hence in $\mathcal{X}'$

$$m_0 \frac{\mathrm{d}\boldsymbol{u}'}{\mathrm{d}t'} = \boldsymbol{f}' \in \mathbb{R}^3 \tag{1.101}$$

also holds exactly. From (1.101), the relativistic equations of motion in an arbitrary reference system may be derived. In (1.101), m_0 is the rest mass and $\boldsymbol{f}'$ the three-dimensional force in $\mathcal{X}'$. Expand the vector $\boldsymbol{f}'$ in (1.101) to a four-vector and call the result $\vec{f}'$:

$$m_0 \frac{\mathrm{d}}{\mathrm{d}t'} \begin{pmatrix} c \\ \boldsymbol{u}' \end{pmatrix} = \begin{pmatrix} 0 \\ \boldsymbol{f}' \end{pmatrix} \stackrel{\text{def}}{=} \vec{f}'. \tag{1.102}$$

Thus $\vec{f}'$ is specified in the resting system $\mathcal{X}'$. In the inertial system $\mathcal{X}$, in which the mass particle moves with the velocity $\boldsymbol{u}$, $\vec{f}$ is obtained by a Lorentz transformation with $\boldsymbol{L}(-\boldsymbol{u})$:

$$\vec{f} = \boldsymbol{L}(-\boldsymbol{u}) \begin{pmatrix} 0 \\ \boldsymbol{f}' \end{pmatrix} = \begin{pmatrix} \frac{\gamma_u}{c} \boldsymbol{u}^\top \boldsymbol{f}' \\ \boldsymbol{A}(\boldsymbol{u}) \boldsymbol{f}' \end{pmatrix} \stackrel{\text{def}}{=} \begin{pmatrix} f_0 \\ \boldsymbol{f} \end{pmatrix}. \tag{1.103}$$

The equation

$$m_0 \gamma \frac{\mathrm{d}}{\mathrm{d}t} \begin{pmatrix} \gamma c \\ \gamma \boldsymbol{u} \end{pmatrix} = \vec{f} = \begin{pmatrix} f_0 \\ \boldsymbol{f} \end{pmatrix},$$

i.e.

$$\boxed{m_0 \vec{a} = \vec{f}} \tag{1.104}$$

possesses all the desired properties! The four-vectors $\vec{a}$ and $\vec{f}$ are Lorentz-invariant and, in the inertial frame of the particle, this equation is reduced to Newton's equation of motion

$$m_0 \begin{pmatrix} 0 \\ \frac{\mathrm{d}\boldsymbol{u}'}{\mathrm{d}t'} \end{pmatrix} = \begin{pmatrix} 0 \\ \boldsymbol{f}' \end{pmatrix}.$$

For the last three components of the equation of motion (1.104),

$$\frac{\mathrm{d}(m_u \boldsymbol{u})}{\mathrm{d}t} = \frac{1}{\gamma_u} \boldsymbol{f} \tag{1.105}$$

with the velocity-dependent mass

$$m_u \stackrel{\text{def}}{=} \gamma_u m_0. \tag{1.106}$$

In the theory of relativity, the time derivative of the momentum $m_u \boldsymbol{u}$ is also interpreted as force. The components f_i of the relativistic equation of motion are thus, according to (1.103) and (1.104),

$$f_0 = \gamma_u \frac{\mathrm{d}}{\mathrm{d}t}(m_u c) = \frac{\gamma_u}{c} \boldsymbol{u}^\top \boldsymbol{f}' \tag{1.107}$$

and

$$\boldsymbol{f} = \gamma_u \frac{\mathrm{d}}{\mathrm{d}t}(m_u \boldsymbol{u}) = \boldsymbol{A}(\boldsymbol{u}) \boldsymbol{f}'. \tag{1.108}$$

1.7.3 Energy and Rest Mass

Equation (1.107) multiplied with c/γ_u provides

$$\frac{\mathrm{d}}{\mathrm{d}t}\left(m_u c^2\right) = \boldsymbol{u}^\top \boldsymbol{f}. \tag{1.109}$$

In (1.109), $\boldsymbol{u}^\top \boldsymbol{f}$ is the instantaneous power. This is the work per unit of time, done by the force $\boldsymbol{f}$. So the left-hand side of (1.109) must be the temporal change of energy, i.e. $m_u c^2 = \gamma_u m_0 c^2$ is energy. We obtain for the *relativistic energy* the most renowned formula of the theory of relativity:

$$\boxed{E = m_u c^2.} \tag{1.110}$$

When $\boldsymbol{u} = \boldsymbol{o}$, i.e. when the particle is at rest, $\gamma_u = 1$, and

$$\boxed{E_0 = m_0 c^2} \tag{1.111}$$

is the "rest energy", Einstein's famous formula. Weinberg says in [36] about the content of this formula: "If some mass is destroyed (as in radioactive decay or fusion, or fission), then very large quantities of kinetic energy will be liberated, with consequences of well-known importance."

The four-dimensional momentum vector $\vec{\boldsymbol{p}}$ is then recognised as the combination of energy and momentum:

$$\vec{\boldsymbol{p}} = \begin{pmatrix} E/c \\ \boldsymbol{p} \end{pmatrix}. \tag{1.112}$$

For the quadratic form in (1.88)

$$\vec{\boldsymbol{p}}^\top \boldsymbol{M} \vec{\boldsymbol{p}} = m_0^2 c^2$$

we now obtain

$$\vec{\boldsymbol{p}}^\top \boldsymbol{M} \vec{\boldsymbol{p}} = \left(E/c, \boldsymbol{p}^\top\right) \begin{pmatrix} E/c \\ -\boldsymbol{p} \end{pmatrix} = E^2/c^2 - p^2 = m_0^2 c^2,$$

i.e.

$$E = \sqrt{(m_0c^2)^2 + p^2c^2}. \tag{1.113}$$

For high velocities the momentum term p^2c^2 dominates here: $E = pc$ as for neutrinos and particles in accelerators (CERN). For small velocities $u \ll c$, we may use the approximation

$$E \approx m_0c^2 + \frac{1}{2}m_0u^2.$$

The first term is the rest energy E_0. The second term is the classical kinetic energy E_{kin}. The relativistic kinetic energy is

$$E_{\text{kin}} = E - m_0c^2 = (\gamma_u - 1)m_0c^2. \tag{1.114}$$

If $v \to c$, then $E_{\text{kin}} \to \infty$. Equation (1.114) gives a reason why mass particles cannot be accelerated up to the velocity of light!

1.7.4 Emission of Energy

A body, at rest in $\mathcal{X}'$ with the rest mass $m_{0,\text{before}}$, radiates at a certain time the energy E'_{Emission} in form of light or heat radiation. This radiation is emitted symmetrically so that the total momentum of the radiated energy in $\mathcal{X}'$ is zero, i.e. the body remains at rest during the radiation process. The energy–momentum vector of the radiation in $\mathcal{X}'$ is therefore

$$\begin{pmatrix} \frac{1}{c}E'_{\text{Emission}} \\ 0 \\ 0 \\ 0 \end{pmatrix}. \tag{1.115}$$

Compared with the inertial system $\mathcal{X}$, the body is moving with the velocity $\boldsymbol{v}$. In this inertial system, the total momentum is equal to the momentum of the body $\boldsymbol{p}_{\text{before}} = \gamma_u m_{0,\text{before}}\boldsymbol{u}$. After radiation, the body has the momentum $\boldsymbol{p}_{\text{after}} = \gamma_u m_{0,\text{after}}\boldsymbol{u}$, and the momentum of radiation is calculated using (1.115) with the Lorentz transformation $\boldsymbol{L}(-\boldsymbol{u})$ from $\mathcal{X}'$ to $\mathcal{X}$:

$$\vec{p} = \boldsymbol{L}(-\boldsymbol{u}) \begin{pmatrix} \frac{1}{c}E'_{\text{Emission}} \\ 0 \\ 0 \\ 0 \end{pmatrix} = \begin{pmatrix} \gamma_u \frac{1}{c}E'_{\text{Emission}} \\ \gamma_u \boldsymbol{u}\frac{1}{c^2}E'_{\text{Emission}} \end{pmatrix}.$$

During the emission of radiation, the body has given away the momentum $\gamma_u \boldsymbol{u}\frac{1}{c^2}E'_{\text{Emission}}$, without changing its speed. This is only possible because the body has changed its rest mass! Because of the momentum conservation law, one has

$$\gamma_u m_{0,\text{before}} \boldsymbol{u} = \gamma_u m_{0,\text{after}} \boldsymbol{u} + \gamma_u \boldsymbol{u} \frac{1}{c^2} E'_{\text{Emission}}.$$

It follows that

$$m_{0,\text{after}} = m_{0,\text{before}} - \frac{1}{c^2} E'_{\text{Emission}}. \tag{1.116}$$

In Einstein's words: "If a body emits the energy E' in the form of radiation, its mass is reduced by E'/c^2."

1.8 Relativistic Electrodynamics

1.8.1 Maxwell's Equations

The magnetic field associated with the induction $\boldsymbol{b}$ and the electric field with the field strength[2] $\boldsymbol{e}$, the sources of the charge q and the current $\boldsymbol{j}$ satisfy the Maxwell's equations:[3]

$$\nabla \times \boldsymbol{b} = \frac{1}{c}\left(\frac{\partial \boldsymbol{e}}{\partial t} + \boldsymbol{j}\right), \tag{1.117}$$

$$\nabla^{\mathsf{T}} \boldsymbol{e} = \rho, \tag{1.118}$$

$$\nabla \times \boldsymbol{e} = -\frac{1}{c}\frac{\partial \boldsymbol{b}}{\partial t}, \tag{1.119}$$

$$\nabla^{\mathsf{T}} \boldsymbol{b} = 0. \tag{1.120}$$

The *Derivative Operator* ∇ is the 3-dimensional column vector

$$\nabla = \begin{pmatrix} \frac{\partial}{\partial x} \\ \frac{\partial}{\partial y} \\ \frac{\partial}{\partial z} \end{pmatrix}$$

and ∇^{T} is the 3-dimensional row-vector

$$\nabla^{\mathsf{T}} = \left(\frac{\partial}{\partial x} \middle| \frac{\partial}{\partial y} \middle| \frac{\partial}{\partial z}\right),$$

[2]Since the magnetic induction and the electric field strength are *vectors*, they are marked by bold small letters $\boldsymbol{b}$ and $\boldsymbol{e}$.

[3]In other symbols, $\nabla \times \boldsymbol{b} = \text{curl}\, \boldsymbol{b}$ and $\nabla^{\mathsf{T}} \boldsymbol{e} = \text{div}\, \boldsymbol{e}$.

so that

$$\nabla^{\mathsf{T}} \boldsymbol{e} = \frac{\partial e_x}{\partial x} + \frac{\partial e_y}{\partial y} + \frac{\partial e_z}{\partial z}, \tag{1.121}$$

and

$$\nabla \times \boldsymbol{b} = -\boldsymbol{B}_\times \nabla \stackrel{\text{def}}{=} \begin{pmatrix} \frac{\partial b_z}{\partial y} - \frac{\partial b_y}{\partial z} \\ -\frac{\partial b_z}{\partial x} + \frac{\partial b_x}{\partial z} \\ \frac{\partial b_y}{\partial x} - \frac{\partial b_x}{\partial y} \end{pmatrix}, \tag{1.122}$$

with[4]

$$\boldsymbol{B}_\times \stackrel{\text{def}}{=} \begin{pmatrix} 0 & -b_z & b_y \\ b_z & 0 & -b_x \\ -b_y & b_x & 0 \end{pmatrix}. \tag{1.123}$$

Equations (1.117) and (1.118) can be summed up after minor changes in the following four equations:

$$\begin{aligned}
& & \frac{\partial e_x}{\partial x} & +\frac{\partial e_y}{\partial y} & +\frac{\partial e_z}{\partial z} & = \rho, \\
-\frac{1}{c}\frac{\partial e_x}{\partial t} & & & -\frac{\partial b_z}{\partial y} & +\frac{\partial b_y}{\partial z} & = \frac{1}{c} j_1, \\
-\frac{1}{c}\frac{\partial e_y}{\partial t} & +\frac{\partial b_z}{\partial x} & & & -\frac{\partial b_x}{\partial z} & = \frac{1}{c} j_2, \\
-\frac{1}{c}\frac{\partial e_z}{\partial t} & -\frac{\partial b_y}{\partial x} & +\frac{\partial b_x}{\partial y} & & & = \frac{1}{c} j_3,
\end{aligned} \tag{1.124}$$

or in matrix form,[5]

$$\underbrace{\begin{pmatrix} 0 & \boldsymbol{e}^{\mathsf{T}} \\ -\boldsymbol{e} & \boldsymbol{B}_\times \end{pmatrix}}_{\stackrel{\text{def}}{=} \boldsymbol{F}_{B,e}} \underbrace{\gamma \begin{pmatrix} -\frac{1}{c}\frac{\partial}{\partial t} \\ \nabla \end{pmatrix}}_{\stackrel{\text{def}}{=} \vec{\nabla}} = \underbrace{\frac{1}{c} \gamma \begin{pmatrix} c\rho \\ -\boldsymbol{j} \end{pmatrix}}_{\stackrel{\text{def}}{=} \vec{j}} \in \mathbb{R}^4. \tag{1.125}$$

The skew-symmetric matrix $\boldsymbol{F}_{B,e}$, composed of $\boldsymbol{B}_\times$ and $\boldsymbol{e}$, is called Faraday's matrix. A similarly structured field strength matrix is obtained if (1.119) and (1.120)

---

[4]An equation $\boldsymbol{c} = \boldsymbol{a} \times \boldsymbol{b}$ cannot in this form be linearly transformed with an invertible transformation matrix $\boldsymbol{T}$. But with $\boldsymbol{A}_\times \stackrel{\text{def}}{=} \begin{pmatrix} 0 & -a_z & a_y \\ a_z & 0 & -a_x \\ -a_y & a_x & 0 \end{pmatrix}$ this is possible in the form of the equation $\boldsymbol{c} = \boldsymbol{A}_\times \boldsymbol{b}$: $(\boldsymbol{T}\boldsymbol{c}) = (\boldsymbol{T}\boldsymbol{A}_\times \boldsymbol{T}^{-1})(\boldsymbol{T}\boldsymbol{b})$.

[5]The factor γ has been added on both sides, so that no difficulties arise later in the invariance of this equation with respect to a Lorentz transformation.

are gathered and summarized in the following equations:

$$
\begin{array}{rrrrl}
 & -\frac{\partial b_x}{\partial x} & -\frac{\partial b_y}{\partial y} & -\frac{\partial b_z}{\partial z} & = 0, \\
+\frac{1}{c}\frac{\partial b_x}{\partial t} & & -\frac{\partial e_z}{\partial y} & +\frac{\partial e_y}{\partial z} & = 0, \\
+\frac{1}{c}\frac{\partial b_y}{\partial t} & +\frac{\partial e_z}{\partial x} & & -\frac{\partial e_x}{\partial z} & = 0, \\
+\frac{1}{c}\frac{\partial b_z}{\partial t} & -\frac{\partial e_y}{\partial x} & +\frac{\partial e_x}{\partial y} & & = 0,
\end{array}
\tag{1.126}
$$

or in matrix form,

$$
\underbrace{\begin{pmatrix} 0 & -\boldsymbol{b}^\top \\ \boldsymbol{b} & \boldsymbol{E}_\times \end{pmatrix}}_{\stackrel{\text{def}}{=} \boldsymbol{F}_{E,b}} \underbrace{\gamma \begin{pmatrix} -\frac{1}{c}\frac{\partial}{\partial t} \\ \nabla \end{pmatrix}}_{\vec{\nabla}} = \boldsymbol{o} \in \mathbb{R}^4. \tag{1.127}
$$

The matrix $\boldsymbol{F}_{E,b}$ is called the Maxwell's matrix.

So (1.125) and (1.127) contain the Maxwell's equations in a new four-dimensional form:

$$
\boxed{\boldsymbol{F}_{B,e}\vec{\nabla} = \frac{1}{c}\vec{j} \quad \text{and} \quad \boldsymbol{F}_{E,b}\vec{\nabla} = \boldsymbol{o}.} \tag{1.128}
$$

This form has a great advantage of being invariant when transitioning to another reference system by a Lorentz transformation, i.e. in every inertial system its keeps the same external form. But the quantities appearing in it take different values in each reference system. This is shown in the following.

1.8.2 Lorentz Transformation of the Maxwell's Equations

Multiplying (1.125) from the left with $\boldsymbol{L}^{-1}$ and inserting $\boldsymbol{L}^{-1}\boldsymbol{L} = \boldsymbol{I}$, we obtain

$$
\underbrace{\boldsymbol{L}^{-1}\boldsymbol{F}_{B,e}\boldsymbol{L}^{-1}}_{\boldsymbol{F}'_{B',e'}} \boldsymbol{L}\vec{\nabla} = \frac{1}{c}\underbrace{\boldsymbol{L}^{-1}\vec{j}}_{\vec{j}'}. \tag{1.129}
$$

For the new Faraday's matrix $\boldsymbol{F}'_{B',e'}$ we get

$$
\begin{aligned}
\boldsymbol{F}'_{B',e'} &= \begin{pmatrix} 0 & \boldsymbol{e}'^\top \\ -\boldsymbol{e}' & \boldsymbol{B}'_\times \end{pmatrix} \\
&= \begin{pmatrix} \gamma_v & \frac{\gamma_v}{c}\boldsymbol{v}^\top \\ \frac{\gamma_v}{c}\boldsymbol{v} & \boldsymbol{A}(\boldsymbol{v}) \end{pmatrix} \begin{pmatrix} 0 & \boldsymbol{e}^\top \\ -\boldsymbol{e} & \boldsymbol{B}_\times \end{pmatrix} \begin{pmatrix} \gamma_v & \frac{\gamma_v}{c}\boldsymbol{v}^\top \\ \frac{\gamma_v}{c}\boldsymbol{v} & \boldsymbol{A}(\boldsymbol{v}) \end{pmatrix}.
\end{aligned} \tag{1.130}
$$

After multiplying the three matrices, we obtain for the bottom left corner vector $\boldsymbol{e}'$ with $\boldsymbol{A} = \boldsymbol{A}(\boldsymbol{v})$ and $\gamma = \gamma_v$:

$$-\boldsymbol{e}' = \frac{\gamma}{c}\boldsymbol{A}\boldsymbol{B}_\times \boldsymbol{v} + \frac{\gamma^2}{c^2}\boldsymbol{v}\boldsymbol{e}^\mathsf{T}\boldsymbol{v} - \gamma\boldsymbol{A}\boldsymbol{e}$$

$$= \frac{\gamma}{c}\left(\boldsymbol{B}_\times\boldsymbol{v} + (\gamma-1)\frac{\boldsymbol{v}\overbrace{\boldsymbol{v}^\mathsf{T}\boldsymbol{B}_\times\boldsymbol{v}}^{0}}{v^2}\right) + \frac{\gamma^2}{c^2}\boldsymbol{v}\boldsymbol{e}^\mathsf{T}\boldsymbol{v} - \gamma\boldsymbol{e} - \frac{\gamma(\gamma-1)}{v^2}\boldsymbol{v}\boldsymbol{v}^\mathsf{T}\boldsymbol{e}$$

$$= \frac{\gamma}{c}\boldsymbol{B}_\times\boldsymbol{v} - \underbrace{\left(-\frac{\gamma^2}{c^2} + \frac{\gamma^2}{v^2} - \frac{\gamma}{v^2}\right)}_{\frac{1}{v^2}}\boldsymbol{v}^\mathsf{T}\boldsymbol{e}\boldsymbol{v} - \gamma\boldsymbol{e},$$

$$\boldsymbol{e}' = \gamma\left(\boldsymbol{e} - \frac{1}{c}\boldsymbol{B}_\times\boldsymbol{v}\right) + \frac{(1-\gamma)}{v^2}\left(\boldsymbol{v}^\mathsf{T}\boldsymbol{e}\right)\boldsymbol{v}, \tag{1.131}$$

i.e.

$$\boxed{\boldsymbol{e}' = \gamma\left(\boldsymbol{e} + \frac{1}{c}\boldsymbol{v}\times\boldsymbol{b}\right) + (1-\gamma)\frac{\boldsymbol{v}^\mathsf{T}\boldsymbol{e}}{v^2}\boldsymbol{v}.} \tag{1.132}$$

In the matrix $\boldsymbol{F}'_{B',e'}$ in (1.130), for the right bottom 3×3 sub-matrix $\boldsymbol{B}'_\times$ after performing the matrix multiplication we obtain

$$\boldsymbol{B}'_\times = \boldsymbol{B}_\times + \frac{(\gamma-1)}{v^2}\left(\boldsymbol{B}_\times\boldsymbol{v}\boldsymbol{v}^\mathsf{T} + \boldsymbol{v}\boldsymbol{v}^\mathsf{T}\boldsymbol{B}_\times\right) + \frac{\gamma}{c}\left(\boldsymbol{v}\boldsymbol{e}^\mathsf{T} - \boldsymbol{e}\boldsymbol{v}^\mathsf{T}\right). \tag{1.133}$$

From this matrix equation, the components of the magnetic induction vector $\boldsymbol{b}'$ can be filtered out with the help of the row or column vectors of the unit matrix ($\boldsymbol{i}_j$ is the jth column of the 3×3 identity matrix $\boldsymbol{I}_3$) as follows:

$$\boldsymbol{B}'_\times \stackrel{\text{def}}{=} \begin{pmatrix} 0 & -b'_z & b'_y \\ b'_z & 0 & -b'_x \\ -b'_y & b'_x & 0 \end{pmatrix}. \tag{1.134}$$

We thus obtain

$$b'_x = \boldsymbol{i}_3^\mathsf{T}\boldsymbol{B}'_\times\boldsymbol{i}_2, \tag{1.135}$$

$$b'_y = \boldsymbol{i}_1^\mathsf{T}\boldsymbol{B}'_\times\boldsymbol{i}_3, \tag{1.136}$$

$$b'_z = \boldsymbol{i}_2^\mathsf{T}\boldsymbol{B}'_\times\boldsymbol{i}_1. \tag{1.137}$$

Using (1.135) for (1.133) results in

$$b'_x = \boldsymbol{i}_3^\mathsf{T}\boldsymbol{B}'_\times\boldsymbol{i}_2 = b_x + \frac{(\gamma-1)}{v^2}\left[(-b_y v_1 + b_x v_2)v_2 + v_3(-b_z v_1 + b_x v_3)\right]$$
$$+ \frac{\gamma}{c}(v_3 e_y - e_z v_2). \tag{1.138}$$

Accordingly, one obtains for the two remaining components

$$b'_y = \boldsymbol{i}_1^\mathsf{T} \boldsymbol{B}'_\times \boldsymbol{i}_3 = b_y + \frac{(\gamma-1)}{v^2}\big[(-b_z v_2 + b_y v_3)v_3 + v_1(b_y v_1 - b_x v_2)\big] \\ + \frac{\gamma}{c}(v_1 e_z - e_x v_3) \tag{1.139}$$

and

$$b'_z = \boldsymbol{i}_2^\mathsf{T} \boldsymbol{B}'_\times \boldsymbol{i}_1 = b_z + \frac{(\gamma-1)}{v^2}\big[(b_z v_1 - b_x v_3)v_1 + v_2(b_z v_2 - b_y v_3)\big] + \frac{\gamma}{c}(v_2 e_x - e_y v_1). \tag{1.140}$$

Summarizing the last terms in (1.138), (1.139) and (1.140) results in the vector product

$$\frac{\gamma}{c}\boldsymbol{e} \times \boldsymbol{v}. \tag{1.141}$$

For the second to last summands one obtains the vector

$$\frac{(\gamma-1)}{v^2}\begin{pmatrix} (-b_y v_1 + b_x v_2)v_2 + v_3(-b_z v_1 + b_x v_3) \\ (-b_z v_2 + b_y v_3)v_3 + v_1(b_y v_1 - b_x v_2) \\ (b_z v_1 - b_x v_3)v_1 + v_2(b_z v_2 - b_y v_3) \end{pmatrix},$$

and using $0 = b_x v_1^2 - b_x v_1^2$ in the first component, $0 = b_y v_2^2 - b_y v_2^2$ in the second, and $0 = b_z v_3^2 - b_z v_3^2$ in the third,

$$\frac{(\gamma-1)}{v^2}\begin{pmatrix} (-b_x v_1^2 - b_y v_1 v_2 - b_z v_1 v_3) + (b_x v_1^2 + b_x v_2^2 + b_x v_3^2) \\ (-b_x v_1 v_2 - b_y v_2^2 - b_z v_2 v_3) + (b_y v_1^2 + b_y v_2^2 + b_y v_3^2) \\ (-b_x v_1 v_3 - b_y v_2 v_3 - b_z v_3^2) + (b_z v_1^2 + b_z v_2^2 + b_z v_3^2) \end{pmatrix}$$

$$= \frac{(\gamma-1)}{v^2}\big(-(\boldsymbol{b}^\mathsf{T}\boldsymbol{v})\boldsymbol{v} + v^2\boldsymbol{b}\big). \tag{1.142}$$

Equations (1.141) and (1.142), together with (1.138), (1.139) and (1.140), yield the final result

$$\boxed{\boldsymbol{b}' = \gamma\left(\boldsymbol{b} - \frac{1}{c}\boldsymbol{v} \times \boldsymbol{e}\right) + (1-\gamma)\frac{\boldsymbol{v}^\mathsf{T}\boldsymbol{b}}{v^2}\boldsymbol{v}.} \tag{1.143}$$

One gets the same result for $\boldsymbol{e}'$ and $\boldsymbol{b}'$ by a Lorentz transformation of $\boldsymbol{F}_{E,b}$!

The formulas (1.132) and (1.143) are simplified considerably for small velocities $v \ll c$ because then $\gamma \approx 1$ and one obtains

$$\boldsymbol{e}' = \boldsymbol{e} + \frac{1}{c}\boldsymbol{v} \times \boldsymbol{b}$$

and

$$b' = b - \frac{1}{c} v \times e.$$

The decomposition of the electromagnetic field in an electric and magnetic field has no absolute significance. If there exists, for example, in a reference system $\mathcal{X}$ a purely electrostatic field with $b = o$, then, due to (1.143), a magnetic field $b' = \frac{\gamma}{c} v \times e \neq o$ exists in a reference system $\mathcal{X}'$ that moves with the speed v relative to the reference system $\mathcal{X}$. Physically this means that all charges rest in $\mathcal{X}$. But these charges move relative to $\mathcal{X}'$ with the velocity v. So there exists a current in $\mathcal{X}'$ which generates a magnetic field in $\mathcal{X}'$.

The equations (1.132) and (1.143) can be summarized in one equation:

$$\begin{pmatrix} b' \\ e' \end{pmatrix} = \underbrace{\left(\begin{array}{c|c} \gamma I + (1-\gamma)\frac{vv^\top}{v^2} & -\frac{\gamma}{c} V_\times \\ \hline \frac{\gamma}{c} V_\times & \gamma I + (1-\gamma)\frac{vv^\top}{v^2} \end{array}\right)}_{P(v)} \begin{pmatrix} b \\ e \end{pmatrix} \in \mathbb{R}^6, \quad (1.144)$$

with

$$V_\times \overset{\text{def}}{=} \begin{pmatrix} 0 & -v_3 & v_2 \\ v_3 & 0 & -v_1 \\ -v_2 & v_1 & 0 \end{pmatrix}.$$

The symmetric 6×6-matrix $P(v)$ occurring in (1.144) has a formal similarity with the Lorentz matrix L!

It is, of course, true that $P(-v) = P^{-1}(v)$, for, with

$$V_\times^2 = vv^\top - v^2 I, \quad (1.145)$$

one obtains easily $P(v)P(-v) = I$.

In the special case when only one component of the velocity vector in the x-direction is nonzero, i.e. $v = [v, 0, 0]^\top$, one obtains the matrix

$$P(v) = \left(\begin{array}{ccc|ccc} 1 & 0 & 0 & 0 & 0 & 0 \\ 0 & \gamma & 0 & 0 & 0 & \frac{\gamma v}{c} \\ 0 & 0 & \gamma & 0 & -\frac{\gamma v}{c} & 0 \\ \hline 0 & 0 & 0 & 1 & 0 & 0 \\ 0 & 0 & -\frac{\gamma v}{c} & 0 & \gamma & 0 \\ 0 & \frac{\gamma v}{c} & 0 & 0 & 0 & \gamma \end{array}\right). \quad (1.146)$$

This special symmetric 6×6-matrix can also be found in [33].

1.8.3 Electromagnetic Invariants

For the electromagnetic field quantities e and b one can form *invariants* which do not change during the transition to another inertial system. For the scalar product of the electric field strength $\boldsymbol{e}$ and the magnetic induction $\boldsymbol{b}$ we obtain

$$\begin{aligned}\boldsymbol{e}'^{\mathsf{T}}\boldsymbol{b}' &= \left(\gamma\left(\boldsymbol{e}^{\mathsf{T}} - \frac{1}{c}(\boldsymbol{v}\times\boldsymbol{b})^{\mathsf{T}}\right) + (1-\gamma)\frac{\boldsymbol{v}^{\mathsf{T}}\boldsymbol{e}}{v^2}\boldsymbol{v}^{\mathsf{T}}\right)\left(\gamma\left(\boldsymbol{b} + \frac{1}{c}\boldsymbol{v}\times\boldsymbol{e}\right)\right. \\ &\quad \left. + (1-\gamma)\frac{\boldsymbol{v}^{\mathsf{T}}\boldsymbol{b}}{v^2}\boldsymbol{v}\right) \\ &= \gamma^2\boldsymbol{e}^{\mathsf{T}}\boldsymbol{b} + \left[2\gamma(1-\gamma) + (1-\gamma^2)\right]\frac{\boldsymbol{e}^{\mathsf{T}}\boldsymbol{v}\boldsymbol{v}^{\mathsf{T}}\boldsymbol{b}}{v^2} - \frac{\gamma^2}{c^2}\underbrace{(\boldsymbol{v}\times\boldsymbol{b})^{\mathsf{T}}(\boldsymbol{v}\times\boldsymbol{e})}_{\boldsymbol{b}^{\mathsf{T}}\boldsymbol{V}_\times^{\mathsf{T}}\boldsymbol{V}_\times\boldsymbol{e}}.\end{aligned}$$

With $\boldsymbol{V}_\times^{\mathsf{T}} = -\boldsymbol{V}_\times$ one obtains due to (1.145)

$$\boldsymbol{V}_\times^{\mathsf{T}}\boldsymbol{V}_\times = -\boldsymbol{V}_\times^2 = v^2\boldsymbol{I} - \boldsymbol{v}\boldsymbol{v}^{\mathsf{T}}.$$

This gives

$$\underline{\underline{\boldsymbol{e}'^{\mathsf{T}}\boldsymbol{b}'}} = \gamma^2\boldsymbol{e}^{\mathsf{T}}\boldsymbol{b} - \frac{\gamma^2}{c^2}\left(\boldsymbol{v}^{\mathsf{T}}\boldsymbol{e}\boldsymbol{v}^{\mathsf{T}}\boldsymbol{b} + v^2\boldsymbol{b}^{\mathsf{T}}\boldsymbol{e} - \boldsymbol{b}^{\mathsf{T}}\boldsymbol{v}\boldsymbol{v}^{\mathsf{T}}\boldsymbol{e}\right) = \left(\gamma^2 - \frac{\gamma^2 v^2}{c^2}\right)\boldsymbol{e}^{\mathsf{T}}\boldsymbol{b} = \underline{\underline{\boldsymbol{e}^{\mathsf{T}}\boldsymbol{b}}},$$

so the scalar product of the electric field intensity $\boldsymbol{e}$ and the magnetic induction $\boldsymbol{b}$ is invariant with respect to a Lorentz transformation!

Using a slightly long calculation, one can show that the difference of the squares of the field intensities is also invariant:

$$\underline{\underline{b'^2 - e'^2 = b^2 - e^2}}. \tag{1.147}$$

You can also arrive at electromagnetic invariants in another way, namely with the help of the invariant Faraday's matrix

$$\boldsymbol{F}_{B,e} = \begin{pmatrix} 0 & \boldsymbol{e}^{\mathsf{T}} \\ -\boldsymbol{e} & \boldsymbol{B}_\times \end{pmatrix}$$

and the invariant Maxwell's matrix

$$\boldsymbol{F}_{E,b} = \begin{pmatrix} 0 & -\boldsymbol{b}^{\mathsf{T}} \\ \boldsymbol{b} & \boldsymbol{E}_\times \end{pmatrix}.$$

With

$$\boldsymbol{F}^*_{B,e} \overset{\text{def}}{=} \boldsymbol{M}\boldsymbol{F}_{B,e}\boldsymbol{M} = \begin{pmatrix} 0 & -\boldsymbol{e}^{\mathsf{T}} \\ \boldsymbol{e} & \boldsymbol{B}_\times \end{pmatrix},$$

one gets, e.g.

$$\boldsymbol{F}^*_{B,e}\boldsymbol{F}_{B,e} = \begin{pmatrix} 0 & -\boldsymbol{e}^{\mathsf{T}} \\ \boldsymbol{e} & \boldsymbol{B}_\times \end{pmatrix}\begin{pmatrix} 0 & \boldsymbol{e}^{\mathsf{T}} \\ -\boldsymbol{e} & \boldsymbol{B}_\times \end{pmatrix} = \begin{pmatrix} \boldsymbol{e}^{\mathsf{T}}\boldsymbol{e} & -\boldsymbol{e}^{\mathsf{T}}\boldsymbol{B}_\times \\ -\boldsymbol{B}_\times\boldsymbol{e} & \boldsymbol{e}\boldsymbol{e}^{\mathsf{T}} + \boldsymbol{B}_\times\boldsymbol{B}_\times \end{pmatrix}$$

$$= \begin{pmatrix} e^2 & s^{\mathsf{T}} \\ s & \begin{pmatrix} e_x^2 - b_z^2 - b_y^2 & \cdots & \cdots \\ \cdots & e_y^2 - b_z^2 - b_x^2 & \cdots \\ \cdots & \cdots & e_z^2 - b_y^2 - b_x^2 \end{pmatrix} \end{pmatrix},$$

with the Poynting vector $\boldsymbol{s} \stackrel{\text{def}}{=} \boldsymbol{e} \times \boldsymbol{b}$. By forming the trace of this matrix product, i.e. taking the sum of the matrix elements on the main diagonal, one obtains

$$\operatorname{trace}\left(\boldsymbol{F}_{B,e}^{*} \boldsymbol{F}_{B,e}\right) = 2e^2 - 2b^2,$$

i.e. the above invariant in the form

$$\boxed{-\frac{1}{2}\operatorname{trace}\left(\boldsymbol{F}_{B,e}^{*} \boldsymbol{F}_{B,e}\right) = b^2 - e^2.} \tag{1.148}$$

One obtains the second invariant $\boldsymbol{e}^{\mathsf{T}}\boldsymbol{b}$ above by taking the trace of the product of the modified Faraday's matrix $\boldsymbol{F}_{B,e}^{*}$ and the Maxwell's matrix $\boldsymbol{F}_{E,b}$:

$$\boxed{-\frac{1}{4}\operatorname{trace}\left(\boldsymbol{F}_{B,e}^{*} \boldsymbol{F}_{E,e}\right) = \boldsymbol{e}^{\mathsf{T}}\boldsymbol{b}.} \tag{1.149}$$

Making invariants of this type later plays a role in the consideration of the singularities of the Schwarzkopf's solution of General Relativity equations.

1.8.4 Electromagnetic Forces

We want to determine the force acting on a charged particle with the charge q, which is in an electromagnetic field moving with the velocity $\boldsymbol{u}$ relative to an inertial frame $\mathcal{X}$. Let $\mathcal{X}'$ be the inertial frame in which the particle rests at the moment. In this system, due to $\boldsymbol{u}' = \boldsymbol{o}$ and $\boldsymbol{u}' \times \boldsymbol{b}' = \boldsymbol{o}$,

$$m_0 \frac{\mathrm{d}\boldsymbol{u}'}{\mathrm{d}t'} = q\boldsymbol{e}' \in \mathbb{R}^3. \tag{1.150}$$

Generally, due to the relativity principle, the Lorentz force in $\mathcal{X}$ is

$$\boldsymbol{f} = q\left(\boldsymbol{e} + \frac{1}{c}\boldsymbol{u} \times \boldsymbol{b}\right) = \frac{q}{c}[\boldsymbol{e} \mid -\boldsymbol{B}_{\times}]\begin{pmatrix} c \\ \boldsymbol{u} \end{pmatrix} \in \mathbb{R}^3. \tag{1.151}$$

The following law expresses how an electromagnetic field acts on a stationary charge q and current $\boldsymbol{j}$:

$$\boldsymbol{f} = q\boldsymbol{e} + \frac{1}{c}\boldsymbol{j} \times \boldsymbol{b} = \frac{1}{c}[\boldsymbol{e} \mid -\boldsymbol{B}_{\times}]\begin{pmatrix} cq \\ \boldsymbol{j} \end{pmatrix} \in \mathbb{R}^3. \tag{1.152}$$

f is completed to a four-vector $\vec{f}$ as in (1.108)

$$\vec{f} \stackrel{\text{def}}{=} \gamma_u \begin{pmatrix} \boldsymbol{u}^\top \boldsymbol{f}/c \\ \boldsymbol{f} \end{pmatrix} \in \mathbb{R}^4. \tag{1.153}$$

Equation (1.151) applied to (1.153) yields (with $\boldsymbol{u}^\top(\boldsymbol{u} \times \boldsymbol{b}) = 0$)

$$\vec{f} = q\gamma_u \begin{pmatrix} \boldsymbol{u}^\top \boldsymbol{e}/c \\ \boldsymbol{e} + \frac{1}{c}\boldsymbol{u} \times \boldsymbol{b} \end{pmatrix} = q\frac{\gamma_u}{c} \underbrace{\begin{pmatrix} 0 & \boldsymbol{e}^\top \\ \boldsymbol{e} & -\boldsymbol{B}_\times \end{pmatrix}}_{\boldsymbol{MF}_{B,e}} \begin{pmatrix} c \\ \boldsymbol{u} \end{pmatrix}$$

$$= \frac{q}{c} \boldsymbol{MF}_{B,e} \underbrace{\gamma_u \begin{pmatrix} c \\ \boldsymbol{u} \end{pmatrix}}_{\vec{u}}, \tag{1.154}$$

so

$$\boxed{\vec{f} = \frac{q}{c} \boldsymbol{MF}_{B,e} \vec{u}.} \tag{1.155}$$

With (1.152) one can write instead of (1.155)

$$\boxed{\vec{f} = \frac{1}{c} \boldsymbol{F}^*_{B,e} \vec{j},} \tag{1.156}$$

in which again

$$\vec{j} = \gamma \begin{pmatrix} cq \\ -\boldsymbol{j} \end{pmatrix} \in \mathbb{R}^4.$$

Subjecting (1.155) to a linear transformation using the Lorentz matrix $\boldsymbol{L}(\boldsymbol{v})$, one obtains

$$\vec{f}' = \boldsymbol{L}(\boldsymbol{v})\vec{f} = \frac{q}{c} \underbrace{\boldsymbol{L}(\boldsymbol{v}) \boldsymbol{MF}_{B,e} \overbrace{\boldsymbol{L}^{-1}(\boldsymbol{v})}^{\boldsymbol{L}(-\boldsymbol{v})}}_{\boldsymbol{MF}'_{B',e'}} \underbrace{\boldsymbol{L}(\boldsymbol{v})\vec{u}}_{\vec{u}'}, \tag{1.157}$$

i.e.

$$\boxed{\vec{f}' = \frac{q}{c} \boldsymbol{MF}'_{B',e'} \vec{u}'.} \tag{1.158}$$

The power equation is also invariant with respect to a Lorentz transformation!

1.9 The Energy–Momentum Matrix

1.9.1 The Electromagnetic Energy–Momentum Matrix

We want to derive one equation which contains the fundamental dynamic equations of the theory of electricity. This equation should also include the energy theorem and the momentum theorems of electrodynamics. We will find again the therein contained energy–momentum matrix in the main equation of the theory of general relativity.

Due to (1.128), one has

$$\boldsymbol{F}_{B,e}\vec{\nabla} = \frac{1}{c}\vec{j}$$

and because of (1.156)

$$\vec{f} = \frac{1}{c}\boldsymbol{F}^*_{B,e}\vec{j}.$$

Is there a matrix such that

$$\vec{f} = \boldsymbol{T}_{b,e}\vec{\nabla}? \tag{1.159}$$

We try $\boldsymbol{T}_{b,e} = \boldsymbol{F}^*_{B,e}\boldsymbol{F}_{B,e}$. Then

$$\boldsymbol{T}_{b,e}\vec{\nabla} = (\boldsymbol{F}^*_{B,e}\boldsymbol{F}_{B,e})\vec{\nabla} = \begin{pmatrix} \boldsymbol{e}^\top\boldsymbol{e} & -\boldsymbol{e}^\top\boldsymbol{B}_\times \\ -\boldsymbol{B}_\times\boldsymbol{e} & \boldsymbol{e}\boldsymbol{e}^\top + \boldsymbol{B}_\times^2 \end{pmatrix}\vec{\nabla}$$

$$= \begin{pmatrix} e^2 & \boldsymbol{s}^\top \\ \boldsymbol{s} & \boldsymbol{e}\boldsymbol{e}^\top + \boldsymbol{b}\boldsymbol{b}^\top - b^2\boldsymbol{I}_3 \end{pmatrix}\vec{\nabla} \overset{!}{=} \vec{f}, \tag{1.160}$$

by considering the Poynting vector

$$\boldsymbol{s} \overset{\text{def}}{=} \boldsymbol{e}\times\boldsymbol{b} \tag{1.161}$$

and the relation $\boldsymbol{B}_\times^2 = \boldsymbol{b}\boldsymbol{b}^\top - b^2\boldsymbol{I}$ in conformance with (1.145). The Poynting vector gives the magnitude and direction of the energy transport in electromagnetic fields. What is the result when differentiating the first line of (1.160)? It is

$$-\frac{1}{c}\frac{\partial(e^2)}{\partial t} + \nabla^\top\boldsymbol{s} \overset{!}{=} \frac{\rho\boldsymbol{u}^\top\boldsymbol{f}}{c}.$$

$\frac{1}{c}\frac{\partial(e^2)}{\partial t}$ is proportional to the temporal change of the energy density of the electromagnetic field when no magnetic field is present. But then also $\boldsymbol{b} = \boldsymbol{o}$, i.e. $\boldsymbol{s} = \boldsymbol{o}$, and thus the whole equation is without a statement. The matter would be different if instead of e^2 in top left corner of the matrix $\boldsymbol{T}_{b,e}$, one would have the expression $(e^2 + b^2)/2$, as then

$$\frac{1}{2c}\frac{\partial(e^2+b^2)}{\partial t} = \frac{1}{c}\frac{\partial w}{\partial t},$$

namely, the temporal change of the energy density $w = (e^2 + b^2)/2$ of the electromagnetic field. By adding $(b^2 - e^2)/2$ to the upper left, there arises, in fact, $(e^2 + b^2)/2$! Therefore, now the

Definition: The *electromagnetic energy–momentum matrix* $\boldsymbol{T}_{b,e}$ has the form

$$\boldsymbol{T}_{b,e} \stackrel{\text{def}}{=} \boldsymbol{F}^*_{B,e}\boldsymbol{F}_{B,e} + \frac{1}{2}(b^2 - e^2)\boldsymbol{I}_4 = \begin{pmatrix} w & \boldsymbol{s}^\top \\ \boldsymbol{s} & (\boldsymbol{e}\boldsymbol{e}^\top + \boldsymbol{b}\boldsymbol{b}^\top - w\boldsymbol{I}_3) \end{pmatrix}, \tag{1.162}$$

with

$$w \stackrel{\text{def}}{=} \frac{1}{2}(e^2 + b^2) \tag{1.163}$$

and

$$\boldsymbol{T}_{b,e} \cdot \vec{\boldsymbol{\nabla}} = \frac{1}{\gamma_v}\vec{\boldsymbol{f}}. \tag{1.164}$$

While doing this, $\vec{\boldsymbol{\nabla}}$ again has the form

$$\vec{\boldsymbol{\nabla}} = \gamma_u \begin{pmatrix} -\frac{1}{c}\frac{\partial}{\partial t} \\ \boldsymbol{\nabla} \end{pmatrix}$$

and

$$\vec{\boldsymbol{f}} = \gamma_u \begin{pmatrix} \frac{\rho}{c}\boldsymbol{u}^\top \boldsymbol{f} \\ \boldsymbol{f} \end{pmatrix}.$$

For the first row of (1.164) now one obtains

$$-\frac{1}{c}\frac{\partial w}{\partial t} + \boldsymbol{s}^\top \boldsymbol{\nabla} = \frac{\rho}{c}\boldsymbol{u}^\top \boldsymbol{f},$$

i.e.

$$c \operatorname{div} \boldsymbol{s} = \frac{\partial w}{\partial t} + \rho \boldsymbol{u}^\top \boldsymbol{f}. \tag{1.165}$$

On the left is the infinitely small volume unit of the entering or exiting energy flow, and it consists of the temporal change of the energy density $\frac{\partial w}{\partial t}$ and the conversion of the electromagnetic energy into mechanical energy per unit time and volume $\rho \boldsymbol{u}^\top \boldsymbol{f}$. So the whole is the *Energy Theorem of Electrodynamics*.

Next we get for the second to fourth components of (1.164):

$$-\frac{1}{c}\frac{\partial \boldsymbol{s}}{\partial t} + (\boldsymbol{e}\boldsymbol{e}^\top)\boldsymbol{\nabla} + (\boldsymbol{b}\boldsymbol{b}^\top)\boldsymbol{\nabla} - (w\boldsymbol{I}_3)\boldsymbol{\nabla} = \boldsymbol{f}. \tag{1.166}$$

With the help of Maxwell's equations, we obtain for the first term on the left-hand side:

$$-\frac{1}{c}\frac{\partial \boldsymbol{s}}{\partial t} = -\frac{1}{c}\frac{\partial \boldsymbol{e} \times \boldsymbol{b}}{\partial t} = -\frac{1}{c}\frac{\partial \boldsymbol{e}}{\partial t} \times \boldsymbol{b} + -\frac{1}{c}\boldsymbol{e} \times \frac{\partial \boldsymbol{b}}{\partial t}$$

$$= \left(\mathbf{curl}\,\boldsymbol{b} - \frac{1}{c}\boldsymbol{j}\right) \times \boldsymbol{b} + \boldsymbol{e} \times (-\mathbf{curl}\,\boldsymbol{e})$$

$$= -\frac{1}{c}\boldsymbol{j} \times \boldsymbol{b} - \boldsymbol{b} \times \mathbf{curl}\,\boldsymbol{b} - \boldsymbol{e} \times \mathbf{curl}\,\boldsymbol{e}. \tag{1.167}$$

For the first component, the x-component of the three-dimensional vector $(\boldsymbol{e}\boldsymbol{e}^\top)\nabla - \frac{1}{c}(e^2\boldsymbol{I}_3)\nabla$ on the left-hand side, one obtains

$$\left[e_x^2 - \frac{1}{2}e^2 | e_x e_y | e_x e_z\right]\nabla$$

$$= 2e_x\frac{\partial e_x}{\partial x} - \left(e_x\frac{\partial e_x}{\partial x} + e_y\frac{\partial e_y}{\partial y} + e_z\frac{\partial e_z}{\partial z}\right) + \frac{\partial e_x}{\partial y}e_y + e_x\frac{\partial e_y}{\partial y} + \frac{\partial e_x}{\partial z}e_z + e_x\frac{\partial e_z}{\partial z}$$

$$= e_x\left(\frac{\partial e_x}{\partial x} + \frac{\partial e_y}{\partial y} + \frac{\partial e_z}{\partial z}\right) + e_z\frac{\partial e_x}{\partial z} + e_y\frac{\partial e_x}{\partial y} - e_z\frac{\partial e_z}{\partial x} - e_y\frac{\partial e_y}{\partial x}$$

$$= e_x \operatorname{div}\boldsymbol{e} - (\boldsymbol{e} \times \mathbf{curl}\,\boldsymbol{e})_x = e_x \cdot \rho - (\boldsymbol{e} \times \mathbf{curl}\,\boldsymbol{e})_x. \tag{1.168}$$

Accordingly, one receives for the y- and z-component together

$$(\boldsymbol{e}\boldsymbol{e}^\top)\nabla - \frac{1}{2}(e^2\boldsymbol{I}_3)\nabla = \boldsymbol{e} \cdot \rho - \boldsymbol{e} \times \mathbf{curl}\,\boldsymbol{e}. \tag{1.169}$$

Also we obtain with $\operatorname{div}\boldsymbol{b} = 0$:

$$(\boldsymbol{b}\boldsymbol{b}^\top)\nabla - \frac{1}{2}(b^2\boldsymbol{I}_3)\nabla = -\boldsymbol{b} \times \mathbf{curl}\,\boldsymbol{b}. \tag{1.170}$$

Overall, (1.165), (1.168) and (1.170) result for the second to fourth rows of (1.164) in

$$\boldsymbol{e} \cdot \rho + \frac{1}{c}\boldsymbol{j} \times \boldsymbol{b} = \boldsymbol{f}. \tag{1.171}$$

But this is the *Momentum Theorem* of Lorentz! Thus the above assertion (1.164) is completely proved.

1.9.2 The Mechanical Energy–Momentum Matrix

A particle at rest at the point $\boldsymbol{x}_0$ has the speed

$$\vec{\boldsymbol{u}} = \frac{\mathrm{d}\vec{x}}{\mathrm{d}\tau} = \frac{\mathrm{d}}{\mathrm{d}\tau}\begin{pmatrix} c\tau \\ \boldsymbol{x}_0 \end{pmatrix} = \begin{pmatrix} c \\ \boldsymbol{o} \end{pmatrix}$$

and the momentum $\vec{\boldsymbol{p}} = m_0\binom{c}{\boldsymbol{o}}$. The zeroth component of $\vec{\boldsymbol{p}}$ is the rest energy of the particle divided by c. For a moving particle, with $E = \gamma_u m_0 c^2$, one gets

$$\vec{\boldsymbol{p}} = \gamma_v m_0 \begin{pmatrix} c \\ \boldsymbol{v} \end{pmatrix} = \begin{pmatrix} E/c \\ \gamma_v \boldsymbol{p} \end{pmatrix}.$$

This equation expresses the fact that in the theory of relativity energy and momentum are the temporal and spatial components of the four-vector $\vec{p}$. They preserve this distinction even after a Lorentz transformation just like the four-vector $\vec{x} = \binom{ct}{x}$: its zeroth component is always the time component and the rest represents the space components.

We now go over to a distributed matter, such as that in a perfect fluid, i.e. a liquid without internal friction, but of quite variable density. It is described by the two scalar fields, density ρ and pressure p, and the velocity vector field $\vec{u}$. The aim of this derivation of the energy–momentum matrix is that this matrix somehow represents the energy content of the liquid and the transition to the curved world of the theory of general relativity, which can serve as a source of the field of gravity. The *continuity equation* describes the conservation of mass. The conservation of mass requires that the change of mass $\mathrm{d}V\delta\rho/\delta t$ in the unit of time, associated with the local compression $\delta\rho/\delta t$, must be equal to the difference between the entering and exiting masses per unit of time. In the x-direction, for this difference the following is valid:

$$\rho u_x(x)\cdot \mathrm{d}y\cdot \mathrm{d}z - \left(\rho u_x + \frac{\partial \rho u_x}{\partial x}\mathrm{d}x\right)\mathrm{d}y\,\mathrm{d}z = -\frac{\partial \rho u_x}{\partial x}\mathrm{d}V.$$

One gets similar expressions for the y- and z-direction, making a total for the conservation of mass

$$\frac{\partial \rho}{\partial t}\mathrm{d}V = -\left(\frac{\partial \rho u_x}{\partial x} + \frac{\partial \rho u_y}{\partial y} + \frac{\partial \rho u_z}{\partial z}\right)\mathrm{d}V.$$

So this is the final continuity equation in differential form:

$$\frac{\partial \rho}{\partial t} + \frac{\partial \rho u_x}{\partial x} + \frac{\partial \rho u_y}{\partial y} + \frac{\partial \rho u_z}{\partial z} = \underline{\underline{\frac{\partial \rho}{\partial t} + \operatorname{div}(\rho \boldsymbol{u}) = 0.}} \tag{1.172}$$

The *dynamic* behaviour is described by the Euler's equation. By using Newton's law on the mass contained in a volume element of a perfect fluid, we obtain the Euler equation of motion, initially only in the x-direction:

$$\mathrm{d}m\frac{\mathrm{d}u_x}{\mathrm{d}t} = \rho\,\mathrm{d}x\,\mathrm{d}y\,\mathrm{d}z\frac{\mathrm{d}u_x}{\mathrm{d}t} = \mathrm{d}x\,\mathrm{d}y\,\mathrm{d}z\, f_{D,x} - \left(\frac{\partial p}{\partial x}\mathrm{d}x\right)\mathrm{d}y\,\mathrm{d}z,$$

which implies

$$\rho\frac{\mathrm{d}u_x}{\mathrm{d}t} = f_{D,x} - \frac{\partial p}{\partial x}, \tag{1.173}$$

where $f_{D,x}$ is the x-component of the force per unit volume (power density), $\boldsymbol{f}_D$, e.g. the gravitational force. Using the total differential of Δu_x,

$$\Delta u_x = \frac{\partial u_x}{\partial t}\Delta t + \frac{\partial u_x}{\partial x}\Delta x + \frac{\partial u_x}{\partial y}\Delta y + \frac{\partial u_x}{\partial z}\Delta z,$$

dividing by Δt and passing to the limit $\Delta t \to 0$, we obtain

$$\frac{\mathrm{d}u_x}{\mathrm{d}t} = \frac{\partial u_x}{\partial t} + \frac{\partial u_x}{\partial x}u_x + \frac{\partial u_x}{\partial y}u_y + \frac{\partial u_x}{\partial z}u_z. \tag{1.174}$$

With the corresponding equations for the y- and z-direction, we obtain in total

$$\rho\left(\frac{\partial u_x}{\partial t} + \frac{\partial u_x}{\partial x}u_x + \frac{\partial u_x}{\partial y}u_y + \frac{\partial x}{\partial z}u_z\right) = f_{D,x} - \frac{\partial p}{\partial x},$$

$$\rho\left(\frac{\partial u_y}{\partial t} + \frac{\partial u_y}{\partial x}u_x + \frac{\partial u_y}{\partial y}u_y + \frac{\partial u_y}{\partial z}u_z\right) = f_{D,y} - \frac{\partial p}{\partial y},$$

$$\rho\left(\frac{\partial u_z}{\partial t} + \frac{\partial u_z}{\partial x}u_x + \frac{\partial u_z}{\partial y}u_y + \frac{\partial u_z}{\partial z}u_z\right) = f_{D,z} - \frac{\partial p}{\partial z},$$

summarized in

$$\underline{\underline{\rho\left(\frac{\partial \boldsymbol{u}}{\partial t} + \frac{\partial \boldsymbol{u}}{\partial \boldsymbol{x}^{\mathsf{T}}}\boldsymbol{u}\right) + \mathbf{grad}\, p = \boldsymbol{f}_D.}} \tag{1.175}$$

This is the Euler's equation in modern form. The first term in the parentheses is called the local and the second the convective change.

The relativistic generalization of the hydrodynamic equations (1.172) and (1.175) will now be established. ρ_0 is the *rest density*, defined as the rest mass per rest volume. With $\vec{\boldsymbol{x}}^{\mathsf{T}} = [ct|\boldsymbol{x}^{\mathsf{T}}]$ and $\vec{\boldsymbol{u}}^{\mathsf{T}} = \gamma_u[c|\boldsymbol{u}^{\mathsf{T}}]$, the Euler's equation (1.175) can also be written as

$$\rho\frac{\partial \boldsymbol{u}}{\partial \vec{\boldsymbol{x}}^{\mathsf{T}}}\vec{\boldsymbol{u}} = \boldsymbol{f}_D - \mathbf{grad}\, p. \tag{1.176}$$

With regard to the subsequent application of the operator $\vec{\nabla}$, we now start for the energy–momentum matrix with

$$\boldsymbol{T}_{\text{mech},1} \stackrel{\text{def}}{=} \rho_0\vec{\boldsymbol{u}}\vec{\boldsymbol{u}}^{\mathsf{T}}. \tag{1.177}$$

This matrix is symmetric and built up from the two values ρ_0 and $\vec{\boldsymbol{u}}$, which describe completely the dynamics of a perfect fluid with the pressure p and the acting external forces $\boldsymbol{f}_D$. For the use of the operator $\vec{\nabla}$ and the further investigation of the result, it is advantageous to first divide the matrix $\boldsymbol{T}_{\text{mech},1}$ similarly as the matrix $\boldsymbol{T}_{b,e}$:

$$\boldsymbol{T}_{\text{mech},1} = \rho_0\gamma_u^2\begin{pmatrix} c^2 & c\boldsymbol{u}^{\mathsf{T}} \\ c\boldsymbol{u} & \boldsymbol{u}\boldsymbol{u}^{\mathsf{T}} \end{pmatrix}. \tag{1.178}$$

In a liquid moving with the velocity $\boldsymbol{u}$, the volume decreases with γ_u while the mass increases with γ_u, thus one obtains the total density $\rho = \gamma_u^2\rho_0$. With this we now define

$$\boldsymbol{T}_{\text{mech},1} \stackrel{\text{def}}{=} \begin{pmatrix} \rho c^2 & \rho c \boldsymbol{u}^\mathsf{T} \\ \rho c \boldsymbol{u} & \rho \boldsymbol{u}\boldsymbol{u}^\mathsf{T} \end{pmatrix}. \tag{1.179}$$

Multiplying the first line in (1.179) from the right with the operator $\vec{\nabla}$ yields

$$-c\frac{\partial \rho}{\partial t} + c(\rho \boldsymbol{u}^\mathsf{T})\nabla.$$

By setting this expression equal to zero, we obtain with $(\rho \boldsymbol{u}^\mathsf{T})\nabla = \operatorname{div}(\rho \boldsymbol{u})$ the classical continuity equation (1.172)!

Multiplying now the second row of the matrix in (1.179) from the right with the operator $\vec{\nabla}$, one obtains

$$\begin{aligned} -\frac{\partial \rho \boldsymbol{u}}{\partial t} + (\rho \boldsymbol{u}\boldsymbol{u}^\mathsf{T})\nabla &= -\rho\frac{\partial \boldsymbol{u}}{\partial t} - \frac{\partial \rho}{\partial t}\boldsymbol{u} + \rho\frac{\partial \boldsymbol{u}}{\partial \boldsymbol{x}^\mathsf{T}}\boldsymbol{u} + \boldsymbol{u}\operatorname{div}(\rho \boldsymbol{u}) \\ &= \left(-\frac{\partial \rho}{\partial t}\boldsymbol{u} + \operatorname{div}(\rho \boldsymbol{u})\right)\boldsymbol{u} + \rho\left(-\frac{\partial \boldsymbol{u}}{\partial t} + \frac{\partial \boldsymbol{u}}{\partial \boldsymbol{x}^\mathsf{T}}\boldsymbol{u}\right). \end{aligned} \tag{1.180}$$

The term inside the first parentheses is equal to zero, due to the continuity equation in the non-relativistic case, and the term in the second parentheses contains the force- and pressure-free Euler's equation!

The pressure p must now be incorporated. Presuppose an isotropic liquid, then the pressure p is direction-independent. In the Euler's equation, the pressure appears in the form of **grad** p which can be written as

$$\textbf{grad}\, p = \begin{pmatrix} p & 0 & 0 \\ 0 & p & 0 \\ 0 & 0 & p \end{pmatrix} \nabla. \tag{1.181}$$

If this is taken into account in the matrix $\boldsymbol{T}_{\text{mech}}$, it must be remembered that (1.181) holds for a reference system $\mathcal{X}'$ moving with the liquid; therefore, this approach makes sense:

$$\boldsymbol{T}'_{\text{mech},2} \stackrel{\text{def}}{=} \begin{pmatrix} 0 & 0 & 0 & 0 \\ 0 & p & 0 & 0 \\ 0 & 0 & p & 0 \\ 0 & 0 & 0 & p \end{pmatrix}. \tag{1.182}$$

This matrix can now with the help of the Lorentz matrix

$$\boldsymbol{L}(-\boldsymbol{u}) = \begin{pmatrix} \gamma_u & \frac{\gamma_u}{c}\boldsymbol{u}^\mathsf{T} \\ \frac{\gamma_u}{c}\boldsymbol{u} & \boldsymbol{I}_3 + (\gamma_u - 1)\frac{\boldsymbol{u}\boldsymbol{u}^\mathsf{T}}{u^2} \end{pmatrix}$$

be transformed back into the resting inertial system $\mathcal{X}$:

$$\boldsymbol{T}_{\text{mech},2} = \boldsymbol{L}(-\boldsymbol{u})\boldsymbol{T}'_{\text{mech},2}\boldsymbol{L}^\mathsf{T}(-\boldsymbol{u}) = p\begin{pmatrix} \frac{\gamma_u^2 u^2}{c^2} & \frac{\gamma_u^2}{c}\boldsymbol{u}^\mathsf{T} \\ \frac{\gamma_u^2}{c}\boldsymbol{u} & \boldsymbol{I}_3 + \frac{\gamma_u^2}{c^2}\boldsymbol{u}\boldsymbol{u}^\mathsf{T} \end{pmatrix}.$$

Considering that $p\gamma^2 u^2/c^2 = p(\gamma^2 - 1)$, this can also be written as

$$\boldsymbol{T}_{\text{mech},2} = \frac{p}{c^2}\vec{\boldsymbol{u}}\vec{\boldsymbol{u}}^\top + p\begin{pmatrix} -1 & \boldsymbol{o}^\top \\ \boldsymbol{o} & \boldsymbol{I}_3 \end{pmatrix}.$$

For the sum of the two matrices $\boldsymbol{T}_{\text{mech},1}$ and $\boldsymbol{T}_{\text{mech},2}$, we finally obtain with the Minkowski matrix $\boldsymbol{M}$

$$\boldsymbol{T}_{\text{mech}} \stackrel{\text{def}}{=} \left(\rho_0 + \frac{p}{c^2}\right)\vec{\boldsymbol{u}}\vec{\boldsymbol{u}}^\top - p\boldsymbol{M}. \tag{1.183}$$

We now summarize everything together to the relativistic generalization of the hydrodynamic equations

$$\boxed{\boldsymbol{T}_{\text{mech}}\vec{\nabla} = \vec{\boldsymbol{f}}_D.} \tag{1.184}$$

By the way, one obtains, when the energy–momentum matrix $\boldsymbol{T}_{\text{mech}}$ from the right is multiplied with $\boldsymbol{M}\vec{\boldsymbol{u}}$, the four-vector momentum density $\rho_0\vec{\boldsymbol{u}}$ multiplied by c^2:

$$\boldsymbol{T}_{\text{mech}}\boldsymbol{M}\vec{\boldsymbol{u}} = \left(\rho_0 + \frac{p}{c^2}\right)\vec{\boldsymbol{u}}\underbrace{\gamma^2(c^2 - v^2)}_{c^2} - p\vec{\boldsymbol{u}} = c^2\rho_0\vec{\boldsymbol{u}}. \tag{1.185}$$

1.9.3 The Total Energy–Momentum Matrix

The derived energy–momentum matrices include the conservation laws of energy and momentum of a *closed* system. For example, if a force density $\boldsymbol{f}_D$ acts from the outside on the fluid, so that an electromagnetic field acts on the electrically charged liquid, then

$$\boldsymbol{f}_D = -\boldsymbol{T}_{b,e}\vec{\nabla} = \boldsymbol{T}_{\text{mech}}\vec{\nabla}, \tag{1.186}$$

or combined,

$$\boldsymbol{T}_{\text{total}} \stackrel{\text{def}}{=} (\boldsymbol{T}_{\text{mech}} + \boldsymbol{T}_{b,e}), \tag{1.187}$$

so

$$\boxed{\boldsymbol{T}_{\text{total}}\vec{\nabla} = \boldsymbol{f}_{\text{total}}.} \tag{1.188}$$

The conservation laws now apply to the whole system, which is fluid plus the electromagnetic field. Since the individual matrices are symmetric, the total energy–momentum matrix $\boldsymbol{T}_{\text{total}}$ is symmetric. If there are other components in the considered system, they can be incorporated in the symmetric total energy–momentum matrix $\boldsymbol{T}_{\text{total}}$ in a similar way as described above, and one obtains (1.188). This form of mathematical representation of the dynamic behaviour of physical systems will play a major role in the main equations of Einstein's theory of General Relativity!

1.10 The Most Important Definitions and Formulas in Special Relativity

For inertial systems, i.e. reference systems that are uniformly moving against each other, the fundamental physical laws are linked with the Lorentz transformation and are invariant. In the Theory of Special Relativity, we derived (see (1.20))

$$\gamma \overset{\text{def}}{=} \frac{1}{\sqrt{1-\frac{v^2}{c^2}}}$$

and the Lorentz transformation matrix (see (1.38))

$$\boldsymbol{L}(\boldsymbol{v}) \overset{\text{def}}{=} \left(\begin{array}{c|c} \gamma & -\frac{\gamma}{c}\boldsymbol{v}^\top \\ \hline -\frac{\gamma}{c}\boldsymbol{v} & \boldsymbol{I} + (\gamma - 1)\frac{\boldsymbol{v}\boldsymbol{v}^\top}{v^2} \end{array}\right)$$

with the following (see (1.39)) equations:

$$ct' = \gamma ct - \frac{\gamma}{c}\boldsymbol{v}^\top \boldsymbol{x}, \qquad \boldsymbol{x}' = \boldsymbol{x} + (\gamma - 1)\frac{\boldsymbol{v}^\top \boldsymbol{x}}{v^2}\boldsymbol{v} - \gamma \boldsymbol{v}t.$$

In (1.65), we derived the relativistic velocity addition:

$$\boldsymbol{w} = \frac{\boldsymbol{v} + \boldsymbol{u} + (\frac{1}{\gamma_v} - 1)(\boldsymbol{u} - \frac{\boldsymbol{v}^\top \boldsymbol{u}}{v^2}\boldsymbol{v})}{1 + \frac{\boldsymbol{v}^\top \boldsymbol{u}}{c^2}}.$$

If the two velocities $\boldsymbol{v}$ and $\boldsymbol{u}$ are parallel, then (see (1.68))

$$\boldsymbol{w} = \frac{\boldsymbol{v} + \boldsymbol{u}}{1 + \frac{\boldsymbol{v}^\top \boldsymbol{u}}{c^2}}.$$

The modified velocity vector in (1.79)

$$\boldsymbol{u} \overset{\text{def}}{=} \frac{\mathrm{d}\boldsymbol{x}}{\mathrm{d}t} \in \mathbb{R}^3, \qquad \gamma_u \overset{\text{def}}{=} \frac{1}{\sqrt{1-\frac{u^2}{c^2}}}, \qquad \vec{\boldsymbol{u}} \overset{\text{def}}{=} \gamma_u \begin{pmatrix} c \\ \boldsymbol{u} \end{pmatrix} \in \mathbb{R}^4$$

is transformed by a Lorentz transformation $\boldsymbol{L}$ into the velocity vector $\vec{\boldsymbol{u}}'$ (see (1.80))

$$\vec{\boldsymbol{u}}' = \boldsymbol{L}\vec{\boldsymbol{u}}.$$

With (1.95),

$$\vec{\boldsymbol{a}} \overset{\text{def}}{=} \gamma_u \cdot \frac{\mathrm{d}}{\mathrm{d}t}\vec{\boldsymbol{u}} \in \mathbb{R}^4$$

is the Lorentz transformation of the acceleration vector $\vec{\boldsymbol{a}}$ (see (1.96)):

$$\boxed{\vec{\boldsymbol{a}}' = \boldsymbol{L}\vec{\boldsymbol{a}}.}$$

Einstein's famous formula (1.111) reads

$$\boxed{E_0 = m_0 c^2.}$$

The invariance of the fundamental equation of mechanics (m_0 is the rest mass) is documented in

$$\boxed{m_o\vec{\boldsymbol{a}} = \vec{\boldsymbol{f}}} \quad \boldsymbol{L} \Rightarrow \; \Leftarrow \boldsymbol{L}^{-1} \quad \boxed{m_o\vec{\boldsymbol{a}}' = \vec{\boldsymbol{f}}'}$$

and in electrodynamics

$$\boxed{\begin{aligned} \boldsymbol{F}_{B,e}\vec{\nabla} &= \tfrac{1}{c}\vec{\boldsymbol{j}} \\ \boldsymbol{F}_{E,b}\vec{\nabla} &= \boldsymbol{o} \end{aligned}} \quad \boldsymbol{L} \Rightarrow \; \Leftarrow \boldsymbol{L}^{-1} \quad \boxed{\begin{aligned} \boldsymbol{F}'_{B',e'}\vec{\nabla}' &= \tfrac{1}{c}\vec{\boldsymbol{j}}' \\ \boldsymbol{F}'_{E',b'}\vec{\nabla}' &= \boldsymbol{o} \end{aligned}}$$

and

$$\boxed{\vec{\boldsymbol{f}} = \frac{q}{c}\boldsymbol{M}\boldsymbol{F}_{B,e}\vec{\boldsymbol{u}}} \quad \boldsymbol{L} \Rightarrow \; \Leftarrow \boldsymbol{L}^{-1} \quad \boxed{\vec{\boldsymbol{f}}' = \frac{q}{c}\boldsymbol{M}\boldsymbol{F}'_{B',e'}\vec{\boldsymbol{u}}'}$$

with

$$\boxed{\boldsymbol{F}_{B,e} \stackrel{\text{def}}{=} \begin{pmatrix} 0 & \boldsymbol{e}^{\mathsf{T}} \\ -\boldsymbol{e} & \boldsymbol{B}_{\times} \end{pmatrix}, \; \boldsymbol{F}_{E,b} \stackrel{\text{def}}{=} \begin{pmatrix} 0 & -\boldsymbol{b}^{\mathsf{T}} \\ \boldsymbol{b} & \boldsymbol{E}_{\times} \end{pmatrix} \text{ and } \vec{\nabla} \stackrel{\text{def}}{=} \begin{pmatrix} -\frac{1}{c}\frac{\partial}{\partial t} \\ \nabla \end{pmatrix}.}$$

With the electromagnetic energy–momentum matrix

$$\boxed{\boldsymbol{T}_{b,e} \stackrel{\text{def}}{=} \begin{pmatrix} w & \boldsymbol{s}^{\mathsf{T}} \\ \boldsymbol{s} & (\boldsymbol{e}\boldsymbol{e}^{\mathsf{T}} + \boldsymbol{b}\boldsymbol{b}^{\mathsf{T}} - w\boldsymbol{I}_3) \end{pmatrix}}$$

where

$$w \stackrel{\text{def}}{=} 1/2(e^2 + b^2),$$

one gets

$$\boxed{\boldsymbol{T}_{b,e} \cdot \vec{\nabla} = \frac{1}{\gamma_v}\vec{\boldsymbol{f}},}$$

and with the mechanical energy–momentum matrix

$$\boxed{\boldsymbol{T}_{\text{mech}} \stackrel{\text{def}}{=} \left(\rho_0 + \frac{p}{c^2}\right)\vec{\boldsymbol{u}}\vec{\boldsymbol{u}}^{\mathsf{T}} - p\boldsymbol{M},}$$

we obtain the relativistic generalization of the hydrodynamic equations

$$\boxed{\boldsymbol{T}_{\text{mech}}\vec{\nabla} = \vec{\boldsymbol{f}}_D.}$$

Remark The operator $\vec{\nabla}$ was somewhat unusual in the above formulas—it was written to the right to the approaching object, e.g. in $\boldsymbol{T}_{\text{mech}}\vec{\nabla}$. This was done so that *column* vectors appear on both sides of the equation. The usual order would be if one transposes the eligible equations. Then there would be *row* vectors on the left- and on the right-hand side, and the operators, though provided with a transpose sign, would have their usual place as, e.g. in

$$\vec{\nabla}^{\mathsf{T}}\boldsymbol{T}_{\text{mech}} = \vec{\boldsymbol{f}}_D^{\mathsf{T}}.$$

(Since the matrix $\boldsymbol{T}_{\text{mech}}$ is symmetric, it need not be transposed.)

Chapter 2
Theory of General Relativity

The chapter begins with the introduction of the metric matrix $\boldsymbol{G}$ and the effect of a homogeneous field of gravitation on a mass particle. Then the motion on geodesic lines in a gravitational field is considered. The general transformation of coordinates leads to the Christoffel matrix and the Riemannian curvature matrix. With the help of the Ricci matrix, the Theory of General Relativity of Einstein can then be formulated.

2.1 General Relativity and Riemannian Geometry

In the theory of general relativity, the invariance of the equations with respect to any coordinate transformation is required. Especially the theorems of physics are to remain unchanged if one transforms them from one coordinate system in another coordinate system by the general transformation equations

$$x_i = x_i\left(x_0', x_1', x_2', x_3'\right) \quad \text{for } i = 0, 1, 2 \text{ and } 3. \tag{2.1}$$

Consider an arbitrary coordinate system $\mathcal{K}$ on the infinitely small neighbourhood of the point P, in which also a field of gravity may be present. In an infinitely small space and an infinitely small time interval, or in other words, in an infinitely small spacetime interval, the coordinate system $\mathcal{K}$ can be replaced by a coordinate system $\mathcal{K}'$ which is accelerated to it and in which no field of gravitation is present. $\mathcal{K}'$ is the *local* spacetime coordinate system in the neighbourhood of a point, and $\mathcal{K}$ the *general* coordinate system. It is now assumed that

For all local coordinate systems $\mathcal{K}'$ the Special Relativity is valid in any infinitely small four-dimensional neighbourhood.

The point P' may be infinitely close to the point P, and have the real coordinates $\mathrm{d}\boldsymbol{x}_0$, $\mathrm{d}\boldsymbol{x}_1$, $\mathrm{d}\boldsymbol{x}_2$ and $\mathrm{d}\boldsymbol{x}_3$ in the Cartesian (rectangular) coordinate system. For a line element, one has

$$\mathrm{d}s^2 = \mathrm{d}x_0^2 - \left(\mathrm{d}x_1^2 + \mathrm{d}x_2^2 + \mathrm{d}x_3^2\right) \tag{2.2}$$

G. Ludyk, *Einstein in Matrix Form*, Graduate Texts in Physics,
DOI 10.1007/978-3-642-35798-5_2,

where x_0 is the time coordinate ct. If $\mathrm{d}s^2$ is positive, then P' is reached from P by a movement with a velocity smaller than the velocity of light. Equation (2.2) can be written using the matrix

$$\boldsymbol{M} = \begin{pmatrix} 1 & 0 & 0 & 0 \\ 0 & -1 & 0 & 0 \\ 0 & 0 & -1 & 0 \\ 0 & 0 & 0 & -1 \end{pmatrix}$$

as a quadratic form

$$\mathrm{d}s^2 = \mathrm{d}\vec{x}^{\mathsf{T}} \boldsymbol{M}\, \mathrm{d}\vec{x}. \tag{2.3}$$

Going on to the coordinate system $\mathcal{K}'$, one gets with

$$\mathrm{d}x_i = \frac{\partial x_i}{\partial x_0'}\,\mathrm{d}x_0' + \frac{\partial x_i}{\partial x_1'}\,\mathrm{d}x_1' + \frac{\partial x_i}{\partial x_2'}\,\mathrm{d}x_2' + \frac{\partial x_i}{\partial x_3'}\,\mathrm{d}x_3'$$

and the Jacobi matrix

$$\boldsymbol{J} \stackrel{\text{def}}{=} \begin{pmatrix} \frac{\partial x_0}{\partial x_0'} & \cdots & \frac{\partial x_0}{\partial x_3'} \\ \vdots & \ddots & \vdots \\ \frac{\partial x_3}{\partial x_0'} & \cdots & \frac{\partial x_3}{\partial x_3'} \end{pmatrix} = \frac{\partial \vec{x}}{\partial \vec{x}'^{\mathsf{T}}} \tag{2.4}$$

the connection

$$\mathrm{d}\vec{x} = \boldsymbol{J}\,\mathrm{d}\vec{x}'. \tag{2.5}$$

This is used in (2.3), yielding

$$\mathrm{d}s^2 = \mathrm{d}\vec{x}'^{\mathsf{T}} \boldsymbol{J}^{\mathsf{T}} \boldsymbol{M} \boldsymbol{J}\,\mathrm{d}\vec{x}' = \mathrm{d}\vec{x}'^{\mathsf{T}} \boldsymbol{G}\,\mathrm{d}\vec{x}'. \tag{2.6}$$

So the *metric matrix* $\boldsymbol{G}$ is defined as

$$\boxed{\boldsymbol{G} \stackrel{\text{def}}{=} \boldsymbol{J}^{\mathsf{T}} \boldsymbol{M} \boldsymbol{J} \in \mathbb{R}^{4\times 4}.} \tag{2.7}$$

The matrix elements g_{ik} are functions of the parameters x_i'. They may change their values from point to point. In Special Relativity, $\boldsymbol{G} = \boldsymbol{M}$ in any finite region. An unforced particle moves in such a region straightforwardly and uniformly.

But if the particle is in a gravitational field, the motion is curvilinear and nonuniform. Depending on the nature of the gravitational field, the g_{ik} are functions of other parameters. At most ten different elements g_{ik} of the symmetric 4×4-matrix $\boldsymbol{G}$ describe the field of gravitation at every point in the coordinate system. In every *local* coordinate system the g_{ik} are constant and can be transformed by a similarity transformation into the form $\boldsymbol{M}$. Because if $\boldsymbol{G} = \boldsymbol{J}^{\mathsf{T}} \boldsymbol{M} \boldsymbol{J}$, then one immediately obtains with the transformation $\mathrm{d}\vec{x}' = \boldsymbol{J}^{-1}\,\mathrm{d}\vec{\xi}$

$$\mathrm{d}\vec{x}'^{\mathsf{T}} \boldsymbol{G}\,\mathrm{d}\vec{x}' = \mathrm{d}\vec{\xi}^{\mathsf{T}} \boldsymbol{J}^{-1\mathsf{T}} \boldsymbol{J}^{\mathsf{T}} \boldsymbol{M} \boldsymbol{J} \boldsymbol{J}^{-1}\,\mathrm{d}\vec{\xi} = \mathrm{d}\vec{\xi}^{\mathsf{T}} \boldsymbol{M}\,\mathrm{d}\vec{\xi}.$$

But this is true only for a point because the matrices $\boldsymbol{J}$ are different from point to point. There is *no* transformation matrix $\boldsymbol{J}$ which is valid globally.

2.2 Motion in a Gravitational Field

What influence does a gravitational field have on a mass particle? Following the relativity principle, in the local inertial system, which means in a coordinate system moving with the particle, the laws of special relativity are valid. For the movement of a particle with no external forces acting on it,

$$\frac{\mathrm{d}^2\vec{\xi}}{\mathrm{d}\tau^2} = \mathbf{0}. \tag{2.8}$$

The proper time τ follows from

$$\mathrm{d}s^2 = c^2\,\mathrm{d}\tau^2 = \mathrm{d}\vec{\xi}^{\mathsf{T}}\,\boldsymbol{M}\,\mathrm{d}\vec{\xi}. \tag{2.9}$$

By integrating (2.8), one gets with the initial position $\vec{\xi}(0)$ and the initial velocity $\dot{\vec{\xi}}(0)$

$$\vec{\xi}(\tau) = \vec{\xi}(0) + \dot{\vec{\xi}}(0)\tau.$$

A photon also moves in a straight line in the local inertial system. But τ is then not the proper time of the photon. A photon has no proper time because for light $\mathrm{d}s = 0 = c\,\mathrm{d}\tau$. Therefore, we introduce the parameter λ such that

$$\frac{\mathrm{d}^2\vec{\xi}}{\mathrm{d}\lambda^2} = \mathbf{0}$$

is the equation of motion of the photon in the local inertial system. Now we move from the local inertial system with the spacetime vector $\vec{\xi}$ to the global inertial system with the spacetime vector $\vec{x}$. With

$$\boldsymbol{J} \stackrel{\text{def}}{=} \frac{\partial\vec{\xi}}{\partial\vec{x}^{\mathsf{T}}}$$

one gets for (2.9)

$$\mathrm{d}s^2 = c^2\,\mathrm{d}\tau^2 = \mathrm{d}\vec{x}^{\mathsf{T}}\,\boldsymbol{J}^{\mathsf{T}}\boldsymbol{M}\boldsymbol{J}\,\mathrm{d}\vec{x} = \mathrm{d}\vec{x}^{\mathsf{T}}\,\boldsymbol{G}\,\mathrm{d}\vec{x}. \tag{2.10}$$

For light this becomes

$$\mathrm{d}\vec{x}^{\mathsf{T}}\,\boldsymbol{G}\,\mathrm{d}\vec{x} = 0. \tag{2.11}$$

2.2.1 First Solution

From (2.8) one obtains for the motion of a particle with

$$\dot{\vec{x}} \stackrel{\text{def}}{=} \frac{\mathrm{d}\vec{x}}{\mathrm{d}\tau},$$

$$\frac{\mathrm{d}}{\mathrm{d}\tau}\left(\frac{\partial \vec{\xi}}{\partial \vec{x}^{\mathsf{T}}}\frac{\mathrm{d}\vec{x}}{\mathrm{d}\tau}\right) = \frac{\mathrm{d}}{\mathrm{d}\tau}(\boldsymbol{J}\dot{\vec{x}}) = \frac{\mathrm{d}}{\mathrm{d}\tau}(\boldsymbol{J})\dot{\vec{x}} + \boldsymbol{J}\ddot{\vec{x}} = \boldsymbol{0}, \tag{2.12}$$

and with (A.90)

$$\frac{\mathrm{d}}{\mathrm{d}\tau}\big(\boldsymbol{J}(\vec{x}(\tau))\big) = \big(\dot{\vec{x}}^{\mathsf{T}} \otimes \boldsymbol{I}_4\big)\frac{\partial \boldsymbol{J}}{\partial \vec{x}}$$

from (2.12)

$$\ddot{\vec{x}} = -\boldsymbol{J}^{-1}\big(\dot{\vec{x}}^{\mathsf{T}} \otimes \boldsymbol{I}_4\big)\frac{\partial \boldsymbol{J}}{\partial \vec{x}}\dot{\vec{x}}, \tag{2.13}$$

or

$$\ddot{\vec{x}} = -\big(\dot{\vec{x}}^{\mathsf{T}} \otimes \boldsymbol{J}^{-1}\big)\frac{\partial \boldsymbol{J}}{\partial \vec{x}}\dot{\vec{x}} = -\big(\dot{\vec{x}}^{\mathsf{T}} \otimes \boldsymbol{I}_4\big)\big(\boldsymbol{I}_4 \otimes \boldsymbol{J}^{-1}\big)\frac{\partial \boldsymbol{J}}{\partial \vec{x}}\dot{\vec{x}},$$

that is,

$$\ddot{\vec{x}} = -\big(\boldsymbol{I}_4 \otimes \dot{\vec{x}}^{\mathsf{T}}\big)\boldsymbol{U}_{4\times 4}\big(\boldsymbol{I}_4 \otimes \boldsymbol{J}^{-1}\big)\frac{\partial \boldsymbol{J}}{\partial \vec{x}}\dot{\vec{x}}. \tag{2.14}$$

With $\boldsymbol{J}_k \stackrel{\text{def}}{=} \frac{\partial \boldsymbol{J}}{\partial x_k} \in \mathbb{R}^{4\times 4}$ and

$$\boxed{\hat{\boldsymbol{\Gamma}} = \begin{pmatrix} \hat{\boldsymbol{\Gamma}}_0 \\ \vdots \\ \hat{\boldsymbol{\Gamma}}_3 \end{pmatrix} \stackrel{\text{def}}{=} \boldsymbol{U}_{4\times 4}\big(\boldsymbol{I}_4 \otimes \boldsymbol{J}^{-1}\big)\frac{\partial \boldsymbol{J}}{\partial \vec{x}} = \boldsymbol{U}_{4\times 4}\begin{pmatrix} \boldsymbol{J}^{-1}\boldsymbol{J}_0 \\ \vdots \\ \boldsymbol{J}^{-1}\boldsymbol{J}_3 \end{pmatrix} \in \mathbb{R}^{16\times 4}} \tag{2.15}$$

one can write in place of (2.14) the compact equation

$$\boxed{\ddot{\vec{x}} = -\big(\boldsymbol{I}_4 \otimes \dot{\vec{x}}^{\mathsf{T}}\big)\hat{\boldsymbol{\Gamma}}\dot{\vec{x}}.} \tag{2.16}$$

With (2.15) and (2.16) one gets for the individual vector components $\ddot{x}_k$ according to the form of the permutation matrix $\boldsymbol{U}_{4\times 4}$ in the appendix ($\boldsymbol{j}_k^{-\mathsf{T}} \in \mathbb{R}^4$ is row k of $\boldsymbol{J}^{-1}$)

$$\ddot{x}_k = -\dot{\vec{x}}^{\mathsf{T}}\big(\boldsymbol{I}_4 \otimes \boldsymbol{j}_k^{-\mathsf{T}}\big)\frac{\partial \boldsymbol{J}}{\partial \vec{x}}\dot{\vec{x}}. \tag{2.17}$$

From (2.15) and (2.17) we can read directly

$$\boxed{\hat{\boldsymbol{\Gamma}}_k = \big(\boldsymbol{I}_4 \otimes \boldsymbol{j}_k^{-\mathsf{T}}\big)\frac{\partial \boldsymbol{J}}{\partial \vec{x}}.} \tag{2.18}$$

The so-called Christoffel matrices $\hat{\boldsymbol{\Gamma}}$ can be calculated directly from the Jacobi matrix $\boldsymbol{J}$, that is, from the transformation matrix for the transition from the local inertial system to the accelerated non-inertial system (the coordinate system with a gravitational field). For the motion of a photon, we obtain in the same manner

$$\boxed{\frac{\mathrm{d}^2\vec{\boldsymbol{x}}}{\mathrm{d}\lambda^2} = -\left(\boldsymbol{I}_4 \otimes \frac{\mathrm{d}\vec{\boldsymbol{x}}}{\mathrm{d}\lambda}^{\mathsf{T}}\right)\hat{\boldsymbol{\Gamma}}\frac{\mathrm{d}\vec{\boldsymbol{x}}}{\mathrm{d}\lambda}.} \tag{2.19}$$

2.2.2 Second Solution

An alternative solution is obtained from the second form in (A.90), namely

$$\frac{\mathrm{d}}{\mathrm{d}\tau}(\boldsymbol{J}) = \frac{\partial \boldsymbol{J}}{\partial \vec{\boldsymbol{x}}^{\mathsf{T}}}(\dot{\vec{\boldsymbol{x}}} \otimes \boldsymbol{I}_4). \tag{2.20}$$

Thus we obtain from (2.12)

$$\ddot{\vec{\boldsymbol{x}}} = -\boldsymbol{J}^{-1}\frac{\partial \boldsymbol{J}}{\partial \vec{\boldsymbol{x}}^{\mathsf{T}}}(\dot{\vec{\boldsymbol{x}}} \otimes \boldsymbol{I}_4)\dot{\vec{\boldsymbol{x}}} = -\boldsymbol{J}^{-1}\frac{\partial \boldsymbol{J}}{\partial \vec{\boldsymbol{x}}^{\mathsf{T}}}(\dot{\vec{\boldsymbol{x}}} \otimes \dot{\vec{\boldsymbol{x}}}), \tag{2.21}$$

that is, with

$$\boxed{\tilde{\boldsymbol{\Gamma}} \stackrel{\text{def}}{=} \boldsymbol{J}^{-1}\frac{\partial \boldsymbol{J}}{\partial \vec{\boldsymbol{x}}^{\mathsf{T}}}} \tag{2.22}$$

written completely as

$$\tilde{\boldsymbol{\Gamma}} = \boldsymbol{J}^{-1}\left[\frac{\partial \boldsymbol{J}}{\partial x_0}\middle|\frac{\partial \boldsymbol{J}}{\partial x_1}\middle|\frac{\partial \boldsymbol{J}}{\partial x_2}\middle|\frac{\partial \boldsymbol{J}}{\partial x_3}\right] \in \mathbb{R}^{4\times 16}, \tag{2.23}$$

so with $\dot{\vec{\boldsymbol{x}}} \otimes \dot{\vec{\boldsymbol{x}}} \in \mathbb{R}^{16}$,

$$\boxed{\ddot{\vec{\boldsymbol{x}}} = -\tilde{\boldsymbol{\Gamma}}(\dot{\vec{\boldsymbol{x}}} \otimes \dot{\vec{\boldsymbol{x}}}).} \tag{2.24}$$

This is an alternative representation of the relation (2.16)!

If one defines

$$\tilde{\boldsymbol{\gamma}}_k^{\mathsf{T}} \stackrel{\text{def}}{=} \boldsymbol{j}_k^{-\mathsf{T}}\frac{\partial \boldsymbol{J}}{\partial \vec{\boldsymbol{x}}^{\mathsf{T}}} \in \mathbb{R}^{16}, \tag{2.25}$$

then for each vector component the following scalar vector product is obtained:

$$\ddot{x}_k = -\tilde{\boldsymbol{\gamma}}_k^{\mathsf{T}} \cdot (\dot{\vec{\boldsymbol{x}}} \otimes \dot{\vec{\boldsymbol{x}}}). \tag{2.26}$$

2.2.3 Relation Between $\tilde{\boldsymbol{\Gamma}}$ and G

Since $\boldsymbol{G} = \boldsymbol{J}^\top \boldsymbol{M} \boldsymbol{J}$ and $\tilde{\boldsymbol{\Gamma}} = \boldsymbol{J}^{-1} \frac{\partial \boldsymbol{J}}{\partial \boldsymbol{x}^\top}$, the matrix $\tilde{\boldsymbol{\Gamma}}$ must depend on $\frac{\partial \boldsymbol{G}}{\partial x}$. This indeed is the case.

On the one hand,

$$\boldsymbol{G}\tilde{\boldsymbol{\Gamma}} = \boldsymbol{J}^\top \boldsymbol{M} \frac{\partial \boldsymbol{J}}{\partial \boldsymbol{x}^\top}, \tag{2.27}$$

and, on the other hand, $g_{\mu\nu} = \boldsymbol{j}_\mu^\top \boldsymbol{M} \boldsymbol{j}_\nu$, so

$$\frac{\partial g_{\mu\nu}}{\partial x_\lambda} = \frac{\partial \boldsymbol{j}_\mu^\top}{\partial x_\lambda} \boldsymbol{M} \boldsymbol{j}_\nu + \boldsymbol{j}_\mu^\top \boldsymbol{M} \frac{\partial \boldsymbol{j}_\nu}{\partial x_\lambda}. \tag{2.28}$$

Furthermore,

$$\frac{\partial g_{\lambda\nu}}{\partial x_\mu} = \frac{\partial \boldsymbol{j}_\lambda^\top}{\partial x_\mu} \boldsymbol{M} \boldsymbol{j}_\nu + \boldsymbol{j}_\lambda^\top \boldsymbol{M} \frac{\partial \boldsymbol{j}_\nu}{\partial x_\mu} \tag{2.29}$$

and

$$\frac{\partial g_{\mu\lambda}}{\partial x_\nu} = \frac{\partial \boldsymbol{j}_\mu^\top}{\partial x_\nu} \boldsymbol{M} \boldsymbol{j}_\lambda + \boldsymbol{j}_\mu^\top \boldsymbol{M} \frac{\partial \boldsymbol{j}_\lambda}{\partial x_\nu}. \tag{2.30}$$

If we add (2.28) and (2.29) and then subtract (2.30), we obtain

$$\frac{\partial g_{\mu\nu}}{\partial x_\lambda} + \frac{\partial g_{\lambda\nu}}{\partial x_\mu} - \frac{\partial g_{\mu\lambda}}{\partial x_\nu} = 2 \frac{\partial \boldsymbol{j}_\mu^\top}{\partial x_\lambda} \boldsymbol{M} \boldsymbol{j}_\nu = 2 \boldsymbol{j}_\nu^\top \boldsymbol{M} \frac{\partial \boldsymbol{j}_\mu}{\partial x_\lambda}. \tag{2.31}$$

Calling $\boldsymbol{G}\tilde{\boldsymbol{\Gamma}} \stackrel{\text{def}}{=} \check{\tilde{\boldsymbol{\Gamma}}}$, then we obtain with (2.27) and (2.31) as the element $\check{\tilde{\Gamma}}^\lambda_{\nu\mu}$ in the νth row and the μth column of $\check{\tilde{\boldsymbol{\Gamma}}}_\lambda$

$$\check{\tilde{\Gamma}}^\lambda_{\nu\mu} = \frac{1}{2} \left(\frac{\partial g_{\mu\nu}}{\partial x_\lambda} + \frac{\partial g_{\lambda\nu}}{\partial x_\mu} - \frac{\partial g_{\mu\lambda}}{\partial x_\nu} \right). \tag{2.32}$$

Since $\tilde{\boldsymbol{\Gamma}} = \boldsymbol{G}^{-1} \boldsymbol{G} \tilde{\boldsymbol{\Gamma}} = \boldsymbol{G}^{-1} \check{\tilde{\boldsymbol{\Gamma}}}$,

$$\tilde{\boldsymbol{\Gamma}}_\lambda = \boldsymbol{G}^{-1} \check{\tilde{\boldsymbol{\Gamma}}}_\lambda = \boldsymbol{G}^{-1} \boldsymbol{J}^\top \boldsymbol{M} \frac{\partial \boldsymbol{J}}{\partial x_\lambda},$$

and one finally obtains with the αth row $\boldsymbol{g}_\alpha^{[-T]}$ of the matrix $\boldsymbol{G}^{-1}$ and the νth element $g_{\alpha\nu}^{[-1]}$ of this row vector the following relation between the elements in the αth row and the μth column of $\tilde{\boldsymbol{\Gamma}}_\lambda$ and the elements of $\boldsymbol{G}$

$$\boxed{\tilde{\Gamma}^\lambda_{\alpha\mu} = \boldsymbol{g}_\alpha^{[-T]} \boldsymbol{J}^\top \boldsymbol{M} \frac{\partial \boldsymbol{j}_\mu}{\partial x_\lambda} = \sum_{\nu=0}^{3} \frac{g_{\alpha\nu}^{[-1]}}{2} \left(\frac{\partial g_{\mu\nu}}{\partial x_\lambda} + \frac{\partial g_{\lambda\nu}}{\partial x_\mu} - \frac{\partial g_{\mu\lambda}}{\partial x_\nu} \right).} \tag{2.33}$$

This is the desired relationship between the matrix elements of $\tilde{\boldsymbol{\Gamma}}$ and $\boldsymbol{G}$!

2.3 Geodesic Lines and Equations of Motion

The motion of photons and particles in a gravitational field will again be considered, but now using the calculus of variations. The same results as in (2.16) are expected. This is to be seen by using the same Christoffel matrix $\boldsymbol{\Gamma}$ in the results. In special relativity, the motion of a photon is given by $c^2t^2 = \boldsymbol{x}^\mathsf{T}\boldsymbol{x}$, that means $s^2 = c^2t^2 - \boldsymbol{x}^\mathsf{T}\boldsymbol{x} = 0$, or $\mathrm{d}s^2 = \mathrm{d}x_0^2 - \mathrm{d}\boldsymbol{x}^\mathsf{T}\,\mathrm{d}\boldsymbol{x} = \mathrm{d}\vec{\boldsymbol{x}}^\mathsf{T}\,\boldsymbol{M}\,\mathrm{d}\vec{\boldsymbol{x}} = 0$ for any small distance $\mathrm{d}\boldsymbol{x}$. If paths were straight lines, that would mean the shortest possible connection between the points P_1 and P_2. Also the theory of general relativity demands that light and particles move on straightest possible paths. These paths are the so-called *geodesic curves* for which the length has an extreme value:

$$\delta \int_{P_1}^{P_2} \mathrm{d}s = 0. \tag{2.34}$$

This produces a system of four differential equations. For the variation of $\mathrm{d}s^2$ we obtain

$$\begin{aligned} \delta\big(\mathrm{d}s^2\big) &= \delta\big(\mathrm{d}\vec{\boldsymbol{x}}^\mathsf{T}\,\boldsymbol{G}\,\mathrm{d}\vec{\boldsymbol{x}}\big),\\ 2(\delta\,\mathrm{d}s)\,\mathrm{d}s &= \big(\delta\,\mathrm{d}\vec{\boldsymbol{x}}^\mathsf{T}\big)\,\boldsymbol{G}\,\mathrm{d}\vec{\boldsymbol{x}} + \mathrm{d}\vec{\boldsymbol{x}}^\mathsf{T}(\delta\boldsymbol{G})\,\mathrm{d}\vec{\boldsymbol{x}} + \mathrm{d}\vec{\boldsymbol{x}}^\mathsf{T}\boldsymbol{G}(\delta\,\mathrm{d}\vec{\boldsymbol{x}}), \end{aligned} \tag{2.35}$$

but since $\boldsymbol{G}$ is symmetric, $\boldsymbol{G} = \boldsymbol{G}^\mathsf{T}$, we get

$$2(\delta\,\mathrm{d}s)\,\mathrm{d}s = 2\,\mathrm{d}\vec{\boldsymbol{x}}^\mathsf{T}\boldsymbol{G}(\delta\,\mathrm{d}\vec{\boldsymbol{x}}) + \mathrm{d}\vec{\boldsymbol{x}}^\mathsf{T}(\delta\boldsymbol{G})\,\mathrm{d}\vec{\boldsymbol{x}}. \tag{2.36}$$

Equation (2.36) divided by $2\,\mathrm{d}s$ results, with $\mathrm{d}(\delta\vec{\boldsymbol{x}}) = \delta\,\mathrm{d}\vec{\boldsymbol{x}}$ and $\frac{\mathrm{d}\vec{\boldsymbol{x}}}{\mathrm{d}s} \stackrel{\text{def}}{=} \dot{\vec{\boldsymbol{x}}}$, in

$$\delta\,\mathrm{d}s = \dot{\vec{\boldsymbol{x}}}^\mathsf{T}\boldsymbol{G}\,\mathrm{d}(\delta\vec{\boldsymbol{x}}) + \frac{1}{2}\dot{\vec{\boldsymbol{x}}}^\mathsf{T}(\delta\boldsymbol{G})\,\mathrm{d}\vec{\boldsymbol{x}}. \tag{2.37}$$

Multiplying the right-hand side by $\mathrm{d}s$ provides

$$\delta\,\mathrm{d}s = \left[\dot{\vec{\boldsymbol{x}}}^\mathsf{T}\boldsymbol{G}\frac{\mathrm{d}(\delta\vec{\boldsymbol{x}})}{\mathrm{d}s} + \frac{1}{2}\dot{\vec{\boldsymbol{x}}}^\mathsf{T}(\delta\boldsymbol{G})\dot{\vec{\boldsymbol{x}}}\right]\mathrm{d}s. \tag{2.38}$$

For the variation of the matrix $\boldsymbol{G}$ we set

$$\delta\boldsymbol{G} = \frac{\partial\boldsymbol{G}}{\partial x_0}\delta x_0 + \frac{\partial\boldsymbol{G}}{\partial x_1}\delta x_1 + \frac{\partial\boldsymbol{G}}{\partial x_2}\delta x_2 + \frac{\partial\boldsymbol{G}}{\partial x_3}\delta x_3. \tag{2.39}$$

This is extended to a quadratic form:

$$\dot{\vec{\boldsymbol{x}}}^\mathsf{T}\delta\boldsymbol{G}\dot{\vec{\boldsymbol{x}}} = \dot{\vec{\boldsymbol{x}}}^\mathsf{T}\frac{\partial\boldsymbol{G}}{\partial x_0}\dot{\vec{\boldsymbol{x}}}\delta x_0 + \dot{\vec{\boldsymbol{x}}}^\mathsf{T}\frac{\partial\boldsymbol{G}}{\partial x_1}\dot{\vec{\boldsymbol{x}}}\delta x_1 + \dot{\vec{\boldsymbol{x}}}^\mathsf{T}\frac{\partial\boldsymbol{G}}{\partial x_2}\dot{\vec{\boldsymbol{x}}}\delta x_2 + \dot{\vec{\boldsymbol{x}}}^\mathsf{T}\frac{\partial\boldsymbol{G}}{\partial x_3}\dot{\vec{\boldsymbol{x}}}\delta x_3 \tag{2.40}$$

and the right-hand side is collected to a vector product, leading to the result

$$\dot{\vec{\boldsymbol{x}}}^\mathsf{T}\delta\boldsymbol{G}\dot{\vec{\boldsymbol{x}}} = \left[\dot{\vec{\boldsymbol{x}}}^\mathsf{T}\frac{\partial\boldsymbol{G}}{\partial x_0}\dot{\vec{\boldsymbol{x}}}, \dot{\vec{\boldsymbol{x}}}^\mathsf{T}\frac{\partial\boldsymbol{G}}{\partial x_1}\dot{\vec{\boldsymbol{x}}}, \dot{\vec{\boldsymbol{x}}}^\mathsf{T}\frac{\partial\boldsymbol{G}}{\partial x_2}\dot{\vec{\boldsymbol{x}}}, \dot{\vec{\boldsymbol{x}}}^\mathsf{T}\frac{\partial\boldsymbol{G}}{\partial x_3}\dot{\vec{\boldsymbol{x}}}\right]\delta\vec{\boldsymbol{x}}. \tag{2.41}$$

The row vector on the right-hand side with

$$\left(\frac{\partial \boldsymbol{G}}{\partial \vec{\boldsymbol{x}}^{\mathsf{T}}}\right) \stackrel{\text{def}}{=} \left[\frac{\partial \boldsymbol{G}}{\partial x_0}, \frac{\partial \boldsymbol{G}}{\partial x_1}, \frac{\partial \boldsymbol{G}}{\partial x_2}, \frac{\partial \boldsymbol{G}}{\partial x_3}\right]$$

and the Kronecker product can be written as

$$\left[\dot{\vec{\boldsymbol{x}}}^{\mathsf{T}} \frac{\partial \boldsymbol{G}}{\partial x_0} \dot{\vec{\boldsymbol{x}}}, \dot{\vec{\boldsymbol{x}}}^{\mathsf{T}} \frac{\partial \boldsymbol{G}}{\partial x_1} \dot{\vec{\boldsymbol{x}}}, \dot{\vec{\boldsymbol{x}}}^{\mathsf{T}} \frac{\partial \boldsymbol{G}}{\partial x_2} \dot{\vec{\boldsymbol{x}}}, \dot{\vec{\boldsymbol{x}}}^{\mathsf{T}} \frac{\partial \boldsymbol{G}}{\partial x_3} \dot{\vec{\boldsymbol{x}}}\right] = \dot{\vec{\boldsymbol{x}}}^{\mathsf{T}} \left(\frac{\partial \boldsymbol{G}}{\partial \vec{\boldsymbol{x}}^{\mathsf{T}}}\right) (\boldsymbol{I}_4 \otimes \dot{\vec{\boldsymbol{x}}}). \tag{2.42}$$

With this relationship we obtain now for the variated integral (2.34)

$$\delta \int_{P_1}^{P_2} \mathrm{d}s = \int_{P_1}^{P_2} \left[\dot{\vec{\boldsymbol{x}}}^{\mathsf{T}} \boldsymbol{G} \frac{\mathrm{d}(\delta \vec{\boldsymbol{x}})}{\mathrm{d}s} + \frac{1}{2} \dot{\vec{\boldsymbol{x}}}^{\mathsf{T}} \left(\frac{\partial \boldsymbol{G}}{\partial \vec{\boldsymbol{x}}^{\mathsf{T}}}\right) (\boldsymbol{I}_4 \otimes \dot{\vec{\boldsymbol{x}}}) \delta \vec{\boldsymbol{x}}\right] \mathrm{d}s. \tag{2.43}$$

Taking in consideration $\delta \vec{\boldsymbol{x}}(P_1) = \delta \vec{\boldsymbol{x}}(P_2) = \boldsymbol{o}$, and performing integration by parts to the left summand in the integral, gives

$$\begin{aligned} \delta \int_{P_1}^{P_2} \mathrm{d}s &= \int_{P_1}^{P_2} \left[-\frac{\mathrm{d}}{\mathrm{d}s} (\dot{\vec{\boldsymbol{x}}}^{\mathsf{T}} \boldsymbol{G}) \delta \vec{\boldsymbol{x}} + \frac{1}{2} \dot{\vec{\boldsymbol{x}}}^{\mathsf{T}} \left(\frac{\partial \boldsymbol{G}}{\partial \vec{\boldsymbol{x}}^{\mathsf{T}}}\right) (\boldsymbol{I}_4 \otimes \dot{\vec{\boldsymbol{x}}}) \delta \vec{\boldsymbol{x}}\right] \mathrm{d}s \\ &= \int_{P_1}^{P_2} \left[-\frac{\mathrm{d}}{\mathrm{d}s} (\dot{\vec{\boldsymbol{x}}}^{\mathsf{T}} \boldsymbol{G}) + \frac{1}{2} \dot{\vec{\boldsymbol{x}}}^{\mathsf{T}} \left(\frac{\partial \boldsymbol{G}}{\partial \vec{\boldsymbol{x}}^{\mathsf{T}}}\right) (\boldsymbol{I}_4 \otimes \dot{\vec{\boldsymbol{x}}})\right] \delta \vec{\boldsymbol{x}} \, \mathrm{d}s = 0. \end{aligned} \tag{2.44}$$

To make the variation of the integral for every arbitrary vector function $\delta \vec{\boldsymbol{x}}(\cdot)$ disappear, in spite of the fundamental theorem of the calculus of variations, the vector function in the brackets must be identically zero:

$$-\frac{\mathrm{d}}{\mathrm{d}s} (\dot{\vec{\boldsymbol{x}}}^{\mathsf{T}} \boldsymbol{G}) + \frac{1}{2} \dot{\vec{\boldsymbol{x}}}^{\mathsf{T}} \left(\frac{\partial \boldsymbol{G}}{\partial \vec{\boldsymbol{x}}^{\mathsf{T}}}\right) (\boldsymbol{I}_4 \otimes \dot{\vec{\boldsymbol{x}}}) = \boldsymbol{o}^{\mathsf{T}}, \tag{2.45}$$

or transposed, where $(\boldsymbol{A} \otimes \boldsymbol{B})^{\mathsf{T}} = (\boldsymbol{A}^{\mathsf{T}} \otimes \boldsymbol{B}^{\mathsf{T}})$ is used,

$$\frac{1}{2} (\boldsymbol{I}_4 \otimes \dot{\vec{\boldsymbol{x}}}^{\mathsf{T}}) \frac{\partial \boldsymbol{G}}{\partial \vec{\boldsymbol{x}}} \dot{\vec{\boldsymbol{x}}} - \frac{\mathrm{d}}{\mathrm{d}s} (\boldsymbol{G} \dot{\vec{\boldsymbol{x}}}) = \boldsymbol{o}. \tag{2.46}$$

For the second term on the left-hand side one gets

$$\frac{\mathrm{d}}{\mathrm{d}s} (\boldsymbol{G} \dot{\vec{\boldsymbol{x}}}) = \boldsymbol{G} \ddot{\vec{\boldsymbol{x}}} + \frac{\mathrm{d}}{\mathrm{d}s} (\boldsymbol{G}) \dot{\vec{\boldsymbol{x}}}. \tag{2.47}$$

And this is

$$\frac{\mathrm{d}}{\mathrm{d}s} (\boldsymbol{G}) = (\dot{\vec{\boldsymbol{x}}}^{\mathsf{T}} \otimes \boldsymbol{I}_4) \frac{\partial \boldsymbol{G}}{\partial \vec{\boldsymbol{x}}}.$$

Inserted into (2.47) this finally yields

$$\boldsymbol{G} \ddot{\vec{\boldsymbol{x}}} = \frac{1}{2} (\boldsymbol{I}_4 \otimes \dot{\vec{\boldsymbol{x}}}^{\mathsf{T}}) \frac{\partial \boldsymbol{G}}{\partial \vec{\boldsymbol{x}}} \dot{\vec{\boldsymbol{x}}} - (\dot{\vec{\boldsymbol{x}}}^{\mathsf{T}} \otimes \boldsymbol{I}_4) \frac{\partial \boldsymbol{G}}{\partial \vec{\boldsymbol{x}}} \dot{\vec{\boldsymbol{x}}} \tag{2.48}$$

or, putting $\ddot{\vec{x}}$ to the left-hand side,

$$\ddot{\vec{x}} = G^{-1}\left[\frac{1}{2}(I_4 \otimes \dot{\vec{x}}^{\mathsf{T}}) - (\dot{\vec{x}}^{\mathsf{T}} \otimes I_4)\right]\frac{\partial G}{\partial \vec{x}}\dot{\vec{x}}. \tag{2.49}$$

Taking into account the lemma in the Appendix "Vectors and Matrices":

$$B \otimes A = U_{s\times p}(A \otimes B)U_{q\times t}, \quad A \in \mathbb{R}^{p\times q}, B \in \mathbb{R}^{s\times t}, \tag{2.50}$$

(2.49) can be converted, with the help of $\dot{\vec{x}}^{\mathsf{T}} \otimes I_4 = (I_4 \otimes \dot{\vec{x}}^{\mathsf{T}})U_{4\times 4}$, into

$$\ddot{\vec{x}} = G^{-1}(I_4 \otimes \dot{\vec{x}}^{\mathsf{T}})\left[\frac{1}{2}I_{16} - U_{4\times 4}\right]\frac{\partial G}{\partial \vec{x}}\dot{\vec{x}}. \tag{2.51}$$

With

$$G^{-1}(I_4 \otimes \dot{\vec{x}}^{\mathsf{T}}) = (G^{-1} \otimes 1)(I_4 \otimes \dot{\vec{x}}^{\mathsf{T}}) = (G^{-1} \otimes \dot{\vec{x}}^{\mathsf{T}}) = (I_4 \otimes \dot{\vec{x}}^{\mathsf{T}})(G^{-1} \otimes I_4)$$

we finally obtain a form in which $\dot{\vec{x}}$ is pulled out to the left and to the right:

$$\boxed{\ddot{\vec{x}} = (I_4 \otimes \dot{\vec{x}}^{\mathsf{T}})(G^{-1} \otimes I_4)\left[\frac{1}{2}I_{16} - U_{4\times 4}\right]\frac{\partial G}{\partial \vec{x}}\dot{\vec{x}}.} \tag{2.52}$$

Summarizing

$$\hat{\Gamma} \stackrel{\text{def}}{=} (G^{-1} \otimes I_4)\left[U_{4\times 4} - \frac{1}{2}I_{16}\right]\frac{\partial G}{\partial \vec{x}} \tag{2.53}$$

$$= U_{4\times 4}(I_4 \otimes G^{-1})\left[\frac{1}{2}I_{16} - U_{4\times 4}\right]\frac{\partial G}{\partial \vec{x}}, \tag{2.54}$$

one gets the compact equation

$$\boxed{\ddot{\vec{x}} = -(I_4 \otimes \dot{\vec{x}}^{\mathsf{T}})\hat{\Gamma}\dot{\vec{x}}} \tag{2.55}$$

which agrees with the equation of motion (2.16), that is, in the language of the calculus of variations, this equation yields an extremal.

In (2.53) with the kth row g_k^{T} of the matrix G is

$$U_{4\times 4}\frac{\partial G}{\partial \vec{x}} = \begin{pmatrix} \frac{\partial g_0^{\mathsf{T}}}{\partial \vec{x}} \\ \vdots \\ \frac{\partial g_3^{\mathsf{T}}}{\partial \vec{x}} \end{pmatrix}.$$

For the four components $\ddot{x}_k$, $k = 0, 1, 2$ and 3, one obtains with the kth row $\boldsymbol{g}_k^{-T}$ of the matrix $\boldsymbol{G}^{-1}$,

$$\ddot{x}_k = \dot{\vec{\boldsymbol{x}}}^{\mathsf{T}}(\boldsymbol{g}_k^{-T} \otimes \boldsymbol{I}_4)\left[\begin{pmatrix} \frac{\partial \boldsymbol{g}_0^{\mathsf{T}}}{\partial \vec{\boldsymbol{x}}} \\ \vdots \\ \frac{\partial \boldsymbol{g}_3^{\mathsf{T}}}{\partial \vec{\boldsymbol{x}}} \end{pmatrix} - \frac{1}{2}\frac{\partial \boldsymbol{G}}{\partial \vec{\boldsymbol{x}}}\right]\dot{\vec{\boldsymbol{x}}}. \tag{2.56}$$

With

$$\boxed{\hat{\boldsymbol{\Gamma}}_k \stackrel{\text{def}}{=} (\boldsymbol{g}_k^{-T} \otimes \boldsymbol{I}_4)\left[\begin{pmatrix} \frac{\partial \boldsymbol{g}_0^{\mathsf{T}}}{\partial \vec{\boldsymbol{x}}} \\ \vdots \\ \frac{\partial \boldsymbol{g}_3^{\mathsf{T}}}{\partial \vec{\boldsymbol{x}}} \end{pmatrix} - \frac{1}{2}\frac{\partial \boldsymbol{G}}{\partial \vec{\boldsymbol{x}}}\right]} \tag{2.57}$$

one can also write for (2.56)

$$\boxed{\ddot{x}_k = -\dot{\vec{\boldsymbol{x}}}^{\mathsf{T}}\hat{\boldsymbol{\Gamma}}_k\dot{\vec{\boldsymbol{x}}}.} \tag{2.58}$$

The so-obtained matrix $\hat{\boldsymbol{\Gamma}}_k$ need not be symmetric. But the value of the quadratic form (2.58) is unchanged, if the matrix $\hat{\boldsymbol{\Gamma}}_k$ in (2.58) is replaced by the symmetric 4×4-matrix

$$\boldsymbol{\Gamma}_k \stackrel{\text{def}}{=} \frac{1}{2}(\hat{\boldsymbol{\Gamma}}_k + \hat{\boldsymbol{\Gamma}}_k^{\mathsf{T}}). \tag{2.59}$$

In this way, the matrix is said to be *symmetrised*. Expanding (2.57) yields

$$\hat{\boldsymbol{\Gamma}}_k = \sum_{i=0}^{3} g_{k,i}^{[-1]}\left(\frac{\partial \boldsymbol{g}_i^{\mathsf{T}}}{\partial \vec{\boldsymbol{x}}} - \frac{1}{2}\frac{\partial \boldsymbol{G}}{\partial x_i}\right),$$

and transposed

$$\hat{\boldsymbol{\Gamma}}_k^{\mathsf{T}} = \sum_{i=0}^{3} g_{k,i}^{[-1]}\left(\frac{\partial \boldsymbol{g}_i}{\partial \vec{\boldsymbol{x}}^{\mathsf{T}}} - \frac{1}{2}\frac{\partial \boldsymbol{G}}{\partial x_i}\right).$$

For this one can also write

$$\hat{\boldsymbol{\Gamma}}_k^{\mathsf{T}} = (\boldsymbol{g}_k^{-T} \otimes \boldsymbol{I}_4)\left[\begin{pmatrix} \frac{\partial \boldsymbol{g}_0}{\partial \vec{\boldsymbol{x}}^{\mathsf{T}}} \\ \vdots \\ \frac{\partial \boldsymbol{g}_3}{\partial \vec{\boldsymbol{x}}^{\mathsf{T}}} \end{pmatrix} - \frac{1}{2}\frac{\partial \boldsymbol{G}}{\partial \vec{\boldsymbol{x}}}\right]. \tag{2.60}$$

Inserting (2.60) into (2.59), we obtain

$$\boldsymbol{\Gamma}_k = \frac{1}{2}(\hat{\boldsymbol{\Gamma}}_k + \hat{\boldsymbol{\Gamma}}_k^{\mathsf{T}}) = \frac{1}{2}(\boldsymbol{g}_k^{-T} \otimes \boldsymbol{I}_4)\left[\begin{pmatrix} \frac{\partial \boldsymbol{g}_0^{\mathsf{T}}}{\partial \vec{\boldsymbol{x}}} \\ \vdots \\ \frac{\partial \boldsymbol{g}_3^{\mathsf{T}}}{\partial \vec{\boldsymbol{x}}} \end{pmatrix} + \begin{pmatrix} \frac{\partial \boldsymbol{g}_0}{\partial \vec{\boldsymbol{x}}^{\mathsf{T}}} \\ \vdots \\ \frac{\partial \boldsymbol{g}_3}{\partial \vec{\boldsymbol{x}}^{\mathsf{T}}} \end{pmatrix} - \frac{\partial \boldsymbol{G}}{\partial \vec{\boldsymbol{x}}}\right]. \tag{2.61}$$

Multiplying out yields the results for the components of the Christoffel matrix $\boldsymbol{\Gamma}_k$, namely the above already derived relationship

$$\boxed{\Gamma^k_{\alpha\beta} = \sum_{i=0}^{3} \frac{g_{ki}^{[-1]}}{2}\left(\frac{\partial g_{\beta i}}{\partial x_\alpha} + \frac{\partial g_{\alpha i}}{\partial x_\beta} - \frac{\partial g_{\alpha\beta}}{\partial x_i}\right).} \tag{2.62}$$

This can with

$$\check{\Gamma}^i_{\alpha\beta} \stackrel{\text{def}}{=} \frac{1}{2}\left(\frac{\partial g_{\beta i}}{\partial x_\alpha} + \frac{\partial g_{\alpha i}}{\partial x_\beta} - \frac{\partial g_{\alpha\beta}}{\partial x_i}\right) \tag{2.63}$$

also be written as

$$\Gamma^k_{\alpha\beta} = \sum_{i=0}^{3} g_{ki}^{[-1]} \check{\Gamma}^i_{\alpha\beta}. \tag{2.64}$$

In addition, from (2.63) the interesting connection follows:

$$\frac{\partial g_{\alpha i}}{\partial x_\beta} = \check{\Gamma}^i_{\alpha\beta} + \check{\Gamma}^\alpha_{i\beta}. \tag{2.65}$$

Assembling the four components $\ddot{x}_k$ of the vector $\ddot{\vec{\boldsymbol{x}}}$ with the matrix $\boldsymbol{\Gamma}_k$ into one vector yields

$$\ddot{\vec{\boldsymbol{x}}} = -\begin{pmatrix} \dot{\vec{\boldsymbol{x}}}^{\mathsf{T}} \boldsymbol{\Gamma}_0 \dot{\vec{\boldsymbol{x}}} \\ \vdots \\ \dot{\vec{\boldsymbol{x}}}^{\mathsf{T}} \boldsymbol{\Gamma}_3 \dot{\vec{\boldsymbol{x}}} \end{pmatrix}, \tag{2.66}$$

respectively,

$$\boxed{\ddot{\vec{\boldsymbol{x}}} = -(\boldsymbol{I}_4 \otimes \dot{\vec{\boldsymbol{x}}}^{\mathsf{T}})\boldsymbol{\Gamma}\dot{\vec{\boldsymbol{x}}},} \tag{2.67}$$

with

$$\boxed{\boldsymbol{\Gamma} \stackrel{\text{def}}{=} \begin{pmatrix} \boldsymbol{\Gamma}_0 \\ \vdots \\ \boldsymbol{\Gamma}_3 \end{pmatrix} = \frac{1}{2}(\boldsymbol{G}^{-1} \otimes \boldsymbol{I}_4)\left[\begin{pmatrix} \frac{\partial \boldsymbol{g}_0^{\mathsf{T}}}{\partial \vec{\boldsymbol{x}}} \\ \vdots \\ \frac{\partial \boldsymbol{g}_3^{\mathsf{T}}}{\partial \vec{\boldsymbol{x}}} \end{pmatrix} + \begin{pmatrix} \frac{\partial \boldsymbol{g}_0}{\partial \vec{\boldsymbol{x}}^{\mathsf{T}}} \\ \vdots \\ \frac{\partial \boldsymbol{g}_3}{\partial \vec{\boldsymbol{x}}^{\mathsf{T}}} \end{pmatrix} - \frac{\partial \boldsymbol{G}}{\partial \vec{\boldsymbol{x}}}\right].} \tag{2.68}$$

This can also be written as

$$\boldsymbol{\Gamma} = \frac{1}{2}\left(\boldsymbol{G}^{-1} \otimes \boldsymbol{I}_4\right)\left[(\boldsymbol{U}_{4\times4} - \boldsymbol{I}_{16})\frac{\partial \boldsymbol{G}}{\partial \vec{\boldsymbol{x}}} + \begin{pmatrix} \frac{\partial \boldsymbol{g}_0}{\partial \vec{\boldsymbol{x}}^\top} \\ \vdots \\ \frac{\partial \boldsymbol{g}_3}{\partial \vec{\boldsymbol{x}}^\top} \end{pmatrix}\right]. \tag{2.69}$$

Introducing the matrix

$$\check{\boldsymbol{\Gamma}} \stackrel{\text{def}}{=} \frac{1}{2}\left[(\boldsymbol{U}_{4\times4} - \boldsymbol{I}_{16})\frac{\partial \boldsymbol{G}}{\partial \vec{\boldsymbol{x}}} + \begin{pmatrix} \frac{\partial \boldsymbol{g}_0}{\partial \vec{\boldsymbol{x}}^\top} \\ \vdots \\ \frac{\partial \boldsymbol{g}_3}{\partial \vec{\boldsymbol{x}}^\top} \end{pmatrix}\right], \tag{2.70}$$

it can also be written as

$$\boldsymbol{\Gamma} = \left(\boldsymbol{G}^{-1} \otimes \boldsymbol{I}_4\right)\check{\boldsymbol{\Gamma}}. \tag{2.71}$$

The matrix difference $\boldsymbol{U}_{4\times4} - \boldsymbol{I}_{16}$, appearing in the matrix $\check{\boldsymbol{\Gamma}}$, has the remarkable property that the first, $(4+2)$th, $(8+3)$th and the 16th row (resp., column) are equal to a zero-row (resp., zero-column)! For the matrix $\check{\boldsymbol{\Gamma}}$ this has the consequence that the corresponding rows consist of $\frac{\partial g_{00}}{\partial \boldsymbol{x}^\top}$, $\frac{\partial g_{11}}{\partial \boldsymbol{x}^\top}$, $\frac{\partial g_{22}}{\partial \boldsymbol{x}^\top}$ and $\frac{\partial g_{33}}{\partial \boldsymbol{x}^\top}$. Furthermore, from (2.71) it follows that

$$\check{\boldsymbol{\Gamma}} = (\boldsymbol{G} \otimes \boldsymbol{I}_4)\boldsymbol{\Gamma}, \tag{2.72}$$

that is,

$$\check{\boldsymbol{\Gamma}}_k = \left(\boldsymbol{g}_k^\top \otimes \boldsymbol{I}_4\right)\boldsymbol{\Gamma} = g_{ko}\boldsymbol{\Gamma}_0 + \cdots + g_{k3}\boldsymbol{\Gamma}_3,$$

therefore,

$$\check{\Gamma}^k_{\alpha\beta} = \sum_{i=0}^{3} g_{ki}\Gamma^i_{\alpha\beta}. \tag{2.73}$$

2.3.1 *Alternative Geodesic Equation of Motion*

Again the equations of motion can be modified as follows: Firstly,

$$\left(\dot{\vec{\boldsymbol{x}}}^\top \otimes \boldsymbol{I}_4\right)\frac{\partial \boldsymbol{G}}{\partial \vec{\boldsymbol{x}}} = \frac{\partial \boldsymbol{G}}{\partial \vec{\boldsymbol{x}}^\top}(\dot{\vec{\boldsymbol{x}}} \otimes \boldsymbol{I}_4), \tag{2.74}$$

and secondly,

$$\left(\boldsymbol{I}_4 \otimes \dot{\vec{\boldsymbol{x}}}^\top\right)\frac{\partial \boldsymbol{G}}{\partial \vec{\boldsymbol{x}}}\dot{\vec{\boldsymbol{x}}} = \begin{pmatrix} \dot{\vec{\boldsymbol{x}}}^\top \boldsymbol{G}_0 \dot{\vec{\boldsymbol{x}}} \\ \vdots \\ \dot{\vec{\boldsymbol{x}}}^\top \boldsymbol{G}_3 \dot{\vec{\boldsymbol{x}}} \end{pmatrix}.$$

Using the $\boldsymbol{vec}$-operator from the Appendix (A.51) to the scalar component $\dot{\vec{x}}^\top \boldsymbol{G}_k \dot{\vec{x}}$ yields

$$\boldsymbol{vec}\big(\dot{\vec{x}}^\top \boldsymbol{G}_k \dot{\vec{x}}\big) = \big(\dot{\vec{x}}^\top \otimes \dot{\vec{x}}^\top\big)\boldsymbol{vec}(\boldsymbol{G}_k) = \big(\boldsymbol{vec}(\boldsymbol{G}_k)\big)^\top (\dot{\vec{x}} \otimes \dot{\vec{x}}),$$

so

$$\underline{\underline{\big(\boldsymbol{I}_4 \otimes \dot{\vec{x}}^\top\big)\frac{\partial \boldsymbol{G}}{\partial \vec{x}}\dot{\vec{x}} = \frac{\overline{\partial \boldsymbol{G}}}{\partial \vec{x}^\top}(\dot{\vec{x}} \otimes \dot{\vec{x}}),}} \tag{2.75}$$

with

$$\frac{\overline{\partial \boldsymbol{G}}}{\partial \vec{x}^\top} \stackrel{\text{def}}{=} \begin{pmatrix} (\boldsymbol{vec}(\boldsymbol{G}_0))^\top \\ \vdots \\ (\boldsymbol{vec}(\boldsymbol{G}_3))^\top \end{pmatrix} = \begin{pmatrix} \boldsymbol{g}_{0,0}^\top & & \boldsymbol{g}_{0,3}^\top \\ \boldsymbol{g}_{1,0}^\top & \cdots & \boldsymbol{g}_{1,3}^\top \\ \boldsymbol{g}_{2,0}^\top & & \boldsymbol{g}_{2,3}^\top \\ \boldsymbol{g}_{3,0}^\top & & \boldsymbol{g}_{3,3}^\top \end{pmatrix} \in \mathbb{R}^{4\times 16}, \tag{2.76}$$

where $\boldsymbol{g}_{i,j}^\top$ is the jth row of $\boldsymbol{G}_i$.

By the "method of careful examination", one can write

$$\frac{\overline{\partial \boldsymbol{G}}}{\partial \vec{x}^\top} = \begin{pmatrix} \boldsymbol{g}_{0,0}^\top \\ \boldsymbol{g}_{1,0}^\top \\ \boldsymbol{g}_{2,0}^\top \\ \boldsymbol{g}_{3,0}^\top \\ \vdots \\ \boldsymbol{g}_{0,3}^\top \\ \boldsymbol{g}_{1,3}^\top \\ \boldsymbol{g}_{2,3}^\top \\ \boldsymbol{g}_{3,3}^\top \end{pmatrix}^B = \left(\boldsymbol{U}_{4\times 4}\frac{\partial \boldsymbol{G}}{\partial \vec{x}}\right)^B, \tag{2.77}$$

where the superscript "B" means the *block-transposition* of the corresponding matrix. The block-transposition of a block matrix is defined as

$$\boldsymbol{A}^B \stackrel{\text{def}}{=} \begin{pmatrix} \boldsymbol{A}_1 \\ \vdots \\ \boldsymbol{A}_n \end{pmatrix}^B = \begin{pmatrix} \boldsymbol{A}_1 & \ldots & \boldsymbol{A}_n \end{pmatrix}.$$

Equation (2.75) used in (2.48) yields

$$\ddot{\vec{x}} = -\boldsymbol{G}^{-1}\left[\frac{\partial \boldsymbol{G}}{\partial \vec{x}^\top} - \frac{1}{2}\frac{\overline{\partial \boldsymbol{G}}}{\partial \vec{x}^\top}\right](\dot{\vec{x}} \otimes \dot{\vec{x}}). \tag{2.78}$$

With

$$\boxed{\tilde{\boldsymbol{\Gamma}} \stackrel{\text{def}}{=} \boldsymbol{G}^{-1}\left[\frac{\partial \boldsymbol{G}}{\partial \vec{x}^\top} - \frac{1}{2}\frac{\overline{\partial \boldsymbol{G}}}{\partial \vec{x}^\top}\right] \in \mathbb{R}^{4\times 16},} \tag{2.79}$$

one finally obtains

$$\boxed{\ddot{\vec{x}} = -\tilde{\boldsymbol{\Gamma}}(\dot{\vec{x}} \otimes \dot{\vec{x}}).} \tag{2.80}$$

One can also reach the form (2.80) as follows: It is $\ddot{x}_k = \dot{\boldsymbol{x}}^\mathsf{T} \boldsymbol{\Gamma}_k \dot{\boldsymbol{x}}$. Applying to this the $\boldsymbol{vec}$-operator (see Appendix), one obtains

$$-\ddot{x}_k = \boldsymbol{vec}\big(\dot{\vec{x}}^\mathsf{T} \boldsymbol{\Gamma}_k \dot{\vec{x}}\big) = \big(\dot{\vec{x}}^\mathsf{T} \otimes \dot{\vec{x}}^\mathsf{T}\big)\boldsymbol{vec}(\boldsymbol{\Gamma}_k) = \big(\boldsymbol{vec}(\boldsymbol{\Gamma}_k)\big)^\mathsf{T}(\dot{\vec{x}} \otimes \dot{\vec{x}}).$$

With

$$\tilde{\boldsymbol{\Gamma}} \overset{\text{def}}{=} \begin{pmatrix} (\boldsymbol{vec}(\boldsymbol{\Gamma}_0))^\mathsf{T} \\ \vdots \\ (\boldsymbol{vec}(\boldsymbol{\Gamma}_3))^\mathsf{T} \end{pmatrix} \tag{2.81}$$

(2.80) is obtained again. Once again, we can write

$$\tilde{\boldsymbol{\Gamma}} = \left(\boldsymbol{U}_{4\times 4}\frac{\partial \boldsymbol{\Gamma}}{\partial \vec{\boldsymbol{x}}}\right)^B. \tag{2.82}$$

Just a word regarding the derivatives with respect to s. If s^2 and $\mathrm{d}s^2$ are positive, then it is a so-called time-like event. Then

$$(\mathrm{d}s)^2 = c^2\,\mathrm{d}t^2 - \mathrm{d}\boldsymbol{x}^\mathsf{T}\,\mathrm{d}\boldsymbol{x} = c^2\,\mathrm{d}t^2 - \dot{\boldsymbol{x}}^\mathsf{T}\dot{\boldsymbol{x}}\,\mathrm{d}t^2,$$
$$\mathrm{d}s = \sqrt{c^2 - v^2}\,\mathrm{d}t,$$

so

$$\gamma\,\mathrm{d}s = c\,\mathrm{d}t.$$

Substituting $\mathrm{d}s = c\,\mathrm{d}\tau$, one obtains

$$\gamma\,\mathrm{d}\tau = \mathrm{d}t.$$

A comparison with the results of special relativity theory provides $\mathrm{d}\tau = \mathrm{d}t'$, i.e. the time that elapses in the moving coordinate system $\mathcal{X}'$. One calls in this context τ as *proper time*.

2.4 Example: Uniformly Rotating Systems

We will consider a fixed inertial frame $\mathcal{X}$ with the coordinates t, x, y and z and a uniformly around the z-axis *rotating* coordinate system $\mathcal{K}$ with the coordinates τ, r, φ and z. Then the transformation equations are

$$\begin{aligned} t &= \tau, \\ x &= r\cos(\varphi + \omega t), \\ y &= r\sin(\varphi + \omega t), \\ z &= z. \end{aligned} \tag{2.83}$$

As Jacobi matrix $\boldsymbol{J}$ one obtains

$$\boldsymbol{J} = \begin{pmatrix} 1 & 0 & 0 & 0 \\ -r\frac{\omega}{c}\sin(\varphi+\omega t) & \cos(\varphi+\omega t) & -r\sin(\varphi+\omega t) & 0 \\ r\frac{\omega}{c}\cos(\varphi+\omega t) & \sin(\varphi+\omega t) & r\cos(\varphi+\omega t) & 0 \\ 0 & 0 & 0 & 1 \end{pmatrix} \tag{2.84}$$

and as the metric matrix

$$\boldsymbol{G} = \boldsymbol{J}^{\mathsf{T}}\boldsymbol{M}\boldsymbol{J} = \begin{pmatrix} 1-r^2\frac{\omega^2}{c^2} & 0 & -r^2\frac{\omega}{c} & 0 \\ 0 & -1 & 0 & 0 \\ -r^2\frac{\omega}{c} & 0 & -r^2 & 0 \\ 0 & 0 & 0 & -1 \end{pmatrix}, \tag{2.85}$$

and from this

$$\boldsymbol{G}^{-1} = \begin{pmatrix} 1 & 0 & -\frac{\omega}{c} & 0 \\ 0 & -1 & 0 & 0 \\ -\frac{\omega}{c} & 0 & \frac{\omega^2}{c^2}-\frac{1}{r^2} & 0 \\ 0 & 0 & 0 & -1 \end{pmatrix}. \tag{2.86}$$

In this case, $\mathrm{d}s^2$ has the value

$$\mathrm{d}s^2 = \left(1-r^2\frac{\omega^2}{c^2}\right)\mathrm{d}\tau^2 - \mathrm{d}r^2 - r\,\mathrm{d}\varphi^2 - 2r^2\frac{\omega}{c}\,\mathrm{d}\varphi\,\mathrm{d}\tau - \mathrm{d}z^2. \tag{2.87}$$

If a clock is in the rotating system at the position (r,θ,z) and one considers two temporally directly adjacent events with $\mathrm{d}r = \mathrm{d}\varphi = \mathrm{d}z = 0$, then one obtains for the proper time $\mathrm{d}s$ in this case the relationship (with $v = r\omega$)

$$\mathrm{d}s = \mathrm{d}\tau\sqrt{1-r^2\omega^2/c^2} = \mathrm{d}\tau\sqrt{1-v^2/c^2} = \mathrm{d}\tau/\gamma.$$

This is the relationship known from the theory of special relativity! For the calculation of the acceleration, the derivatives of the metric matrix are needed. In this case, $\boldsymbol{G}_0 = \boldsymbol{G}_2 = \boldsymbol{G}_3 = \boldsymbol{0}$, but

$$\boldsymbol{G}_1 = \frac{\partial \boldsymbol{G}}{\partial r} = \begin{pmatrix} -2r\frac{\omega^2}{c^2} & 0 & -2r\frac{\omega}{c} & 0 \\ 0 & 0 & 0 & 0 \\ -2r\frac{\omega}{c} & 0 & -2r & 0 \\ 0 & 0 & 0 & 0 \end{pmatrix}.$$

Since only the matrix $\boldsymbol{G}_1 \neq \boldsymbol{0}$, in this case (2.56) is simplified to

$$\ddot{x}_k = \left[\frac{1}{2}\left(g^{k1}\dot{\boldsymbol{x}}^{\mathsf{T}}\right) - \left(\dot{x}_1 \boldsymbol{g}_k^{-\mathsf{T}}\right)\right]\boldsymbol{G}_1\dot{\boldsymbol{x}}.$$

In particular, with $\dot{\vec{x}}^\mathsf{T} = [c|\dot{r}|\dot{\varphi}|\dot{z}]$ we obtain

$$\begin{aligned}\ddot{r} &= -\frac{1}{2}\dot{\vec{x}}^\mathsf{T}\boldsymbol{G}_1\dot{\vec{x}} \\ &= -\frac{1}{2}\big[-2r\omega^2/c - \dot{\varphi}2r\omega/c \,|\, 0 \,|\, -2r\omega - 2r\dot{\varphi} \,|\, 0\big]\dot{\boldsymbol{x}} = r(\omega+\dot{\varphi})^2\end{aligned} \tag{2.88}$$

and

$$\ddot{\varphi} = -\dot{r}\big[-\omega/c \,|\, 0 \,|\, \omega^2/c^2 - 1/r^2 \,|\, 0\big]\boldsymbol{G}_1\dot{\vec{x}} = -2\dot{r}\omega/r - 2\dot{r}\dot{\varphi}/r, \tag{2.89}$$

or

$$r\ddot{\varphi} = -2\dot{r}(\dot{\varphi}+\omega). \tag{2.90}$$

Equation (2.88) multiplied by the mass m represents the centrifugal force, and (2.90) multiplied by the mass m is the so-called Coriolis force! The accelerations occurring in this rotating system are determined by the elements $g_{ij} = g_{ij}(\vec{x})$ of the coordinate-dependent metric matrix $\boldsymbol{G}(\vec{x})$. For a local reference system, one can always specify a coordinate transformation (with $\boldsymbol{J}^{-1}(\vec{x})$), so that the transformed system is obviously an inertial frame. In general, for an accelerated or non-uniformly moving (e.g. rotating) system no *globally valid* transformation matrix $\boldsymbol{J}$ can be specified. The given space is *curved*!

It should now be shown that the Christoffel matrices are the same for the rotating system by applying the formula (2.18). According to (2.84),

$$\boldsymbol{J} = \begin{pmatrix} 1 & 0 & 0 & 0 \\ -r\frac{\omega}{c}\sin(\varphi+\omega t) & \cos(\varphi+\omega t) & -r\sin(\varphi+\omega t) & 0 \\ r\frac{\omega}{c}\cos(\varphi+\omega t) & \sin(\varphi+\omega t) & r\cos(\varphi+\omega t) & 0 \\ 0 & 0 & 0 & 1 \end{pmatrix}, \tag{2.91}$$

i.e.

$$\boldsymbol{J}^{-1} = \begin{pmatrix} 1 & 0 & 0 & 0 \\ 0 & \cos(\varphi+\omega t) & \sin(\varphi+\omega t) & 0 \\ -\frac{\omega}{c} & \frac{1}{r}\sin(\varphi+\omega t) & \frac{1}{r}\cos(\varphi+\omega t) & 0 \\ 0 & 0 & 0 & 1 \end{pmatrix}. \tag{2.92}$$

Equations (2.88) and (2.89) can be used for the Christoffel matrices. The non-zero matrix elements are:

$$\Gamma^1_{00} = -r\frac{\omega^2}{c^2}, \qquad \Gamma^1_{02} = -r\frac{\omega}{c}, \qquad \Gamma^1_{03} = -r\frac{\omega}{c}, \qquad \Gamma^1_{22} = -1,$$

$$\Gamma^2_{01} = \frac{4\omega}{rc}, \quad \text{and} \quad \Gamma^2_{12} = \frac{4}{r}.$$

Accordingly, we obtain with (2.18), e.g.

$$\Gamma^1_{00} = \left[0 \mid \cos(\varphi+\omega t) \mid \sin(\varphi+\omega t) \mid 0\right] \begin{pmatrix} 0 \\ -r\frac{\omega^2}{c^2}\cos(\varphi+\omega t) \\ -r\frac{\omega^2}{c^2}\sin(\varphi+\omega t) \\ 0 \end{pmatrix}$$

$$= -r\frac{\omega^2}{c^2}.$$

2.5 General Coordinate Transformations

In the theory of general relativity, the invariance of the mathematical descriptions is demanded for the general laws of nature with respect to each other arbitrarily moving coordinate systems. Even more generally:

The invariance of the mathematical descriptions with respect to arbitrary coordinate transformations is required.

2.5.1 *Absolute Derivatives*

First, we must clarify how the derivatives eventually have to be modified so that the derived expressions are invariant under coordinate transformations.

Suppose a vector field $\boldsymbol{a}(\lambda)$ is defined along a curve whose parametric representation is given by $\vec{\boldsymbol{x}}(\lambda)$. Going on to another coordinate system $\mathcal{K}'$ with $\boldsymbol{a}'$, for the mathematical description of dynamic processes one is especially interested in how the derivative $\mathrm{d}\boldsymbol{a}/\mathrm{d}\lambda$ is transformed into $\mathrm{d}\boldsymbol{a}'/\mathrm{d}\lambda$. Because $\boldsymbol{T} = \boldsymbol{T}(\vec{\boldsymbol{x}}(\lambda))$, it follows that

$$\frac{\mathrm{d}\boldsymbol{a}'}{\mathrm{d}\lambda} = \frac{\mathrm{d}(\boldsymbol{T}\boldsymbol{a})}{\mathrm{d}\lambda} = \boldsymbol{T}\frac{\mathrm{d}\boldsymbol{a}}{\mathrm{d}\lambda} + \left(\frac{\mathrm{d}\vec{\boldsymbol{x}}^{\mathsf{T}}}{\mathrm{d}\lambda} \otimes \boldsymbol{I}_4\right)\frac{\partial \boldsymbol{T}}{\partial \vec{\boldsymbol{x}}}\boldsymbol{a}, \tag{2.93}$$

i.e. $\mathrm{d}\boldsymbol{a}/\mathrm{d}\lambda$ is not transformed into $\mathrm{d}\boldsymbol{a}'/\mathrm{d}\lambda$ as was $\boldsymbol{a}$ by a simple multiplication with the transformation matrix $\boldsymbol{T}$. For this the reason, the definition of the derivative is

$$\frac{\mathrm{d}\boldsymbol{a}}{\mathrm{d}\lambda} = \lim_{\delta\lambda\to 0} \frac{\boldsymbol{a}(\lambda+\delta\lambda) - \boldsymbol{a}(\lambda)}{\delta\lambda},$$

where the difference of the vectors is formed at different places on the curve γ, to which transformation matrices $\boldsymbol{T}(\lambda) \neq \boldsymbol{T}(\lambda+\delta\lambda)$ generally belong.

Thus in order to be always able to take the same transformation matrix, the difference of two vectors must be taken at the *same* place of the curve. It is true that

$$\delta\boldsymbol{a} \approx \frac{\mathrm{d}\boldsymbol{a}}{\mathrm{d}\lambda}\delta\lambda \tag{2.94}$$

and, if the shift of $\boldsymbol{a}$ is taken along a geodesic line,

$$\frac{\mathrm{d}\boldsymbol{a}}{\mathrm{d}\lambda} + (\boldsymbol{I}_4 \otimes \boldsymbol{a}^{\mathsf{T}})\boldsymbol{\Gamma}\frac{\mathrm{d}\vec{\boldsymbol{x}}}{\mathrm{d}\lambda} = \boldsymbol{0},$$

so

$$\frac{\mathrm{d}\boldsymbol{a}}{\mathrm{d}\lambda} = -(\boldsymbol{I}_4 \otimes \boldsymbol{a}^{\mathsf{T}})\boldsymbol{\Gamma}\frac{\mathrm{d}\vec{\boldsymbol{x}}}{\mathrm{d}\lambda}. \tag{2.95}$$

Here $\frac{\mathrm{d}\vec{\boldsymbol{x}}}{\mathrm{d}\lambda}$ is the tangent vector to the geodesic curve γ, and $\vec{\boldsymbol{x}}(\lambda)$ is the parametric representation of γ. Multiplying (2.95) with $\delta\lambda$, we obtain

$$\boxed{\delta\boldsymbol{a} = -(\boldsymbol{I}_4 \otimes \boldsymbol{a}^{\mathsf{T}})\boldsymbol{\Gamma}\delta\vec{\boldsymbol{x}}.} \tag{2.96}$$

Moving the vector $\boldsymbol{a}(\lambda)$ from the position $\vec{\boldsymbol{x}}(\lambda)$ parallel to the position $\vec{\boldsymbol{x}}(\lambda+\delta\lambda)$, the vector

$$\overline{\boldsymbol{a}} \stackrel{\text{def}}{=} \boldsymbol{a}(\lambda) + \delta\boldsymbol{a}$$

is obtained, or by (2.96),

$$\overline{\boldsymbol{a}} \approx \boldsymbol{a}(\lambda) - (\boldsymbol{I}_4 \otimes \boldsymbol{a}^{\mathsf{T}})\boldsymbol{\Gamma}\delta\vec{\boldsymbol{x}}. \tag{2.97}$$

On the other hand, $\boldsymbol{a}(\lambda+\delta\lambda) - \overline{\boldsymbol{a}}$ is a vector at the position $\gamma(\lambda+\delta\lambda)$, as is $(\boldsymbol{a}(\lambda+\delta\lambda) - \overline{\boldsymbol{a}})/\delta\lambda$. As $\delta\lambda \to 0$ the quotient is always a vector at the same location that varies, however.

The limit of this ratio is called *absolute derivative* $\frac{\mathrm{D}\boldsymbol{a}}{\mathrm{d}\lambda}$ of $\boldsymbol{a}(\lambda)$ along the curve γ:

$$\lim_{\delta\lambda\to 0}\frac{\boldsymbol{a}(\lambda+\delta\lambda) - \overline{\boldsymbol{a}}}{\delta\lambda} \approx \frac{\mathrm{d}\boldsymbol{a}}{\mathrm{d}\lambda} + \lim_{\delta\lambda\to 0}(\boldsymbol{I}_4 \otimes \boldsymbol{a}^{\mathsf{T}})\boldsymbol{\Gamma}\frac{\delta\vec{\boldsymbol{x}}}{\delta\lambda}$$

which is obtained with (2.97), and therefore one defines the absolute derivative as

$$\boxed{\frac{\mathrm{D}\boldsymbol{a}}{\mathrm{d}\lambda} \stackrel{\text{def}}{=} \frac{\mathrm{d}\boldsymbol{a}}{\mathrm{d}\lambda} + (\boldsymbol{I}_4 \otimes \boldsymbol{a}^{\mathsf{T}})\boldsymbol{\Gamma}\frac{\mathrm{d}\vec{\boldsymbol{x}}}{\mathrm{d}\lambda} = \dot{\boldsymbol{a}} + (\boldsymbol{I}_4 \otimes \boldsymbol{a}^{\mathsf{T}})\boldsymbol{\Gamma}\dot{\vec{\boldsymbol{x}}}.} \tag{2.98}$$

The differentiated vector can be decomposed into

$$\dot{\boldsymbol{a}} = \frac{\partial\boldsymbol{a}}{\partial\vec{\boldsymbol{x}}^{\mathsf{T}}}\dot{\vec{\boldsymbol{x}}}, \tag{2.99}$$

so that in (2.98) one can extract $\dot{\vec{\boldsymbol{x}}}$ to the right:

$$\boxed{\frac{\mathrm{D}\boldsymbol{a}}{\mathrm{d}\lambda} = \left[\frac{\partial\boldsymbol{a}}{\partial\vec{\boldsymbol{x}}^{\mathsf{T}}} + (\boldsymbol{I}_4 \otimes \boldsymbol{a}^{\mathsf{T}})\boldsymbol{\Gamma}\right]\dot{\vec{\boldsymbol{x}}}.} \tag{2.100}$$

The expression appearing in brackets is called the *covariant derivative of* $\boldsymbol{a}$ and written $\boldsymbol{a}_{||\vec{x}^{\mathsf{T}}} (\in \mathbb{R}^{4\times 4})$:

$$\boxed{\boldsymbol{a}_{||\vec{x}^{\mathsf{T}}} \stackrel{\text{def}}{=} \frac{\partial \boldsymbol{a}}{\partial \vec{x}^{\mathsf{T}}} + (\boldsymbol{I}_4 \otimes \boldsymbol{a}^{\mathsf{T}})\boldsymbol{\Gamma}.} \tag{2.101}$$

This covariant derivative $\boldsymbol{a}_{||\vec{x}^{\mathsf{T}}}$ becomes the normal partial derivative $\frac{a}{\vec{x}^{\mathsf{T}}}$, if $\boldsymbol{\Gamma} = \boldsymbol{0}$, that means, there is no gravitational field.

2.5.2 Transformation of the Christoffel Matrix $\tilde{\boldsymbol{\Gamma}}$

In (2.22), the Christoffel matrix is defined:

$$\tilde{\boldsymbol{\Gamma}} \stackrel{\text{def}}{=} \boldsymbol{J}^{-1}\frac{\partial \boldsymbol{J}}{\partial \vec{x}^{\mathsf{T}}} = \frac{\partial \vec{x}}{\partial \vec{\xi}^{\mathsf{T}}} \cdot \frac{\partial^2 \vec{\xi}}{\partial \vec{x}^{\mathsf{T}} \partial \vec{x}^{\mathsf{T}}}. \tag{2.102}$$

If one changes over from the coordinate system with $\vec{x}$ to the coordinate system with $\vec{x}'$, one obtains with the transformation matrices

$$\boldsymbol{T} \stackrel{\text{def}}{=} \frac{\partial \vec{x}'}{\partial \vec{x}^{\mathsf{T}}} \tag{2.103}$$

and

$$\bar{\boldsymbol{T}} \stackrel{\text{def}}{=} \frac{\partial \vec{x}}{\partial \vec{x}'^{\mathsf{T}}}, \tag{2.104}$$

the Christoffel matrix $\tilde{\boldsymbol{\Gamma}}'$ in the coordinate system with $\vec{x}'$

$$\begin{aligned}
\tilde{\boldsymbol{\Gamma}}' \stackrel{\text{def}}{=} \frac{\partial \vec{x}'}{\partial \vec{\xi}^{\mathsf{T}}} \cdot \frac{\partial^2 \vec{\xi}}{\partial \vec{x}'^{\mathsf{T}} \partial \vec{x}'^{\mathsf{T}}} &= \frac{\partial \vec{x}'}{\partial \vec{x}^{\mathsf{T}}} \frac{\partial \vec{x}}{\partial \vec{\xi}^{\mathsf{T}}} \frac{\partial}{\partial \vec{x}'^{\mathsf{T}}} \left(\frac{\partial \vec{\xi}}{\partial \vec{x}'^{\mathsf{T}}} \right) \\
&= \boldsymbol{T} \cdot \underbrace{\frac{\partial \vec{x}}{\partial \vec{\xi}^{\mathsf{T}}}}_{\boldsymbol{J}^{-1}} \frac{\partial}{\partial \vec{x}'^{\mathsf{T}}} \left(\underbrace{\frac{\partial \vec{\xi}}{\partial \vec{x}^{\mathsf{T}}}}_{\boldsymbol{J}} \cdot \underbrace{\frac{\partial \vec{x}}{\partial \vec{x}'^{\mathsf{T}}}}_{\bar{\boldsymbol{T}}} \right).
\end{aligned} \tag{2.105}$$

According to the product and chain rules, one obtains

$$\begin{aligned}
\frac{\partial}{\partial \vec{x}'^{\mathsf{T}}} (\boldsymbol{J} \cdot \bar{\boldsymbol{T}}) &= \frac{\partial \boldsymbol{J}}{\partial \vec{x}'^{\mathsf{T}}} (\boldsymbol{I}_4 \otimes \bar{\boldsymbol{T}}) + \boldsymbol{J} \frac{\partial \bar{\boldsymbol{T}}}{\partial \vec{x}'^{\mathsf{T}}} \\
&= \frac{\partial \boldsymbol{J}}{\partial \vec{x}^{\mathsf{T}}} (\bar{\boldsymbol{T}} \otimes \boldsymbol{I}_4)(\boldsymbol{I}_4 \otimes \bar{\boldsymbol{T}}) + \boldsymbol{J} \frac{\partial \bar{\boldsymbol{T}}}{\partial \vec{x}^{\mathsf{T}}} (\bar{\boldsymbol{T}} \otimes \boldsymbol{I}_4) \\
&= \frac{\partial \boldsymbol{J}}{\partial \vec{x}^{\mathsf{T}}} (\bar{\boldsymbol{T}} \otimes \bar{\boldsymbol{T}}) + \boldsymbol{J} \frac{\partial \bar{\boldsymbol{T}}}{\partial \vec{x}^{\mathsf{T}}} (\bar{\boldsymbol{T}} \otimes \boldsymbol{I}_4).
\end{aligned} \tag{2.106}$$

Inserting (2.106) in (2.105) reveals

$$\boxed{\tilde{\boldsymbol{\Gamma}}' = \boldsymbol{T}\tilde{\boldsymbol{\Gamma}}(\bar{\boldsymbol{T}} \otimes \bar{\boldsymbol{T}}) + \boldsymbol{T}\frac{\partial \bar{\boldsymbol{T}}}{\partial \vec{\boldsymbol{x}}^{\mathsf{T}}}(\bar{\boldsymbol{T}} \otimes \boldsymbol{I}_4).} \tag{2.107}$$

The second term on the right-hand side shows the coordinate dependence of the transformation matrix $\boldsymbol{T}$.

Another important characteristic is obtained as follows. Differentiating $\boldsymbol{I}_4 = \boldsymbol{T}\bar{\boldsymbol{T}}$ with respect to $\vec{\boldsymbol{x}}^{\mathsf{T}}$ yields

$$\boldsymbol{O} = \frac{\partial \boldsymbol{T}}{\partial \vec{\boldsymbol{x}}^{\mathsf{T}}}(\boldsymbol{I}_4 \otimes \bar{\boldsymbol{T}}) + \boldsymbol{T}\frac{\partial \bar{\boldsymbol{T}}}{\partial \vec{\boldsymbol{x}}^{\mathsf{T}}},$$

i.e.

$$\boldsymbol{T}\frac{\partial \bar{\boldsymbol{T}}}{\partial \vec{\boldsymbol{x}}^{\mathsf{T}}} = -\frac{\partial \boldsymbol{T}}{\partial \vec{\boldsymbol{x}}^{\mathsf{T}}}(\boldsymbol{I}_4 \otimes \bar{\boldsymbol{T}}). \tag{2.108}$$

Putting this in (2.107), one gets another form of the transformed Christoffel matrix, namely

$$\tilde{\boldsymbol{\Gamma}}' = \boldsymbol{T}\tilde{\boldsymbol{\Gamma}}(\bar{\boldsymbol{T}} \otimes \bar{\boldsymbol{T}}) - \frac{\partial \boldsymbol{T}}{\partial \vec{\boldsymbol{x}}^{\mathsf{T}}}(\bar{\boldsymbol{T}} \otimes \bar{\boldsymbol{T}}). \tag{2.109}$$

Moreover,

$$\frac{\mathrm{d}^2\vec{\boldsymbol{x}}'}{\mathrm{d}\tau^2} = \frac{\mathrm{d}}{\mathrm{d}\tau}\Bigg(\underbrace{\frac{\partial \vec{\boldsymbol{x}}'}{\partial \vec{\boldsymbol{x}}^{\mathsf{T}}}}_{\boldsymbol{T}} \cdot \frac{\mathrm{d}\vec{\boldsymbol{x}}}{\mathrm{d}\tau}\Bigg) = \boldsymbol{T}\frac{\mathrm{d}^2\vec{\boldsymbol{x}}}{\mathrm{d}\tau^2} + \frac{\partial \boldsymbol{T}}{\partial \vec{\boldsymbol{x}}^{\mathsf{T}}}\underbrace{\left(\frac{\mathrm{d}\vec{\boldsymbol{x}}}{\mathrm{d}\tau} \otimes \boldsymbol{I}_4\right)\frac{\mathrm{d}\vec{\boldsymbol{x}}}{\mathrm{d}\tau}}_{(\frac{\mathrm{d}\vec{x}}{\mathrm{d}\tau} \otimes \frac{\mathrm{d}\vec{x}}{\mathrm{d}\tau})}. \tag{2.110}$$

Multiplying (2.109) by the vector $(\frac{\mathrm{d}\vec{x}'}{\mathrm{d}\tau} \otimes \frac{\mathrm{d}\vec{x}'}{\mathrm{d}\tau})$ from the right yields

$$\begin{aligned}\tilde{\boldsymbol{\Gamma}}'\left(\frac{\mathrm{d}\vec{\boldsymbol{x}}'}{\mathrm{d}\tau} \otimes \frac{\mathrm{d}\vec{\boldsymbol{x}}'}{\mathrm{d}\tau}\right) &= \boldsymbol{T}\tilde{\boldsymbol{\Gamma}}\,\underbrace{(\bar{\boldsymbol{T}} \otimes \bar{\boldsymbol{T}})\left(\frac{\mathrm{d}\vec{\boldsymbol{x}}'}{\mathrm{d}\tau} \otimes \frac{\mathrm{d}\vec{\boldsymbol{x}}'}{\mathrm{d}\tau}\right)}_{(\frac{\mathrm{d}\vec{x}}{\mathrm{d}\tau} \otimes \frac{\mathrm{d}\vec{x}}{\mathrm{d}\tau})} \\ &\quad - \frac{\partial \boldsymbol{T}}{\partial \vec{\boldsymbol{x}}^{\mathsf{T}}}\underbrace{(\bar{\boldsymbol{T}} \otimes \bar{\boldsymbol{T}})\left(\frac{\mathrm{d}\vec{\boldsymbol{x}}'}{\mathrm{d}\tau} \otimes \frac{\mathrm{d}\vec{\boldsymbol{x}}'}{\mathrm{d}\tau}\right)}_{(\frac{\mathrm{d}\vec{x}}{\mathrm{d}\tau} \otimes \frac{\mathrm{d}\vec{x}}{\mathrm{d}\tau})}.\end{aligned} \tag{2.111}$$

Adding the two equations (2.110) and (2.111), we finally obtain

$$\boxed{\frac{\mathrm{d}^2\vec{\boldsymbol{x}}'}{\mathrm{d}\tau^2} + \tilde{\boldsymbol{\Gamma}}'\left(\frac{\mathrm{d}\vec{\boldsymbol{x}}'}{\mathrm{d}\tau} \otimes \frac{\mathrm{d}\vec{\boldsymbol{x}}'}{\mathrm{d}\tau}\right) = \boldsymbol{T}\left[\frac{\mathrm{d}^2\vec{\boldsymbol{x}}}{\mathrm{d}\tau^2} + \tilde{\boldsymbol{\Gamma}}\left(\frac{\mathrm{d}\vec{\boldsymbol{x}}}{\mathrm{d}\tau} \otimes \frac{\mathrm{d}\vec{\boldsymbol{x}}}{\mathrm{d}\tau}\right)\right].} \tag{2.112}$$

The vector in the brackets in (2.112) is transformed as a vector in general! The equation of motion is invariant.

2.5.3 Transformation of the Christoffel Matrix $\hat{\boldsymbol{\Gamma}}$

By (2.15),

$$\hat{\boldsymbol{\Gamma}} = \boldsymbol{U}_{4\times4}\begin{pmatrix} \boldsymbol{J}^{-1}\boldsymbol{J}_0 \\ \vdots \\ \boldsymbol{J}^{-1}\boldsymbol{J}_3 \end{pmatrix} = \boldsymbol{U}_{4\times4}\big(\boldsymbol{I}_4 \otimes \boldsymbol{J}^{-1}\big)\frac{\partial \boldsymbol{J}}{\partial \vec{x}}, \tag{2.113}$$

where

$$\boldsymbol{J} = \frac{\partial \vec{\xi}}{\partial \vec{x}^{\mathsf{T}}}.$$

Defining furthermore

$$\boldsymbol{J}' = \frac{\partial \vec{\xi}}{\partial \vec{x}'^{\mathsf{T}}}, \tag{2.114}$$

we obtain with the transformation matrices $\boldsymbol{T}$ and $\bar{\boldsymbol{T}}$ the relation

$$\boldsymbol{J} = \frac{\partial \vec{\xi}}{\partial \vec{x}^{\mathsf{T}}} = \frac{\partial \vec{\xi}}{\partial \vec{x}'^{\mathsf{T}}}\frac{\partial \vec{x}'}{\partial \vec{x}^{\mathsf{T}}} = \frac{\partial \vec{\xi}}{\partial \vec{x}'^{\mathsf{T}}}\boldsymbol{T}, \tag{2.115}$$

i.e. with (2.114)

$$\boldsymbol{J} = \boldsymbol{J}'\boldsymbol{T}, \tag{2.116}$$

or

$$\boldsymbol{J}' = \boldsymbol{J}\bar{\boldsymbol{T}}. \tag{2.117}$$

Hence

$$\hat{\boldsymbol{\Gamma}}' = \boldsymbol{U}_{4\times4}\big(\boldsymbol{I}_4 \otimes \boldsymbol{J}'^{-1}\big)\frac{\partial \boldsymbol{J}'}{\partial \vec{x}'}. \tag{2.118}$$

With

$$\boldsymbol{J}'^{-1} = \boldsymbol{T}\boldsymbol{J}^{-1}$$

and

$$\frac{\partial \boldsymbol{J}'}{\partial \vec{x}'} = \frac{\partial \boldsymbol{J}\bar{\boldsymbol{T}}}{\partial \vec{x}'} = \frac{\partial \boldsymbol{J}}{\partial \vec{x}'}\bar{\boldsymbol{T}} + (\boldsymbol{I}_4 \otimes \boldsymbol{J})\frac{\partial \bar{\boldsymbol{T}}}{\partial \boldsymbol{x}'},$$

we then obtain

$$\begin{aligned} \hat{\boldsymbol{\Gamma}}' &= \boldsymbol{U}_{4\times4}(\boldsymbol{I}_4 \otimes \boldsymbol{T})\big(\boldsymbol{I}_4 \otimes \boldsymbol{J}^{-1}\big)\left(\frac{\partial \boldsymbol{J}}{\partial \vec{x}'}\bar{\boldsymbol{T}} + (\boldsymbol{I}_4 \otimes \boldsymbol{J})\frac{\partial \bar{\boldsymbol{T}}}{\partial \boldsymbol{x}'}\right) \\ &= \boldsymbol{U}_{4\times4}(\boldsymbol{I}_4 \otimes \boldsymbol{T})\left(\big(\boldsymbol{I}_4 \otimes \boldsymbol{J}^{-1}\big)\frac{\partial \boldsymbol{J}}{\partial \vec{x}'}\bar{\boldsymbol{T}} + \frac{\partial \bar{\boldsymbol{T}}}{\partial \boldsymbol{x}'}\right). \end{aligned} \tag{2.119}$$

Furthermore,

$$\frac{\partial \boldsymbol{J}}{\partial \vec{\boldsymbol{x}}'} = \left(\frac{\partial \vec{\boldsymbol{x}}^{\mathsf{T}}}{\partial \vec{\boldsymbol{x}}'} \otimes \boldsymbol{I}_4\right)\frac{\partial \boldsymbol{J}}{\partial \vec{\boldsymbol{x}}} = (\bar{\boldsymbol{T}}^{\mathsf{T}} \otimes \boldsymbol{I}_4)\frac{\partial \boldsymbol{J}}{\partial \vec{\boldsymbol{x}}}, \tag{2.120}$$

so

$$\begin{aligned}\hat{\boldsymbol{\Gamma}}' &= \boldsymbol{U}_{4\times4}\left[(\bar{\boldsymbol{T}}^{\mathsf{T}} \otimes \boldsymbol{T}\boldsymbol{J}^{-1})\frac{\partial \boldsymbol{J}}{\partial \vec{\boldsymbol{x}}}\bar{\boldsymbol{T}} + (\boldsymbol{I}_4 \otimes \boldsymbol{T})\frac{\partial \bar{\boldsymbol{T}}}{\partial \vec{\boldsymbol{x}}'}\right] \\ &= \underbrace{\boldsymbol{U}_{4\times4}(\bar{\boldsymbol{T}}^{\mathsf{T}} \otimes \boldsymbol{T})\boldsymbol{U}_{4\times4}}_{\boldsymbol{T}\otimes\bar{\boldsymbol{T}}^{\mathsf{T}}}\underbrace{\boldsymbol{U}_{4\times4}(\boldsymbol{I}_4 \otimes \boldsymbol{J}^{-1})\frac{\partial \boldsymbol{J}}{\partial \vec{\boldsymbol{x}}}}_{\hat{\boldsymbol{\Gamma}}}\bar{\boldsymbol{T}} + \boldsymbol{U}_{4\times4}(\boldsymbol{I}_4 \otimes \boldsymbol{T})\frac{\partial \bar{\boldsymbol{T}}}{\partial \vec{\boldsymbol{x}}'},\end{aligned}$$

and with (2.113)

$$\boxed{\hat{\boldsymbol{\Gamma}}' = (\boldsymbol{T} \otimes \bar{\boldsymbol{T}}^{\mathsf{T}})\hat{\boldsymbol{\Gamma}}\bar{\boldsymbol{T}} + \boldsymbol{U}_{4\times4}(\boldsymbol{I}_4 \otimes \boldsymbol{T})\frac{\partial \bar{\boldsymbol{T}}}{\partial \vec{\boldsymbol{x}}'}.} \tag{2.121}$$

The second term on the right-hand side expresses again the dependence on the transformation matrix $\boldsymbol{T}$.

2.5.4 Coordinate Transformation and Covariant Derivative

It is true that

$$\boldsymbol{T}\bar{\boldsymbol{T}} = \boldsymbol{I},$$

so

$$\frac{\partial}{\partial \vec{\boldsymbol{x}}'}(\boldsymbol{T}\bar{\boldsymbol{T}}) = \boldsymbol{0} = \frac{\partial \boldsymbol{T}}{\partial \vec{\boldsymbol{x}}'}\bar{\boldsymbol{T}} + (\boldsymbol{I}_4 \otimes \boldsymbol{T})\frac{\partial \bar{\boldsymbol{T}}}{\partial \vec{\boldsymbol{x}}'},$$

or

$$(\boldsymbol{I}_4 \otimes \boldsymbol{T})\frac{\partial \bar{\boldsymbol{T}}}{\partial \vec{\boldsymbol{x}}'} = -\frac{\partial \boldsymbol{T}}{\partial \vec{\boldsymbol{x}}'}\bar{\boldsymbol{T}}. \tag{2.122}$$

Look now at the coordinate transformation

$$\boldsymbol{a}' = \boldsymbol{T}\boldsymbol{a} \tag{2.123}$$

and its partial derivative with respect to $\vec{\boldsymbol{x}}'$:

$$\frac{\partial \boldsymbol{a}'}{\partial \vec{\boldsymbol{x}}'} = \frac{\partial}{\partial \vec{\boldsymbol{x}}'}(\boldsymbol{T}\boldsymbol{a}) = \frac{\partial \boldsymbol{T}}{\partial \vec{\boldsymbol{x}}'}\boldsymbol{a} + (\boldsymbol{I}_4 \otimes \boldsymbol{T})\frac{\partial \boldsymbol{a}}{\partial \vec{\boldsymbol{x}}'}. \tag{2.124}$$

On the other hand,

$$\frac{\partial \boldsymbol{a}}{\partial \vec{\boldsymbol{x}}'} = \left(\frac{\partial \vec{\boldsymbol{x}}^{\mathsf{T}}}{\partial \vec{\boldsymbol{x}}'} \otimes \boldsymbol{I}_4\right)\frac{\partial \boldsymbol{a}}{\partial \vec{\boldsymbol{x}}} = (\bar{\boldsymbol{T}}^{\mathsf{T}} \otimes \boldsymbol{I}_4)\frac{\partial \boldsymbol{a}}{\partial \vec{\boldsymbol{x}}}.$$

Inserting this into (2.124) yields

$$\frac{\partial a'}{\partial \vec{x}'} = \left(\bar{T}^{\mathsf{T}} \otimes T\right)\frac{\partial a}{\partial \vec{x}} + \frac{\partial T}{\partial \vec{x}'}a. \tag{2.125}$$

How does the product of the matrix Γ and the vector a transform? We have

$$\Gamma' a' = \left(T \otimes \bar{T}^{\mathsf{T}}\right)\Gamma \bar{T} T a + U_{4\times 4}(I_4 \otimes T)\frac{\partial T^{-1}}{\partial \vec{x}'}T a, \tag{2.126}$$

and hence obtain

$$\frac{\partial T^{-1}}{\partial \vec{x}'} = -\left(I_4 \otimes T^{-1}\right)\frac{\partial T}{\partial \vec{x}'}T^{-1}.$$

This in (2.126) yields

$$\Gamma' a' = \left(T \otimes \bar{T}^{\mathsf{T}}\right)\Gamma a - U_{4\times 4}\frac{\partial T}{\partial \vec{x}'}a. \tag{2.127}$$

After adding (2.125) and (2.127), multiplied from the left by $U_{4\times 4}$, we get

$$\frac{\partial a'}{\partial \vec{x}'} + U_{4\times 4}\Gamma' a' = \left(\bar{T}^{\mathsf{T}} \otimes T\right)\frac{\partial a}{\partial \vec{x}} + U_{4\times 4}\left(T \otimes \bar{T}^{\mathsf{T}}\right)U_{4\times 4}U_{4\times 4}\Gamma a,$$

so

$$\frac{\partial a'}{\partial \vec{x}'} + U_{4\times 4}\Gamma' a' = \left(\bar{T}^{\mathsf{T}} \otimes T\right)\left[\frac{\partial a}{\partial \vec{x}} + U_{4\times 4}\Gamma a\right]. \tag{2.128}$$

With the definition

$$\Gamma^* \stackrel{\text{def}}{=} U_{4\times 4}\Gamma = \frac{1}{2}\left(I \otimes G^{-1}\right)\left[(I_{16} - U_{4\times 4})\frac{\partial G}{\partial \vec{x}} + U_{4\times 4}\begin{pmatrix} \frac{\partial g_0}{\partial \vec{x}^{\mathsf{T}}} \\ \vdots \\ \frac{\partial g_3}{\partial \vec{x}^{\mathsf{T}}} \end{pmatrix}\right] \tag{2.129}$$

the following compact relation is obtained:

$$\boxed{\frac{\partial a'}{\partial \vec{x}'} + \Gamma^{*\prime} a' = \left(\bar{T}^{\mathsf{T}} \otimes T\right)\left[\frac{\partial a}{\partial \vec{x}} + \Gamma^* a\right].} \tag{2.130}$$

On the right-hand side of this equation, a vector from $\mathbb{R}^{16}$ is multiplied by a 16×16-matrix! We introduce the abbreviations

$$a_{|\vec{x}} \stackrel{\text{def}}{=} \frac{\partial a}{\partial \vec{x}} \tag{2.131}$$

and

$$\underline{\underline{a_{\|\vec{x}} \stackrel{\text{def}}{=} a_{|\vec{x}} + \Gamma^* a,}} \tag{2.132}$$

and call $\boldsymbol{a}_{\|\vec{x}}$ the *covariant derivative of* $\boldsymbol{a}$ *with respect to* $\vec{x}$. Now (2.130) is written as

$$\boxed{\boldsymbol{a}'_{\|\vec{x}'} = (\bar{\boldsymbol{T}}^\mathsf{T} \otimes \boldsymbol{T})\boldsymbol{a}_{\|\vec{x}}.} \tag{2.133}$$

The form on the right-hand side of (2.130) reminds us of the right-hand side of the expression used in the $\boldsymbol{vec}$-operator (see Appendix), that is,

$$\boldsymbol{vec}(\boldsymbol{ABC}) = \left(\boldsymbol{C}^\mathsf{T} \otimes \boldsymbol{A}\right)\boldsymbol{vec}(\boldsymbol{B}). \tag{2.134}$$

To use this lemma, (2.130) is first written in some detail. With the new matrix

$$\boldsymbol{\Gamma}_k^* \stackrel{\text{def}}{=} \left(\boldsymbol{i}_k^\mathsf{T} \otimes \boldsymbol{G}^{-1}\right)\left[\frac{1}{2}\boldsymbol{I}_{16} - \boldsymbol{U}_{4\times 4}\right]\frac{\partial \boldsymbol{G}}{\partial \vec{\boldsymbol{x}}}, \tag{2.135}$$

where $\boldsymbol{i}_k^\mathsf{T}$ is the kth row of the unit matrix $\boldsymbol{I}_4$, one obtains for (2.130)

$$\begin{aligned}
\begin{pmatrix} \frac{\partial \boldsymbol{a}'}{\partial x_0'} \\ \vdots \\ \frac{\partial \boldsymbol{a}'}{\partial x_3'} \end{pmatrix} + \begin{pmatrix} \boldsymbol{\Gamma}_0^{*\prime}\boldsymbol{a}' \\ \vdots \\ \boldsymbol{\Gamma}_3^{*\prime}\boldsymbol{a}' \end{pmatrix} &= (\boldsymbol{I}_4^\mathsf{T} \otimes \boldsymbol{I}_4)\left[\begin{pmatrix} \frac{\partial \boldsymbol{a}'}{\partial x_0'} \\ \vdots \\ \frac{\partial \boldsymbol{a}'}{\partial x_3'} \end{pmatrix} + \begin{pmatrix} \boldsymbol{\Gamma}_0^{*\prime}\boldsymbol{a}' \\ \vdots \\ \boldsymbol{\Gamma}_3^{*\prime}\boldsymbol{a}' \end{pmatrix}\right] \\
&= (\bar{\boldsymbol{T}}^\mathsf{T} \otimes \boldsymbol{T})\left[\begin{pmatrix} \frac{\partial \boldsymbol{a}}{\partial x_0} \\ \vdots \\ \frac{\partial \boldsymbol{a}}{\partial x_3} \end{pmatrix} + \begin{pmatrix} \boldsymbol{\Gamma}_0^{*}\boldsymbol{a} \\ \vdots \\ \boldsymbol{\Gamma}_3^{*}\boldsymbol{a} \end{pmatrix}\right].
\end{aligned} \tag{2.136}$$

Now Lemma (2.134) applied to this equation shows that

$$\begin{aligned}
&\left[\frac{\partial \boldsymbol{a}'}{\partial x_0'}\middle|\cdots\middle|\frac{\partial \boldsymbol{a}'}{\partial x_3'}\right] + \left[\boldsymbol{\Gamma}_0^{*\prime}\boldsymbol{a}'|\ldots|\boldsymbol{\Gamma}_3^{*\prime}\boldsymbol{a}'\right] \\
&\quad = \boldsymbol{T}\left\{\left[\frac{\partial \boldsymbol{a}}{\partial x_0}\middle|\cdots\middle|\frac{\partial \boldsymbol{a}}{\partial x_3}\right] + \left[\boldsymbol{\Gamma}_0^{*}\boldsymbol{a}|\ldots|\boldsymbol{\Gamma}_3^{*}\boldsymbol{a}\right]\right\}\boldsymbol{T}^{-1}.
\end{aligned} \tag{2.137}$$

With

$$\frac{\partial \boldsymbol{a}}{\partial \vec{\boldsymbol{x}}^\mathsf{T}} \stackrel{\text{def}}{=} \left[\frac{\partial \boldsymbol{a}}{\partial x_0}\middle|\cdots\middle|\frac{\partial \boldsymbol{a}}{\partial x_3}\right] \in \mathbb{R}^{4\times 4} \quad \text{and} \quad \bar{\boldsymbol{\Gamma}} \stackrel{\text{def}}{=} \left[\boldsymbol{\Gamma}_0^*|\ldots|\boldsymbol{\Gamma}_3^*\right] \in \mathbb{R}^{4\times 16}$$

one obtains

$$\boxed{\frac{\partial \boldsymbol{a}'}{\partial \vec{\boldsymbol{x}}'^\mathsf{T}} + \bar{\boldsymbol{\Gamma}}'\left(\boldsymbol{I}_4 \otimes \boldsymbol{a}'\right) = \boldsymbol{T}\left[\frac{\partial \boldsymbol{a}}{\partial \vec{\boldsymbol{x}}^\mathsf{T}} + \bar{\boldsymbol{\Gamma}}(\boldsymbol{I}_4 \otimes \boldsymbol{a})\right]\boldsymbol{T}^{-1}.} \tag{2.138}$$

The sum of matrices

$$\frac{\partial \boldsymbol{a}}{\partial \vec{\boldsymbol{x}}^\mathsf{T}} + \bar{\boldsymbol{\Gamma}}(\boldsymbol{I}_4 \otimes \boldsymbol{a})$$

is therefore transformed by a normal similarity transformation in to the following sum of matrices

$$\frac{\partial \boldsymbol{a}'}{\partial \vec{\boldsymbol{x}}'^{\mathsf{T}}} + \bar{\boldsymbol{\Gamma}}'\left(\boldsymbol{I}_4 \otimes \boldsymbol{a}'\right).$$

And again some abbreviations are introduced (now the vector $\vec{x}$ is transposed!):

$$\boldsymbol{a}_{|\vec{x}^{\mathsf{T}}} \stackrel{\text{def}}{=} \frac{\partial \boldsymbol{a}}{\partial \vec{\boldsymbol{x}}^{\mathsf{T}}} \tag{2.139}$$

and

$$\boxed{\boldsymbol{a}_{\|\vec{x}^{\mathsf{T}}} \stackrel{\text{def}}{=} \boldsymbol{a}_{|\vec{x}^{\mathsf{T}}} + \bar{\boldsymbol{\Gamma}}(\boldsymbol{I}_4 \otimes \boldsymbol{a}) \in \mathbb{R}^{4\times 4}.} \tag{2.140}$$

$\boldsymbol{a}_{\|\vec{x}^{\mathsf{T}}}$ is called again the *covariant derivative* of $\boldsymbol{a}$ and it is

$$\boxed{\boldsymbol{a}'_{\|\vec{x}'^{\mathsf{T}}} = \boldsymbol{T}\boldsymbol{a}_{\|\vec{x}^{\mathsf{T}}}\boldsymbol{T}^{-1}.} \tag{2.141}$$

Important conclusion:

The formulas in the theory of general relativity are invariant with respect to coordinate transformations if in formulas of special relativity ordinary derivatives $\frac{\partial a}{\partial x^{\mathsf{T}}}$ are replaced by covariant derivatives $\boldsymbol{a}_{\|x^{\mathsf{T}}}$!

The covariant derivative defined in (2.140) differs from the derivative defined in (2.100) by the summand $\bar{\boldsymbol{\Gamma}}(\boldsymbol{I}_4 \otimes \boldsymbol{a})$; there one has instead $(\boldsymbol{I}_4 \otimes \boldsymbol{a}^{\mathsf{T}})\boldsymbol{\Gamma}$. But it is indeed true that

$$\bar{\boldsymbol{\Gamma}}(\boldsymbol{I}_4 \otimes \boldsymbol{a}) = \left(\boldsymbol{I}_4 \otimes \boldsymbol{a}^{\mathsf{T}}\right)\boldsymbol{\Gamma}. \tag{2.142}$$

In fact, if we rename the jth row of the sub-matrix $\boldsymbol{\Gamma}_i$ by $\boldsymbol{\gamma}^{i\,\mathsf{T}}_{j}$, then the matrix $\boldsymbol{\Gamma}^*$ is composed as

$$\boldsymbol{\Gamma}^* = \boldsymbol{U}_{4\times 4}\boldsymbol{\Gamma} = \begin{pmatrix} \boldsymbol{\gamma}_0^{0\mathsf{T}} \\ \boldsymbol{\gamma}_0^{1\mathsf{T}} \\ \boldsymbol{\gamma}_0^{2\mathsf{T}} \\ \boldsymbol{\gamma}_0^{3\mathsf{T}} \\ \hline \vdots \\ \hline \boldsymbol{\gamma}_3^{0\mathsf{T}} \\ \boldsymbol{\gamma}_3^{1\mathsf{T}} \\ \boldsymbol{\gamma}_3^{2\mathsf{T}} \\ \boldsymbol{\gamma}_3^{3\mathsf{T}} \end{pmatrix},$$

i.e.

$$\bar{\boldsymbol{\Gamma}} = \begin{pmatrix} \boldsymbol{\gamma}_0^{0\mathsf{T}} & & \boldsymbol{\gamma}_3^{0\mathsf{T}} \\ \boldsymbol{\gamma}_0^{1\mathsf{T}} & & \boldsymbol{\gamma}_3^{1\mathsf{T}} \\ \boldsymbol{\gamma}_0^{2\mathsf{T}} & \dots & \boldsymbol{\gamma}_3^{2\mathsf{T}} \\ \boldsymbol{\gamma}_0^{3\mathsf{T}} & & \boldsymbol{\gamma}_3^{3\mathsf{T}} \end{pmatrix}. \tag{2.143}$$

With $\boldsymbol{\Gamma}_i = \boldsymbol{\Gamma}_i^{\mathsf{T}}$ one obtains:

$$\underline{\underline{\bar{\boldsymbol{\Gamma}}(\boldsymbol{I}_4 \otimes \boldsymbol{a})}} = \begin{pmatrix} \boldsymbol{\gamma}_0^{0\mathsf{T}}\boldsymbol{a} & & \boldsymbol{\gamma}_3^{0\mathsf{T}}\boldsymbol{a} \\ \boldsymbol{\gamma}_0^{1\mathsf{T}}\boldsymbol{a} & & \boldsymbol{\gamma}_3^{1\mathsf{T}}\boldsymbol{a} \\ \boldsymbol{\gamma}_0^{2\mathsf{T}}\boldsymbol{a} & \dots & \boldsymbol{\gamma}_3^{2\mathsf{T}}\boldsymbol{a} \\ \boldsymbol{\gamma}_0^{3\mathsf{T}}\boldsymbol{a} & & \boldsymbol{\gamma}_3^{3\mathsf{T}}\boldsymbol{a} \end{pmatrix} = \begin{pmatrix} \boldsymbol{a}^{\mathsf{T}}\boldsymbol{\gamma}_0^{0} & & \boldsymbol{a}^{\mathsf{T}}\boldsymbol{\gamma}_3^{0} \\ \boldsymbol{a}^{\mathsf{T}}\boldsymbol{\gamma}_0^{1} & & \boldsymbol{a}^{\mathsf{T}}\boldsymbol{\gamma}_3^{1} \\ \boldsymbol{a}^{\mathsf{T}}\boldsymbol{\gamma}_0^{2} & \dots & \boldsymbol{a}^{\mathsf{T}}\boldsymbol{\gamma}_3^{2} \\ \boldsymbol{a}^{\mathsf{T}}\boldsymbol{\gamma}_0^{3} & & \boldsymbol{a}^{\mathsf{T}}\boldsymbol{\gamma}_3^{3} \end{pmatrix}$$

$$= \begin{pmatrix} \boldsymbol{a}^{\mathsf{T}}\boldsymbol{\Gamma}_0^{\mathsf{T}} \\ \vdots \\ \boldsymbol{a}^{\mathsf{T}}\boldsymbol{\Gamma}_3^{\mathsf{T}} \end{pmatrix} = \begin{pmatrix} \boldsymbol{a}^{\mathsf{T}}\boldsymbol{\Gamma}_0 \\ \vdots \\ \boldsymbol{a}^{\mathsf{T}}\boldsymbol{\Gamma}_3 \end{pmatrix} = \underline{\underline{(\boldsymbol{I}_4 \otimes \boldsymbol{a}^{\mathsf{T}})\boldsymbol{\Gamma}}}, \tag{2.144}$$

so we can instead of (2.140) finally write the same as in (2.100)

$$\boxed{\boldsymbol{a}_{\|\vec{x}^{\mathsf{T}}} = \boldsymbol{a}_{|\vec{x}^{\mathsf{T}}} + (\boldsymbol{I}_4 \otimes \boldsymbol{a}^{\mathsf{T}})\boldsymbol{\Gamma} \in \mathbb{R}^{4\times 4}.} \tag{2.145}$$

2.6 Incidental Remark

If one starts with an equation which is valid in the presence of gravity in general relativity, then this equation must for $v^2 \ll c^2$ pass over to Newton's equation. The force of interaction of two discrete masses m and m_1 is proportional to the product of the two known masses and inversely proportional to the square of the distance of the two centres of gravity:

$$\boldsymbol{f} = G\frac{mm_1}{|\boldsymbol{x}-\boldsymbol{x}_1|^2}\frac{\boldsymbol{x}-\boldsymbol{x}_1}{|\boldsymbol{x}-\boldsymbol{x}_1|},$$

or

$$\boldsymbol{f} = m \cdot G\frac{m_1}{|\boldsymbol{x}-\boldsymbol{x}_1|^3}(\boldsymbol{x}-\boldsymbol{x}_1).$$

For several discrete masses m_i, this attraction is obtained as

$$\boldsymbol{f} = m \cdot G\sum_i \frac{m_i}{|\boldsymbol{x}-\boldsymbol{x}_i|^3}(\boldsymbol{x}-\boldsymbol{x}_i),$$

and for a distributed mass with the mass density ρ

$$\boldsymbol{f} = m \cdot G\int_V \rho(\boldsymbol{x}_i)\frac{\boldsymbol{x}-\boldsymbol{x}_i}{|\boldsymbol{x}-\boldsymbol{x}_i|^3}\,\mathrm{d}V.$$

In an electric field of strength $\boldsymbol{e}$, which together with a charge q produces the force $\boldsymbol{f}=q\boldsymbol{e}$, we define the gravitational field strength $\boldsymbol{e}_G$, producing the force $\boldsymbol{f}=m\boldsymbol{e}_G$ acting on the mass m. For several discrete masses, the gravitational field strength is

$$\boldsymbol{e}_G = G\sum_i \frac{m_i}{|\boldsymbol{x}-\boldsymbol{x}_i|^3}(\boldsymbol{x}-\boldsymbol{x}_i).$$

So we can divide the analysis of the problem into two steps. In the first step, the gravitational field generated by several masses m_i at the point $\boldsymbol{x}$ is determined, and in the second step, the force acting on the mass m at the point $\boldsymbol{x}$ is determined.

The potential energy is the integral of force times distance, so

$$U = -\int \boldsymbol{f}^{\mathrm{T}}\,\mathrm{d}\boldsymbol{s} = -m\int \boldsymbol{e}_G^{\mathrm{T}}\,\mathrm{d}\boldsymbol{s} \stackrel{\text{def}}{=} m\phi.$$

If a mass m is displaced by a small distance Δx, the work done is equal to the change in potential energy

$$\Delta W = -\Delta U = f_x \Delta x.$$

Dividing this equation by Δx, this force in the x-direction is given as

$$f_x = -\frac{\Delta U}{\Delta x}.$$

Dividing by the mass m, one gets the x-component of the strength of the gravitational field

$$e_x = -\frac{\Delta\phi}{\Delta x}.$$

Whence it follows generally with $\Delta\boldsymbol{x}\to\boldsymbol{0}$, that

$$\boldsymbol{e} = -\nabla\phi,$$

so

$$\boldsymbol{f} = -m\nabla\phi,$$

or

$$\frac{\mathrm{d}^2\boldsymbol{x}}{\mathrm{d}t^2} = -\nabla\phi(\boldsymbol{x}), \tag{2.146}$$

where the gravitational potential ϕ, which is a scalar function of the position $\boldsymbol{x}$, is obtained from the second-order linear partial differential equation, the Poisson equation

$$\Delta\phi(\boldsymbol{x}) = 4\pi G\rho(\boldsymbol{x}), \tag{2.147}$$

with the gravitational constant G and the mass density $\rho(\boldsymbol{x})$. This equation shows the relationship between gravitational potential and matter in the Newtonian Physics. The above two steps are thus:

Step 1. Find the solution $\phi(\boldsymbol{x})$ of the Poisson equation (2.147).
Step 2. Use the solution of (2.147) to find $\boldsymbol{x}(t)$.

This is the approach in the classical Newtonian Physics. How must we do or modify the two steps in the theory of general relativity, i.e. how does one generally get the g_{ik}'s and how does one establish the dynamic equations? In the case when only the element g_{00} depends on $\boldsymbol{x}$, i.e. only the 00-element of the sub-matrix $\boldsymbol{\Gamma}_i$ is different from zero, then

$$\boldsymbol{\Gamma}_i = \begin{pmatrix} \frac{\partial g_{00}}{\partial x_i} & 0 & 0 & 0 \\ 0 & 0 & 0 & 0 \\ 0 & 0 & 0 & 0 \\ 0 & 0 & 0 & 0 \end{pmatrix}.$$

According to (2.146), the acceleration is proportional to the partial derivative of the gravitational potential ϕ with respect to the coordinates x_i. Looking at the equation

$$\ddot{\boldsymbol{x}} = -(\boldsymbol{I}_4 \otimes \dot{\boldsymbol{x}}^\top)\boldsymbol{\Gamma}\dot{\boldsymbol{x}}, \tag{2.148}$$

on the left-hand side of the equation one has acceleration and on the right-hand side there is the matrix $\boldsymbol{\Gamma}$ comprising partial derivatives of the g_{ij}'s with respect to the coordinates x_ℓ. The g_{ij}'s apparently play in General Relativity the same role as the gravitational potential ϕ does in classical Physics. There the gravitational potential ϕ was determined by the Poisson equation whose form is, for the most part, given by the Laplace operator Δ:

$$\Delta\phi = \frac{\partial^2\phi}{\partial x_1^2} + \frac{\partial^2\phi}{\partial x_2^2} + \frac{\partial^2\phi}{\partial x_3^2} = 4\pi G\rho(\boldsymbol{x}).$$

So one is now looking for a mathematical expression involving the second derivatives of the g_{ij}'s with respect to the four spacetime coordinates x_i. Such a term appears, in fact, in the differential geometry of Gauss and Riemann, to be precise, in the investigation of the curvature of surfaces or hyper-surfaces in three- or n-dimensional spaces, where the surfaces are described by quadratic forms with the g_{ij}'s as the corresponding matrix elements. Therefore, the appendix "Some Differential Geometry" deals with the theory of curvature of surfaces in three- and n-dimensional spaces.

2.7 Parallel Transport

For further considerations, we need the definition of *parallel transport*:

1. The parallel displacement of a vector $\boldsymbol{a}$, which is tangential to the curved surface and runs along a geodesic of this surface, is defined as follows: The origin point of the vector moves along the geodesic and the vector itself moves continuously so that its angle with the geodesic and its length remains constant. It then changes by a parallel transport along $\delta\boldsymbol{x}$ according to (2.96) with $\delta\boldsymbol{a} = -(\boldsymbol{I}_4 \otimes \boldsymbol{a}^\top)\boldsymbol{\Gamma}\delta\boldsymbol{x}$.

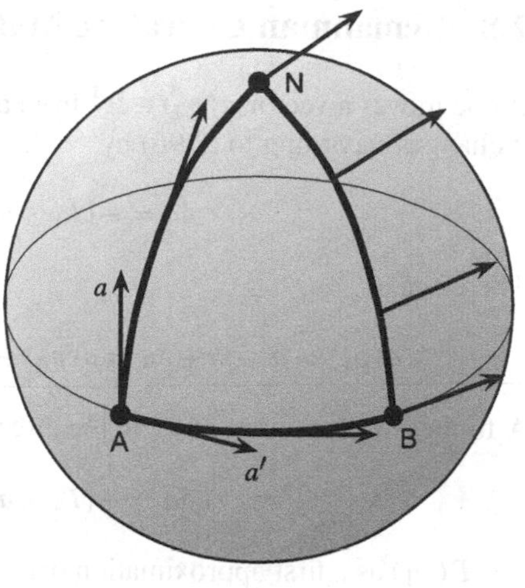

Fig. 2.1 Parallel transport

2. The parallel displacement of a vector on a surface along a broken line, consisting of some geodesic pieces, is done so that from the first corner to the second corner the vector is moved along the first geodesic arc, then along the second arc, etc.
3. Finally, the parallel translation of a vector along a smooth curve is described by the limiting process, in which the curve is approximated by broken lines consisting of geodesic pieces.

If a vector $\boldsymbol{a}$ in a *flat* space, where $\boldsymbol{\Gamma} = \mathbf{0}$, moves in parallel along a closed loop, it comes back to the starting point with the same length and direction. But if a modified vector comes back, so it must be true that $\boldsymbol{\Gamma} \neq \mathbf{0}$—there exists curvature.

Example If you move a vector $\boldsymbol{a}$ on a sphere (Fig. 2.1), beginning at the equator at a point A, along a meridian to the north pole N, then along another meridian back to the equator at a point B, and finally back along the equator to the starting point A, then the incoming vector $\boldsymbol{a}'$ has a direction other than the initial vector $\boldsymbol{a}$. Calling the difference between the starting and end vector

$$\Delta \boldsymbol{a} \stackrel{\text{def}}{=} \boldsymbol{a}' - \boldsymbol{a},$$

the question is: What happens to $\Delta \boldsymbol{a}$ when the circulated area is smaller and smaller? Of course, $\Delta \boldsymbol{a}$ approaches the zero vector, but not the ratio $\Delta \boldsymbol{a}/(\text{circulated area})$.

We now replace the spherical triangle of the example by a differentially small square and consider not the difference vector $\Delta \boldsymbol{a}$ of a complete circulation. Instead, the initial vector $\boldsymbol{a}$ is moved the half way around the square in one direction. Then the same vector $\boldsymbol{a}$ is shifted half way in the other direction. At the meeting point the difference $\Delta \boldsymbol{a}$ arises. This difference vector is deduced and considered more precisely in the following section.

2.8 Riemannian Curvature Matrix

If one moves a vector $\boldsymbol{a}(p_0) \in \mathbb{R}^4$ from a point p_0 by $\delta\boldsymbol{x} \in \mathbb{R}^4$ to the point p_1, then it changes according to (2.96) by

$$\delta\boldsymbol{a} = -\big(\boldsymbol{I}_4 \otimes \boldsymbol{a}(p_0)^\mathsf{T}\big)\boldsymbol{\Gamma}(p_0)\delta\boldsymbol{x}. \tag{2.149}$$

Therefore,

$$\boldsymbol{a}(p_1) = \boldsymbol{a}(p_0) + \delta\boldsymbol{a} = \boldsymbol{a}(p_0) - \big(\boldsymbol{I}_4 \otimes \boldsymbol{a}(p_0)^\mathsf{T}\big)\boldsymbol{\Gamma}(p_0)\delta\boldsymbol{x}. \tag{2.150}$$

A further displacement of p_1 in the direction $\delta\bar{\boldsymbol{x}}$ to p_2 gives the change

$$\delta\bar{\boldsymbol{a}} = -\big(\boldsymbol{I}_4 \otimes \boldsymbol{a}(p_1)^\mathsf{T}\big)\boldsymbol{\Gamma}(p_1)\delta\bar{\boldsymbol{x}}. \tag{2.151}$$

For $\boldsymbol{\Gamma}(p_1)$ as a first approximation one can write

$$\boldsymbol{\Gamma}(p_1) \approx \boldsymbol{\Gamma}(p_0) + \sum_{\nu=0}^{3} \frac{\partial \boldsymbol{\Gamma}}{\partial x_\nu}\delta x_\nu = \boldsymbol{\Gamma}(p_0) + \frac{\partial \boldsymbol{\Gamma}}{\partial \boldsymbol{x}^\mathsf{T}}(\delta\boldsymbol{x} \otimes \boldsymbol{I}_4). \tag{2.152}$$

Equations (2.150) and (2.152) used in (2.151) yield

$$\begin{aligned}
\delta\bar{\boldsymbol{a}} &= -\Big(\boldsymbol{I}_4 \otimes \big[\boldsymbol{a}(p_0) - \big(\boldsymbol{I}_4 \otimes \boldsymbol{a}(p_0)^\mathsf{T}\big)\boldsymbol{\Gamma}(p_0)\delta\boldsymbol{x}\big]^\mathsf{T}\Big)\left(\boldsymbol{\Gamma}(p_0) + \frac{\partial \boldsymbol{\Gamma}}{\partial \boldsymbol{x}^\mathsf{T}}(\delta\boldsymbol{x} \otimes \boldsymbol{I}_4)\right)\delta\bar{\boldsymbol{x}} \\
&= -\big(\boldsymbol{I}_4 \otimes \boldsymbol{a}(p_0)^\mathsf{T}\big)\boldsymbol{\Gamma}(p_0)\delta\bar{\boldsymbol{x}} - \big(\boldsymbol{I}_4 \otimes \boldsymbol{a}(p_0)^\mathsf{T}\big)\frac{\partial \boldsymbol{\Gamma}}{\partial \boldsymbol{x}^\mathsf{T}}(\delta\boldsymbol{x} \otimes \boldsymbol{I}_4)\delta\bar{\boldsymbol{x}} \\
&\quad + \Big(\boldsymbol{I}_4 \otimes \big[\big(\boldsymbol{I}_4 \otimes \boldsymbol{a}(p_0)^\mathsf{T}\big)\boldsymbol{\Gamma}\delta\boldsymbol{x}\big]^\mathsf{T}\Big)\boldsymbol{\Gamma}\delta\bar{\boldsymbol{x}} + \mathcal{O}\big(\delta\bar{x} \cdot (\mathrm{d}x^2)\big) \\
&= \bigg[-\big(\boldsymbol{I}_4 \otimes \boldsymbol{a}^\mathsf{T}\big)\boldsymbol{\Gamma} - \big(\boldsymbol{I}_4 \otimes \boldsymbol{a}^\mathsf{T}\big)\frac{\partial \boldsymbol{\Gamma}}{\partial \boldsymbol{x}^\mathsf{T}}(\delta\boldsymbol{x} \otimes \boldsymbol{I}_4) \\
&\quad + \Big(\boldsymbol{I}_4 \otimes \big[\big(\boldsymbol{I}_4 \otimes \boldsymbol{a}(p_0)^\mathsf{T}\big)\boldsymbol{\Gamma}\delta\boldsymbol{x}\big]^\mathsf{T}\Big)\boldsymbol{\Gamma}\bigg]\delta\bar{\boldsymbol{x}}.
\end{aligned} \tag{2.153}$$

The third term in the brackets can be transformed into the following form:

$$\begin{aligned}
\Big(\boldsymbol{I}_4 \otimes \big[\big(\boldsymbol{I}_4 \otimes \boldsymbol{a}^\mathsf{T}\big)\boldsymbol{\Gamma}\delta\boldsymbol{x}\big]^\mathsf{T}\Big)\boldsymbol{\Gamma} &= \overline{\boldsymbol{\Gamma}}\big(\boldsymbol{I}_4 \otimes \big(\boldsymbol{I}_4 \otimes \boldsymbol{a}^\mathsf{T}\big)\boldsymbol{\Gamma}\delta\boldsymbol{x}\big) \\
&= \overline{\boldsymbol{\Gamma}}\big(\boldsymbol{I}_{16} \otimes \boldsymbol{a}^\mathsf{T}\big)(\boldsymbol{I}_4 \otimes \boldsymbol{\Gamma}\delta\boldsymbol{x}) \\
&= \big(\overline{\boldsymbol{\Gamma}} \otimes \boldsymbol{a}^\mathsf{T}\big)(\boldsymbol{I}_4 \otimes \boldsymbol{\Gamma})(\boldsymbol{I}_4 \otimes \delta\boldsymbol{x}) \\
&= \big(\boldsymbol{I}_4 \otimes \boldsymbol{a}^\mathsf{T}\big)(\overline{\boldsymbol{\Gamma}} \otimes \boldsymbol{I}_4)(\boldsymbol{I}_4 \otimes \boldsymbol{\Gamma})(\boldsymbol{I}_4 \otimes \delta\boldsymbol{x}).
\end{aligned} \tag{2.154}$$

Using this in (2.153) results in

$$\delta\bar{\boldsymbol{a}} = -\big(\boldsymbol{I}_4 \otimes \boldsymbol{a}^\mathsf{T}\big)\bigg[\boldsymbol{\Gamma} + \frac{\partial \boldsymbol{\Gamma}}{\partial \boldsymbol{x}^\mathsf{T}}(\delta\boldsymbol{x} \otimes \boldsymbol{I}_4) - (\overline{\boldsymbol{\Gamma}} \otimes \boldsymbol{I}_4)(\boldsymbol{I}_4 \otimes \boldsymbol{\Gamma})(\boldsymbol{I}_4 \otimes \delta\boldsymbol{x})\bigg]\delta\bar{\boldsymbol{x}}. \tag{2.155}$$

If we now first go in the direction $\delta\bar{x}$ and then in the direction δx, we obtain accordingly

$$\delta\bar{\bar{a}} = -\left(I_4 \otimes a^\top\right)\left[\Gamma + \frac{\partial \Gamma}{\partial x^\top}(\delta\bar{x} \otimes I_4) - (\overline{\Gamma} \otimes I_4)(I_4 \otimes \Gamma)(I_4 \otimes \delta\bar{x})\right]\delta x. \quad (2.156)$$

For the final product in the third summand of (2.155) we can also write

$$(I_4 \otimes \delta x)\delta\bar{x} = (I_4 \otimes \delta x)(\delta\bar{x} \otimes 1) = (\delta\bar{x} \otimes \delta x) = U_{4\times 4}(\delta x \otimes \delta\bar{x}). \quad (2.157)$$

It is true that

$$\Delta a = (a + \delta\bar{a}) - (a + \delta\bar{\bar{a}}) = \delta\bar{a} - \delta\bar{\bar{a}}, \quad (2.158)$$

so with (2.155), (2.156) and (2.157),

$$\Delta a = \left(I_4 \otimes a^\top\right)\Bigg(\Gamma(\delta x - \delta\bar{x}) + \left[\frac{\partial \Gamma}{\partial x^\top} + (\overline{\Gamma} \otimes I_4)(I_4 \otimes \Gamma)\right](U_{4\times 4} - I_{16})(\delta x \otimes \delta\bar{x})\Bigg). \quad (2.159)$$

With the 16×16 Riemannian *curvature matrix*

$$R \stackrel{\text{def}}{=} \left[\frac{\partial \Gamma}{\partial x^\top} + (\overline{\Gamma} \otimes I_4)(I_4 \otimes \Gamma)\right](U_{4\times 4} - I_{16}) \in \mathbb{R}^{16\times 16}, \quad (2.160)$$

we have

$$\Delta a = \left(I_4 \otimes a^\top\right)\left[\Gamma(\delta x - \delta\bar{x}) + R(\delta x \otimes \delta\bar{x})\right] \in \mathbb{R}^4. \quad (2.161)$$

We also define a slightly modified curvature matrix:

$$\check{R} \stackrel{\text{def}}{=} (G \otimes I_4)R. \quad (2.162)$$

2.9 Properties of the Riemannian Curvature Matrix

2.9.1 Composition of R and $\check{R}$

Which form do the elements of the Riemannian curvature matrix

$$R = \left[\frac{\partial \Gamma}{\partial x^\top} + (\overline{\Gamma} \otimes I_4)(I_4 \otimes \Gamma)\right](U_{4\times 4} - I_{16}) \in \mathbb{R}^{16\times 16}$$

have? The property, already mentioned earlier, that in the occurring matrix difference $U_{4\times 4} - I_{16}$ the first, $(4+2)$th, $(8+3)$th and the 16-th row/column are equal to the zero row/column, has for the Riemannian curvature matrix the consequence that its corresponding columns are zero columns!

Next we obtain with (2.143)

$$\bar{\boldsymbol{\Gamma}} = \begin{pmatrix} \boldsymbol{\gamma}_0^{0\mathsf{T}} & & \boldsymbol{\gamma}_3^{0\mathsf{T}} \\ \boldsymbol{\gamma}_0^{1\mathsf{T}} & & \boldsymbol{\gamma}_3^{1\mathsf{T}} \\ \boldsymbol{\gamma}_0^{2\mathsf{T}} & \cdots & \boldsymbol{\gamma}_3^{2\mathsf{T}} \\ \boldsymbol{\gamma}_0^{3\mathsf{T}} & & \boldsymbol{\gamma}_3^{3\mathsf{T}} \end{pmatrix} = [\bar{\boldsymbol{\Gamma}}_0, \ldots, \bar{\boldsymbol{\Gamma}}_3]$$

for

$$(\overline{\boldsymbol{\Gamma}} \otimes \boldsymbol{I}_4)(\boldsymbol{I}_4 \otimes \boldsymbol{\Gamma}) = \left[(\overline{\boldsymbol{\Gamma}}_0 \otimes \boldsymbol{I}_4), \ldots, (\overline{\boldsymbol{\Gamma}}_3 \otimes \boldsymbol{I}_4)\right] \begin{pmatrix} \boldsymbol{\Gamma} & 0 & 0 & 0 \\ 0 & \boldsymbol{\Gamma} & 0 & 0 \\ 0 & 0 & \boldsymbol{\Gamma} & 0 \\ 0 & 0 & 0 & \boldsymbol{\Gamma} \end{pmatrix}$$

$$= \left[(\overline{\boldsymbol{\Gamma}}_0 \otimes \boldsymbol{I}_4)\boldsymbol{\Gamma}, \ldots, (\overline{\boldsymbol{\Gamma}}_3 \otimes \boldsymbol{I}_4)\boldsymbol{\Gamma}\right].$$

This matrix product contributes to the matrix element $R^{\gamma\delta}_{\alpha\beta}$ the sum

$$\left[(\boldsymbol{\gamma}_\delta^{\gamma\mathsf{T}} \otimes \boldsymbol{I}_4)\boldsymbol{\Gamma}\right]_{\alpha\beta} = \left[\Gamma^{\gamma}_{\delta 0}\boldsymbol{\Gamma}_0 + \cdots + \Gamma^{\gamma}_{\delta 3}\boldsymbol{\Gamma}_3\right]_{\alpha\beta} = \underline{\underline{\sum_\nu \Gamma^{\gamma}_{\delta\nu}\Gamma^{\nu}_{\alpha\beta}}}, \qquad (2.163)$$

where $\boldsymbol{\gamma}_\delta^{\gamma\mathsf{T}} \in \mathbb{R}^4$ is the γth row of the sub-matrix $\overline{\boldsymbol{\Gamma}}_\delta$, i.e. the δth row of the sub-matrix $\boldsymbol{\Gamma}_\gamma$.

Furthermore,

$$(\overline{\boldsymbol{\Gamma}} \otimes \boldsymbol{I}_4)(\boldsymbol{I}_4 \otimes \boldsymbol{\Gamma})\boldsymbol{U}_{4\times 4}$$

$$= (\overline{\boldsymbol{\Gamma}} \otimes \boldsymbol{I}_4) \begin{pmatrix} \boldsymbol{\Gamma} & 0 & 0 & 0 \\ 0 & \boldsymbol{\Gamma} & 0 & 0 \\ 0 & 0 & \boldsymbol{\Gamma} & 0 \\ 0 & 0 & 0 & \boldsymbol{\Gamma} \end{pmatrix} \boldsymbol{U}_{4\times 4}$$

$$= (\overline{\boldsymbol{\Gamma}} \otimes \boldsymbol{I}_4) \left(\begin{array}{cc|cc|cc|cc} \boldsymbol{\gamma}_0 & & \boldsymbol{\gamma}_1 & & \boldsymbol{\gamma}_2 & & \boldsymbol{\gamma}_3 & \\ & \ddots & & \ddots & & \ddots & & \ddots \\ & \boldsymbol{\gamma}_0 & & \boldsymbol{\gamma}_1 & & \boldsymbol{\gamma}_2 & & \boldsymbol{\gamma}_3 \end{array}\right).$$

This matrix product ($\boldsymbol{\gamma}_\delta \in \mathbb{R}^{16}$ is the δth column of $\boldsymbol{\Gamma}$) contributes to the matrix element $R^{\gamma\delta}_{\alpha\beta}$ the sum

$$\left[(\boldsymbol{\gamma}^{\gamma\mathsf{T}} \otimes \boldsymbol{I}_4) \begin{pmatrix} \boldsymbol{\gamma}_\delta & & \\ & \ddots & \\ & & \boldsymbol{\gamma}_\delta \end{pmatrix}\right]_{\alpha\beta}$$

$$= \left[[\boldsymbol{\gamma}_0^{\gamma\mathsf{T}} \otimes \boldsymbol{I}_4, \ldots, \boldsymbol{\gamma}_0^{\gamma\mathsf{T}} \otimes \boldsymbol{I}_4] \begin{pmatrix} \boldsymbol{\gamma}_\delta & & \\ & \ddots & \\ & & \boldsymbol{\gamma}_\delta \end{pmatrix} \right]_{\alpha\beta}$$

$$= [(\boldsymbol{\gamma}_0^{\gamma\mathsf{T}} \otimes \boldsymbol{I}_4)\boldsymbol{\gamma}_\delta, \ldots, (\boldsymbol{\gamma}_3^{\gamma\mathsf{T}} \otimes \boldsymbol{I}_4)\boldsymbol{\gamma}_\delta]_{\alpha\beta} = [(\boldsymbol{\gamma}_\beta^{\gamma\mathsf{T}} \otimes \boldsymbol{I}_4)\boldsymbol{\gamma}_\delta]_\alpha$$

$$= [\Gamma_{\beta 0}^{\gamma} \boldsymbol{\gamma}_\delta^0 + \cdots + \Gamma_{\beta 3}^{\gamma} \boldsymbol{\gamma}_\delta^3]_\alpha = \sum_\nu \Gamma_{\beta\nu}^{\gamma} \Gamma_{\delta\alpha}^{\nu}. \tag{2.164}$$

In accordance with (2.160), one gets finally with (2.163) and (2.164)

$$\boxed{R_{\alpha\beta}^{\gamma\delta} = \frac{\partial}{\partial x_\beta} \Gamma_{\alpha\delta}^{\gamma} - \frac{\partial}{\partial x_\delta} \Gamma_{\alpha\beta}^{\gamma} + \sum_\nu \Gamma_{\beta\nu}^{\gamma} \Gamma_{\delta\alpha}^{\nu} - \sum_\nu \Gamma_{\delta\nu}^{\gamma} \Gamma_{\alpha\beta}^{\nu}.} \tag{2.165}$$

From this form, one can immediately read off the property

$$R_{\alpha\beta}^{\gamma\delta} = -R_{\alpha\delta}^{\gamma\beta}. \tag{2.166}$$

With the help of (2.165) one can also verify the so-called *cyclic identity*:

$$\boxed{R_{\alpha\beta}^{\gamma\delta} + R_{\beta\delta}^{\gamma\alpha} + R_{\delta\alpha}^{\gamma\beta} = 0.} \tag{2.167}$$

From (2.165), using (2.63), (2.73) and (2.65), one can also derive a closed form for $\check{R}_{\alpha\beta}^{\gamma\delta}$ in (2.160) as follows:

$$\check{R}_{\alpha\beta}^{\gamma\delta} = \sum_l g_{\gamma i} R_{\alpha\beta}^{i\delta} = \sum_i g_{\gamma i} \left(\frac{\partial}{\partial x_\beta} \Gamma_{\alpha\delta}^{i} - \frac{\partial}{\partial x_\delta} \Gamma_{\alpha\beta}^{i} + \sum_\nu \Gamma_{\beta\nu}^{i} \Gamma_{\delta\alpha}^{\nu} - \sum_\nu \Gamma_{\delta\nu}^{i} \Gamma_{\alpha\beta}^{\nu} \right)$$

$$= \left(\frac{\partial}{\partial x_\beta} \check{\Gamma}_{\alpha\delta}^{\gamma} - \sum_i \Gamma_{\alpha\delta}^{i} \frac{\partial g_{\gamma i}}{\partial x_\beta} \right) - \left(\frac{\partial}{\partial x_\delta} \check{\Gamma}_{\alpha\beta}^{\gamma} - \sum_i \Gamma_{\alpha\beta}^{i} \frac{\partial g_{\gamma i}}{\partial x_\delta} \right)$$

$$+ \sum_\nu \check{\Gamma}_{\beta\nu}^{\gamma} \Gamma_{\delta\alpha}^{\nu} - \sum_\nu \check{\Gamma}_{\delta\nu}^{\gamma} \Gamma_{\alpha\beta}^{\nu}$$

$$= \left(\frac{\partial}{\partial x_\beta} \check{\Gamma}_{\alpha\delta}^{\gamma} - \sum_i \Gamma_{\alpha\delta}^{i} (\check{\Gamma}_{\gamma\beta}^{i} + \check{\Gamma}_{i\beta}^{\gamma}) \right) - \left(\frac{\partial}{\partial x_\delta} \check{\Gamma}_{\alpha\beta}^{\gamma} \right.$$

$$\left. - \sum_i \Gamma_{\alpha\beta}^{i} (\check{\Gamma}_{\gamma\delta}^{i} + \check{\Gamma}_{i\delta}^{\gamma}) \right) + \sum_\nu \check{\Gamma}_{\beta\nu}^{\gamma} \Gamma_{\delta\alpha}^{\nu} - \sum_\nu \check{\Gamma}_{\delta\nu}^{\gamma} \Gamma_{\alpha\beta}^{\nu}$$

$$= \frac{1}{2} \left(\frac{\partial}{\partial x_\beta} \left(\frac{\partial g_{\delta\gamma}}{\partial x_\alpha} + \frac{\partial g_{\alpha\gamma}}{\partial x_\delta} - \frac{\partial g_{\alpha\delta}}{\partial x_\gamma} \right) - \frac{\partial}{\partial x_\delta} \left(\frac{\partial g_{\beta\gamma}}{\partial x_\alpha} + \frac{\partial g_{\alpha\gamma}}{\partial x_\beta} - \frac{\partial g_{\alpha\beta}}{\partial x_\gamma} \right) \right)$$

$$-\sum_i \Gamma^i_{\alpha\delta}\check{\Gamma}^i_{\gamma\beta} - \sum_i \Gamma^i_{\alpha\delta}\check{\Gamma}^\gamma_{i\beta} + \sum_i \Gamma^i_{\alpha\beta}\check{\Gamma}^i_{\gamma\delta} + \sum_i \Gamma^i_{\alpha\beta}\check{\Gamma}^\gamma_{i\delta}$$
$$+\sum_\nu \check{\Gamma}^\gamma_{\beta\nu}\Gamma^\nu_{\delta\alpha} - \sum_\nu \check{\Gamma}^\gamma_{\delta\nu}\Gamma^\nu_{\alpha\beta}.$$

After cancelling out some of the terms, we finally get the closed form

$$\boxed{\check{R}^{\gamma\delta}_{\alpha\beta} = \frac{\partial}{\partial x_\beta}\check{\Gamma}^\gamma_{\alpha\delta} - \frac{\partial}{\partial x_\delta}\check{\Gamma}^\gamma_{\alpha\beta} + \sum_i \Gamma^i_{\alpha\beta}\check{\Gamma}^i_{\gamma\delta} - \sum_i \Gamma^i_{\alpha\delta}\check{\Gamma}^i_{\gamma\beta}} \tag{2.168}$$

$$\boxed{\begin{aligned} &= \frac{1}{2}\left(\frac{\partial^2 g_{\delta\gamma}}{\partial x_\alpha \partial x_\beta} - \frac{\partial^2 g_{\alpha\delta}}{\partial x_\gamma \partial x_\beta} - \frac{\partial^2 g_{\beta\gamma}}{\partial x_\alpha \partial x_\delta} + \frac{\partial^2 g_{\alpha\beta}}{\partial x_\gamma \partial x_\delta}\right) \\ &\quad + \sum_i \Gamma^i_{\alpha\beta}\check{\Gamma}^i_{\gamma\delta} - \sum_i \Gamma^i_{\alpha\delta}\check{\Gamma}^i_{\gamma\beta}, \end{aligned}} \tag{2.169}$$

or

$$\boxed{\begin{aligned} \check{R}^{\gamma\delta}_{\alpha\beta} &= \frac{1}{2}\left(\frac{\partial^2 g_{\delta\gamma}}{\partial x_\alpha \partial x_\beta} - \frac{\partial^2 g_{\alpha\delta}}{\partial x_\gamma \partial x_\beta} - \frac{\partial^2 g_{\beta\gamma}}{\partial x_\alpha \partial x_\delta} + \frac{\partial^2 g_{\alpha\beta}}{\partial x_\gamma \partial x_\delta}\right) \\ &\quad + \sum_i \check{\Gamma}^i_{\gamma\delta}\sum_\nu g^{(-1)}_{i\nu}\check{\Gamma}^\nu_{\alpha\beta} - \sum_i \check{\Gamma}^i_{\gamma\beta}\sum_\nu g^{(-1)}_{i\nu}\check{\Gamma}^\nu_{\alpha\delta}. \end{aligned}} \tag{2.170}$$

From (2.170) the following identities are obtained directly by comparing the corresponding forms:

$$\underline{\underline{\check{R}^{\gamma\delta}_{\alpha\beta} = -\check{R}^{\alpha\delta}_{\gamma\beta},}} \tag{2.171}$$

$$\underline{\underline{\check{R}^{\gamma\delta}_{\alpha\beta} = -\check{R}^{\gamma\beta}_{\alpha\delta},}} \tag{2.172}$$

$$\underline{\underline{\check{R}^{\gamma\delta}_{\alpha\beta} = \check{R}^{\alpha\beta}_{\gamma\delta}.}} \tag{2.173}$$

Also in this case, a *cyclic identity* is valid:

$$\boxed{\check{R}^{\gamma\delta}_{\alpha\beta} + \check{R}^{\gamma\alpha}_{\beta\delta} + \check{R}^{\gamma\beta}_{\delta\alpha} = 0.} \tag{2.174}$$

If the two vectors $\mathrm{d}\boldsymbol{x}$ and $\mathrm{d}\bar{\boldsymbol{x}}$ are perpendicular to each other, then the area of the formed rectangle is equal to $|\mathrm{d}\boldsymbol{x}| \cdot |\mathrm{d}\bar{\boldsymbol{x}}|$. In the *differential geometry*, the limit of the ratio of $\Delta\boldsymbol{a}$ to the area is now called the *curvature* κ:

$$\kappa \overset{\text{def}}{=} \lim_{|\mathrm{d}\boldsymbol{x}|,|\mathrm{d}\bar{\boldsymbol{x}}|\to 0} \frac{|\Delta\boldsymbol{a}(\mathrm{d}\boldsymbol{x}, \mathrm{d}\bar{\boldsymbol{x}})|}{|\mathrm{d}\boldsymbol{x}| \cdot |\mathrm{d}\bar{\boldsymbol{x}}|}$$

or with $\boldsymbol{n} \stackrel{\text{def}}{=} \frac{\mathrm{d}\boldsymbol{x}}{|\mathrm{d}\boldsymbol{x}|}$ and $\bar{\boldsymbol{n}} \stackrel{\text{def}}{=} \frac{\mathrm{d}\bar{\boldsymbol{x}}}{|\mathrm{d}\bar{\boldsymbol{x}}|}$,

$$\kappa \stackrel{\text{def}}{=} \lim_{\epsilon \to 0} \frac{|\Delta \boldsymbol{a}(\epsilon \boldsymbol{n}, \epsilon \bar{\boldsymbol{n}})|}{\epsilon^2}.$$

With the help of (2.161) we then obtain

$$\kappa = \left|\left(\boldsymbol{I}_4 \otimes \boldsymbol{a}^\mathsf{T}\right)\boldsymbol{R}(\boldsymbol{n} \otimes \bar{\boldsymbol{n}})\right|. \tag{2.175}$$

Riemannian Coordinate System For the study of the properties of the Riemannian curvature matrix, it is advantageous to first perform a coordinate transformation, so that in the new coordinate system, the Christoffel matrices have the property $\boldsymbol{\Gamma} = \boldsymbol{0}$. Such a coordinate transformation is in the case of a curved spacetime, where $\boldsymbol{G}$ depends on $\vec{\boldsymbol{x}}$, only locally possible, but then for each $\vec{\boldsymbol{x}}$! By the inverse transformation of the obtained statements, they are again globally valid. So, we look for a local coordinate transformation such that in the new coordinate system $\boldsymbol{\Gamma} = \boldsymbol{0}$ is valid. For geodesic lines the following applies to the four coordinates:

$$\frac{\mathrm{d}^2 x_k}{\mathrm{d}s^2} + \left(\frac{\mathrm{d}\vec{\boldsymbol{x}}}{\mathrm{d}s}\right)^\mathsf{T} \boldsymbol{\Gamma}_k \frac{\mathrm{d}\vec{\boldsymbol{x}}}{\mathrm{d}s} = 0. \tag{2.176}$$

The x_k's belong to an arbitrary coordinate system in which the geodesic is described by $x_k = x_k(s)$ and s is the arc length along the curve. At a fixed point $\mathcal{P}$ with the coordinate $\vec{\boldsymbol{x}}^{(0)}$, any coordinate can be developed in a power series:

$$x_k = x_k^{(0)} + \zeta_k s + \frac{1}{2}\left(\frac{\mathrm{d}^2 x_k}{\mathrm{d}s^2}\right)_{\mathcal{P}} s^2 + \frac{1}{3!}\left(\frac{\mathrm{d}^3 x_k}{\mathrm{d}s^3}\right)_{\mathcal{P}} s^3 + \cdots, \tag{2.177}$$

where ζ_k is the kth component of the tangent vector

$$\boldsymbol{\zeta} \stackrel{\text{def}}{=} \left(\frac{\mathrm{d}\vec{\boldsymbol{x}}}{\mathrm{d}s}\right)_{\mathcal{P}}$$

to the geodesic at the point $\mathcal{P}$. Then, however, according to (2.176),

$$\frac{\mathrm{d}^2 x_k}{\mathrm{d}s^2} = -\boldsymbol{\zeta}^\mathsf{T}(\boldsymbol{\Gamma}_k)_{\mathcal{P}}\boldsymbol{\zeta}. \tag{2.178}$$

Plugging this into (2.177), considering a small neighbourhood of $\mathcal{P}$, a small $x_k - x_k^{(0)}$, and neglecting powers higher than two gives

$$x_k = x_k^{(0)} + \zeta_k s - \frac{1}{2}\boldsymbol{\zeta}^\mathsf{T}(\boldsymbol{\Gamma}_k)_{\mathcal{P}}\boldsymbol{\zeta} s^2. \tag{2.179}$$

Now, calling $\boldsymbol{\zeta} s = \vec{\boldsymbol{x}}'$, the following is obtained from (2.179):

$$x_k = x_k^{(0)} + x_k' - \frac{1}{2}\vec{\boldsymbol{x}}'^\mathsf{T}(\boldsymbol{\Gamma}_k)_{\mathcal{P}}\vec{\boldsymbol{x}}'.$$

This relation suggests the following coordinate transformation from $\vec{x}$ to $\vec{x}'$:

$$x'_k = x_k - x_k^{(0)} + \frac{1}{2}(\vec{x} - \vec{x}^{(0)})^{\mathsf{T}}(\boldsymbol{\Gamma}_k)_{\mathcal{P}}(\vec{x} - \vec{x}^{(0)}). \tag{2.180}$$

What is the corresponding metric matrix $\boldsymbol{G}'$ in

$$\mathrm{d}s^2 = \vec{x}'^{\mathsf{T}}\boldsymbol{G}'\vec{x}'? \tag{2.181}$$

In this coordinate system, the geodesic has the equation

$$\frac{\mathrm{d}^2\vec{x}'}{\mathrm{d}s^2} + \left(\boldsymbol{I}_4 \otimes \left(\frac{\mathrm{d}\vec{x}'}{\mathrm{d}s}\right)^{\mathsf{T}}\right)\boldsymbol{\Gamma}'(\vec{x}')\left(\frac{\mathrm{d}\vec{x}'}{\mathrm{d}s}\right) = \mathbf{0}. \tag{2.182}$$

But since in the new coordinate system the geodesics are straight lines of the form $\vec{x}' = \boldsymbol{\zeta}s$, in (2.182) the expression $(\boldsymbol{I}_4 \otimes \boldsymbol{\zeta}^{\mathsf{T}})\boldsymbol{\Gamma}'(\boldsymbol{\zeta}s)\boldsymbol{\zeta}$ must be equal to the zero vector. Since $\boldsymbol{\zeta}$ are arbitrary vectors, the following must be valid for $s = 0$:

$$\underline{\underline{\boldsymbol{\Gamma}'(0) = \mathbf{0}.}} \tag{2.183}$$

Implications for the Riemannian Curvature Matrix If at a point $\mathcal{P}$ for a special coordinate system one has $\boldsymbol{\Gamma}_{\mathcal{P}} = \mathbf{0}$, then naturally any $\check{\Gamma}^{\gamma}_{\alpha\beta}$ defined in (2.63) is equal to zero. But since

$$\underline{\underline{\check{\Gamma}^{i}_{k\ell} + \check{\Gamma}^{\ell}_{ki} = \frac{1}{2}\left(\frac{\partial g_{\ell i}}{\partial x_k} + \frac{\partial g_{ki}}{\partial x_\ell} - \frac{\partial g_{k\ell}}{\partial x_i}\right) + \frac{1}{2}\left(\frac{\partial g_{i\ell}}{\partial x_k} + \frac{\partial g_{k\ell}}{\partial x_i} - \frac{\partial g_{ki}}{\partial x_\ell}\right) = \frac{\partial g_{i\ell}}{\partial x_k}}} \tag{2.184}$$

also all first partial derivatives of the elements of the metric matrix are zero, i.e.

$$\underline{\underline{\left.\frac{\partial \boldsymbol{G}}{\partial \vec{x}}\right|_{\mathcal{P}} = \mathbf{0}.}} \tag{2.185}$$

In the local coordinate system with $\boldsymbol{\Gamma}_{\mathcal{P}} = \mathbf{0}$, the Riemannian curvature matrix has the form

$$\boldsymbol{R}_{\mathcal{P}} = \left.\frac{\partial \boldsymbol{\Gamma}}{\partial \vec{x}^{\mathsf{T}}}\right|_{\mathcal{P}}(\boldsymbol{U}_{4\times 4} - \boldsymbol{I}_{16}). \tag{2.186}$$

The structure is

$$\boldsymbol{R} = \begin{pmatrix} \boldsymbol{R}^{00} & \boldsymbol{R}^{01} & \boldsymbol{R}^{02} & \boldsymbol{R}^{03} \\ \boldsymbol{R}^{10} & \boldsymbol{R}^{11} & \boldsymbol{R}^{12} & \boldsymbol{R}^{13} \\ \boldsymbol{R}^{20} & \boldsymbol{R}^{21} & \boldsymbol{R}^{22} & \boldsymbol{R}^{23} \\ \boldsymbol{R}^{30} & \boldsymbol{R}^{31} & \boldsymbol{R}^{32} & \boldsymbol{R}^{33} \end{pmatrix}. \tag{2.187}$$

Each sub-matrix $\boldsymbol{R}^{\gamma\delta}$ has the form

$$\boldsymbol{R}^{\gamma\delta} = \begin{pmatrix} R^{\gamma\delta}_{00} & R^{\gamma\delta}_{01} & R^{\gamma\delta}_{02} & R^{\gamma\delta}_{03} \\ R^{\gamma\delta}_{10} & R^{\gamma\delta}_{11} & R^{\gamma\delta}_{12} & R^{\gamma\delta}_{13} \\ R^{\gamma\delta}_{20} & R^{\gamma\delta}_{21} & R^{\gamma\delta}_{22} & R^{\gamma\delta}_{23} \\ R^{\gamma\delta}_{30} & R^{\gamma\delta}_{31} & R^{\gamma\delta}_{32} & R^{\gamma\delta}_{33} \end{pmatrix}. \tag{2.188}$$

$\boldsymbol{R}^{\gamma\delta}$ is thus the sub-matrix of $\boldsymbol{R}$ at the intersection of the γth row and δth column. Furthermore, $R^{\gamma\delta}_{\alpha\beta}$ is the matrix element of the sub-matrix $\boldsymbol{R}^{\gamma\delta}$ at the intersection of the αth row and βth column.

With (2.68) and (2.185) one has

$$\begin{aligned} \left.\frac{\partial \boldsymbol{\Gamma}}{\partial \vec{\boldsymbol{x}}^{\mathsf{T}}}\right|_{\mathcal{P}} &= \frac{1}{2}\frac{\partial}{\partial \vec{\boldsymbol{x}}^{\mathsf{T}}}\left((\boldsymbol{G}^{-1}\otimes \boldsymbol{I}_4)\left[\begin{pmatrix} \frac{\partial \boldsymbol{g}_0^{\mathsf{T}}}{\partial \vec{\boldsymbol{x}}} \\ \vdots \\ \frac{\partial \boldsymbol{g}_3^{\mathsf{T}}}{\partial \vec{\boldsymbol{x}}} \end{pmatrix} + \begin{pmatrix} \frac{\partial \boldsymbol{g}_0}{\partial \vec{\boldsymbol{x}}^{\mathsf{T}}} \\ \vdots \\ \frac{\partial \boldsymbol{g}_3}{\partial \vec{\boldsymbol{x}}^{\mathsf{T}}} \end{pmatrix} - \frac{\partial \boldsymbol{G}}{\partial \vec{\boldsymbol{x}}}\right]\right) \\ &= \frac{1}{2}\frac{\partial}{\partial \vec{\boldsymbol{x}}^{\mathsf{T}}}(\boldsymbol{G}^{-1}\otimes \boldsymbol{I}_4)[\boldsymbol{0}+\boldsymbol{0}-\boldsymbol{0}] \\ &\quad + \frac{1}{2}(\boldsymbol{G}^{-1}\otimes \boldsymbol{I}_4)\left[\begin{pmatrix} \frac{\partial^2 \boldsymbol{g}_0^{\mathsf{T}}}{\partial \vec{\boldsymbol{x}}^{\mathsf{T}}\partial \vec{\boldsymbol{x}}} \\ \vdots \\ \frac{\partial^2 \boldsymbol{g}_3^{\mathsf{T}}}{\partial \vec{\boldsymbol{x}}^{\mathsf{T}}\partial \vec{\boldsymbol{x}}} \end{pmatrix} + \begin{pmatrix} \frac{\partial^2 \boldsymbol{g}_0}{\partial \vec{\boldsymbol{x}}^{\mathsf{T}}\partial \vec{\boldsymbol{x}}^{\mathsf{T}}} \\ \vdots \\ \frac{\partial^2 \boldsymbol{g}_3}{\partial \vec{\boldsymbol{x}}^{\mathsf{T}}\partial \vec{\boldsymbol{x}}^{\mathsf{T}}} \end{pmatrix} - \frac{\partial^2 \boldsymbol{G}}{\partial \vec{\boldsymbol{x}}^{\mathsf{T}}\partial \vec{\boldsymbol{x}}}\right]. \end{aligned} \tag{2.189}$$

For

$$\check{\boldsymbol{R}} = (\boldsymbol{G}\otimes \boldsymbol{I}_4)\boldsymbol{R} \tag{2.190}$$

with

$$\check{\boldsymbol{\Gamma}} \stackrel{\text{def}}{=} (\boldsymbol{G}\otimes \boldsymbol{I}_4)\boldsymbol{\Gamma} \tag{2.191}$$

one finally obtains

$$\check{\boldsymbol{R}}_{\mathcal{P}} \stackrel{\text{def}}{=} \frac{\partial \check{\boldsymbol{\Gamma}}}{\partial \vec{\boldsymbol{x}}^{\mathsf{T}}}(\boldsymbol{U}_{4\times 4} - \boldsymbol{I}_{16}). \tag{2.192}$$

Naturally, because of the in $\boldsymbol{R}$ occurring matrix difference $\boldsymbol{U}_{4\times 4} - \boldsymbol{I}_{16}$, the first, $(4+2)$th, $(8+3)$th and the 16th column of $\check{\boldsymbol{R}}$ and of $\check{\boldsymbol{R}}_{\mathcal{P}}$ are equal to the zero column. In

$$\check{\boldsymbol{\Gamma}} = \begin{pmatrix} \check{\boldsymbol{\Gamma}}_0 \\ \check{\boldsymbol{\Gamma}}_1 \\ \check{\boldsymbol{\Gamma}}_2 \\ \check{\boldsymbol{\Gamma}}_3 \end{pmatrix},$$

$\check{\boldsymbol{\Gamma}}_\nu$ has the form

$$\check{\boldsymbol{\Gamma}}_\nu = \frac{1}{2}\left(\frac{\partial \boldsymbol{g}_\nu^\mathsf{T}}{\partial \boldsymbol{x}} + \frac{\partial \boldsymbol{g}_\nu}{\partial \boldsymbol{x}^\mathsf{T}} - \frac{\partial \boldsymbol{G}}{\partial x_\nu}\right),$$

i.e. the elements of $\check{\boldsymbol{\Gamma}}_\nu$ are (see also (2.63))

$$\check{\Gamma}^\nu_{\alpha\beta} = \frac{1}{2}\left(\frac{\partial g_{\beta\nu}}{\partial x_\alpha} + \frac{\partial g_{\alpha\nu}}{\partial x_\beta} - \frac{\partial g_{\alpha\beta}}{\partial x_\nu}\right). \tag{2.193}$$

For $(\check{R}^{\gamma\delta}_{\alpha\beta})_\mathcal{P}$ from (2.165) one obtains

$$\underline{\underline{(\check{R}^{\gamma\delta}_{\alpha\beta})_\mathcal{P} = \frac{1}{2}\left(\frac{\partial^2 g_{\gamma\delta}}{\partial x_\alpha \partial x_\beta} - \frac{\partial^2 g_{\alpha\delta}}{\partial x_\gamma \partial x_\beta} - \frac{\partial^2 g_{\beta\gamma}}{\partial x_\alpha \partial x_\delta} + \frac{\partial^2 g_{\alpha\beta}}{\partial x_\gamma \partial x_\delta}\right).}} \tag{2.194}$$

From (2.194), by comparing the corresponding forms, follows

$$\underline{\underline{(\check{R}^{\gamma\delta}_{\alpha\beta})_\mathcal{P} = (\check{R}^{\delta\gamma}_{\alpha\beta})_\mathcal{P},}} \tag{2.195}$$

and the so-called *cyclic identity*

$$\underline{\underline{(\check{R}^{\gamma\delta}_{\alpha\beta})_\mathcal{P} + (\check{R}^{\delta\beta}_{\alpha\gamma})_\mathcal{P} + (\check{R}^{\beta\gamma}_{\alpha\delta})_\mathcal{P} = 0.}} \tag{2.196}$$

From (2.165), at a point $\mathcal{P}$, we get

$$\underline{\underline{(R^{\gamma\delta}_{\alpha\beta})_\mathcal{P} = \left(\frac{\partial}{\partial x_\beta}\Gamma^\gamma_{\alpha\delta} - \frac{\partial}{\partial x_\delta}\Gamma^\gamma_{\alpha\beta}\right)_\mathcal{P}.}} \tag{2.197}$$

Partially differentiating (2.197) yields

$$\underline{\underline{\left(\frac{\partial}{\partial x_\kappa}R^{\gamma\delta}_{\alpha\beta}\right)_\mathcal{P} = \left(\frac{\partial^2}{\partial x_\kappa \partial x_\beta}\Gamma^\gamma_{\alpha\delta} - \frac{\partial^2}{\partial x_\kappa \partial x_\delta}\Gamma^\gamma_{\alpha\beta}\right)_\mathcal{P}.}} \tag{2.198}$$

With the help of (2.198), by substituting the corresponding terms, the so-called *Bianchi identity* is obtained:

$$\underline{\underline{\frac{\partial}{\partial x_\kappa}R^{\gamma\delta}_{\alpha\beta} + \frac{\partial}{\partial x_\beta}R^{\gamma\kappa}_{\alpha\delta} + \frac{\partial}{\partial x_\delta}R^{\gamma\beta}_{\alpha\kappa} = 0.}} \tag{2.199}$$

Since this is valid in *any* event $\mathcal{P}$, it is valid *anywhere*.

2.10 The Ricci Matrix and Its Properties

The reason for considering the Riemannian curvature theory is to find with its help a way to determine the components of the Christoffel matrix $\boldsymbol{\Gamma}$, which are needed to calculate the solution of the equations

$$\ddot{\boldsymbol{x}} = -\left(\boldsymbol{I}_4 \otimes \dot{\boldsymbol{x}}^{\mathsf{T}}\right)\boldsymbol{\Gamma}\dot{\boldsymbol{x}},$$

describing the dynamic behaviour of particles in a gravitational field. But the Riemannian curvature matrix has, being a 16×16-matrix, 256 components, so, it would provide 256 equations in extreme cases. Considering, however, that there are 4 rows and 4 columns of $\boldsymbol{R}$ with all zeros, only $12 \cdot 12 = 144$ equations remain—still too many. The number of equations can be reduced significantly, however, by reducing, using clever addition of matrix elements, the number of components of the newly formed matrix. Setting up in this way a *symmetric* 4×4-matrix, we would get exactly 10 independent equations for the determination of the 10 independent components of $\boldsymbol{G}$. Such a matrix is called the Ricci matrix, which can be obtained in two ways. One way is through the sum of sub-matrices on the diagonal of $\boldsymbol{R}$; it is described in the Appendix. The second way is as follows:

The Ricci matrix $\boldsymbol{R}_{\text{Ric}}$ consists of the traces of sub-matrices $\boldsymbol{R}^{\gamma\delta}$ of $\boldsymbol{R}$

$$R_{\text{Ric},\gamma\delta} \stackrel{\text{def}}{=} \operatorname{trace}\left(\boldsymbol{R}^{\gamma\delta}\right) = \sum_{\nu=0}^{3} R_{\nu\nu}^{\gamma\delta}. \tag{2.200}$$

Accordingly, the components of the new matrix $\check{\boldsymbol{R}}_{\text{Ric}}$ are defined as

$$\check{R}_{\text{Ric},\gamma\delta} \stackrel{\text{def}}{=} \sum_{\nu=0}^{3} \check{R}_{\nu\nu}^{\gamma\delta}. \tag{2.201}$$

From (2.194) one may read immediately that the Ricci matrix $\check{\boldsymbol{R}}_{\text{Ric}}$ is *symmetric* because $\check{R}_{\nu\nu}^{\gamma\delta} = \check{R}_{\nu\nu}^{\delta\gamma}$.

Moreover, because of (2.162),

$$\boldsymbol{R} = \left(\boldsymbol{G}^{-1} \otimes \boldsymbol{I}_4\right)\check{\boldsymbol{R}}, \tag{2.202}$$

so

$$\boldsymbol{R}^{\gamma\delta} = \left(\boldsymbol{g}_\gamma^{-T} \otimes \boldsymbol{I}_4\right)\check{\boldsymbol{R}}^{\delta} = \sum_{\mu=0}^{3} g_{\gamma\mu}^{[-1]}\check{\boldsymbol{R}}^{\mu\delta}, \tag{2.203}$$

where $\boldsymbol{g}_\gamma^{-T}$ is the γth row of $\boldsymbol{G}^{-1}$ and $\check{\boldsymbol{R}}^{\delta}$ the 16×4-matrix consisting of the four sub-matrices in the δth block column of $\check{\boldsymbol{R}}$, i.e. for the matrix elements one has

$$R_{\alpha\beta}^{\gamma\delta} = \sum_{\mu=0}^{3} g_{\gamma\mu}^{[-1]}\check{R}_{\alpha\beta}^{\mu\delta}. \tag{2.204}$$

With the help of (2.200), for the Ricci matrix components one gets

$$R_{\text{Ric},\gamma\delta} = \sum_{\nu} R_{\nu\nu}^{\gamma\delta} = \sum_{\nu}\sum_{\mu} g_{\gamma\mu}^{[-1]}\check{R}_{\nu\nu}^{\mu\delta}, \tag{2.205}$$

or, due to (2.173),

$$R_{\mathrm{Ric},\gamma\delta} = \sum_{\nu}\sum_{\mu} g_{\gamma\mu}^{[-1]} \check{R}_{\mu\delta}^{\nu\nu}. \tag{2.206}$$

The *curvature scalar* R is obtained by taking the trace of the Ricci matrix:

$$\begin{aligned} R &\stackrel{\text{def}}{=} \operatorname{trace}(\boldsymbol{R}_{\mathrm{Ric}}) = \sum_{\alpha}\sum_{\nu} R_{\nu\nu}^{\alpha\alpha} \\ &= \sum_{\alpha}\sum_{\nu}\sum_{\mu} g_{\alpha\mu}^{[-1]} \check{R}_{\nu\nu}^{\mu\alpha} = \sum_{\alpha}\sum_{\mu} g_{\alpha\mu}^{[-1]} \check{R}_{\mathrm{Ric},\mu\alpha}. \end{aligned} \tag{2.207}$$

Conversely, one obtains

$$\check{R}_{\alpha\beta}^{\gamma\delta} = \sum_{\mu=0}^{3} g_{\gamma\mu} R_{\alpha\beta}^{\mu\delta}. \tag{2.208}$$

Because of (2.165) it follows directly that

$$R_{\mathrm{Ric},\gamma\delta} = \sum_{\nu=0}^{3}\left(\frac{\partial}{\partial x_\delta}\Gamma_{\gamma\nu}^{\nu} - \frac{\partial}{\partial x_\nu}\Gamma_{\gamma\delta}^{\nu} + \sum_{\mu=0}^{3}\Gamma_{\delta\mu}^{\nu}\Gamma_{\nu\gamma}^{\mu} - \sum_{\mu=0}^{3}\Gamma_{\nu\mu}^{\mu}\Gamma_{\gamma\delta}^{\mu}\right) \tag{2.209}$$

and from (2.168)

$$\check{R}_{\mathrm{Ric},\gamma\delta} = \sum_{\nu=0}^{3}\left(\frac{\partial}{\partial x_\delta}\check{\Gamma}_{\gamma\nu}^{\nu} - \frac{\partial}{\partial x_\nu}\check{\Gamma}_{\gamma\delta}^{\nu} + \sum_{\mu=0}^{3}\Gamma_{\gamma\delta}^{\mu}\check{\Gamma}_{\nu\nu}^{\mu} - \sum_{\mu=0}^{3}\Gamma_{\gamma\nu}^{\mu}\check{\Gamma}_{\nu\delta}^{\mu}\right). \tag{2.210}$$

2.10.1 Symmetry of the Ricci Matrix $\boldsymbol{R}_{\mathrm{Ric}}$

Even if the Riemannian matrix $\boldsymbol{R}$ itself is not symmetric, the derived Ricci matrix $\boldsymbol{R}_{\mathrm{Ric}}$ is nevertheless symmetric; and that will be shown below. The symmetry will follow from the component equations (2.209) of the Ricci matrix. One sees immediately that the second and fourth summand are symmetric in γ and δ. Looking at $\sum_{\nu=0}^{3}\frac{\partial}{\partial x_\delta}\Gamma_{\gamma\nu}^{\nu}$, it is not directly seen that this term is symmetric in γ and δ. This can be seen using the Laplace expansion theorem for determinants.[1] Developing the determinant of $\boldsymbol{G}$ along the νth row, one gets

$$g \stackrel{\text{def}}{=} \det(\boldsymbol{G}) = g_{\nu 1}A_{\nu 1} + \cdots + g_{\nu\delta}A_{\nu\delta} + \cdots + g_{\nu n}A_{\nu n},$$

[1]The sum of the products of all elements of a row (or column) with its adjuncts is equal to the determinant's value: $\det A = \sum_{j=1}^{n}(-1)^{i+j}\cdot a_{ij}\cdot\det A_{ij}$ (development along the ith row)

where $A_{\nu\delta}$ is the element in the νth row and δth column of the adjoint of $\boldsymbol{G}$. If $g_{\delta\nu}^{[-1]}$ is the $(\delta\nu)$-element of the inverse matrix $\boldsymbol{G}^{-1}$, then $g_{\delta\nu}^{[-1]} = \frac{1}{g} A_{\nu\delta}$, so $A_{\nu\delta} = g g_{\delta\nu}^{[-1]}$. Thus we obtain

$$\frac{\partial g}{\partial g_{\nu\delta}} = A_{\nu\delta} = g g_{\delta\nu}^{[-1]},$$

or

$$\delta g = g g_{\delta\nu}^{[-1]} \delta g_{\nu\delta},$$

and

$$\frac{\partial g}{\partial x_\gamma} = g g_{\delta\nu}^{[-1]} \frac{\partial g_{\nu\delta}}{\partial x_\gamma},$$

i.e.

$$\frac{1}{g}\frac{\partial g}{\partial x_\gamma} = g_{\delta\nu}^{[-1]} \frac{\partial g_{\nu\delta}}{\partial x_\gamma}. \tag{2.211}$$

Due to (2.62), on the other hand, one has

$$\sum_{\nu=0}^{3} \Gamma_{\gamma\nu}^{\nu} = \sum_{\nu=0}^{3}\sum_{\delta=0}^{3} \frac{g_{\delta\nu}^{[-1]}}{2}\left(\frac{\partial g_{\nu\delta}}{\partial x_\gamma} + \frac{\partial g_{\gamma\delta}}{\partial x_\nu} - \frac{\partial g_{\gamma\nu}}{\partial x_\delta}\right),$$

i.e. the last two summands cancel out, and the following remains:

$$\sum_{\nu=0}^{3} \Gamma_{\gamma\nu}^{\nu} = \sum_{\nu=0}^{3}\sum_{\delta=0}^{3} \frac{g_{\delta\nu}^{[-1]}}{2} \frac{\partial g_{\nu\delta}}{\partial x_\gamma}.$$

Then it follows by (2.211) that

$$\underline{\underline{\sum_{\nu=0}^{3} \frac{\partial}{\partial x_\delta} \Gamma_{\gamma\nu}^{\nu} = \sum_{\nu=0}^{3}\sum_{\delta=0}^{3} \frac{1}{\sqrt{|g|}} \frac{\partial^2 \sqrt{|g|}}{\partial x_\gamma \partial x_\delta}}}. \tag{2.212}$$

From this form the symmetry in γ and δ can immediately be observed.

Now remains to show that the third term

$$\sum_{\nu=0}^{3}\sum_{\mu=0}^{3} \Gamma_{\delta\mu}^{\nu} \Gamma_{\nu\gamma}^{\mu}$$

in (2.209) is symmetric. But this can spotted from the following equalities:

$$\sum_{\nu,\mu=0}^{3} \Gamma_{\delta\mu}^{\nu} \Gamma_{\nu\gamma}^{\mu} = \sum_{\nu,\mu=0}^{3} \Gamma_{\mu\delta}^{\nu} \Gamma_{\gamma\nu}^{\mu} = \sum_{\nu,\mu=0}^{3} \Gamma_{\nu\delta}^{\mu} \Gamma_{\gamma\mu}^{\nu}.$$

Combining everything together demonstrates that the Ricci matrix $\boldsymbol{R}_{\mathrm{Ric}}$ is, in fact, symmetric!

2.10.2 The Divergence of the Ricci Matrix

Multiplying the Bianchi identities (2.199) in the form

$$\frac{\partial}{\partial x_\kappa} R^{\nu\delta}_{\alpha\beta} + \frac{\partial}{\partial x_\beta} R^{\nu\kappa}_{\alpha\delta} + \frac{\partial}{\partial x_\delta} R^{\nu\beta}_{\alpha\kappa} = 0$$

with $g_{\gamma\nu}$ and summing over ν, we obtain at a point $\mathcal{P}$, where $\frac{\partial \boldsymbol{G}}{\partial \boldsymbol{x}} = \boldsymbol{0}$,

$$\frac{\partial}{\partial x_\kappa} \sum_{\nu=0}^{3} g_{\gamma\nu} R^{\nu\delta}_{\alpha\beta} + \frac{\partial}{\partial x_\beta} \sum_{\nu=0}^{3} g_{\gamma\nu} R^{\nu\kappa}_{\alpha\delta} + \frac{\partial}{\partial x_\delta} \sum_{\nu=0}^{3} g_{\gamma\nu} R^{\nu\beta}_{\alpha\kappa} = 0.$$

With (2.208) we obtain the following modified Bianchi identity:

$$\frac{\partial}{\partial x_\kappa} \check{R}^{\gamma\delta}_{\alpha\beta} + \frac{\partial}{\partial x_\beta} \check{R}^{\gamma\kappa}_{\alpha\delta} + \frac{\partial}{\partial x_\delta} \check{R}^{\gamma\beta}_{\alpha\kappa} = 0. \tag{2.213}$$

For the third term we can, with respect to (2.172), also write

$$-\frac{\partial}{\partial x_\delta} \check{R}^{\gamma\kappa}_{\alpha\beta}.$$

Substituting now $\alpha = \beta$ and summing over α, we obtain

$$\frac{\partial}{\partial x_\kappa} \check{R}_{\mathrm{Ric},\gamma\delta} + \sum_{\alpha=0}^{3} \frac{\partial}{\partial x_\alpha} \check{R}^{\gamma\kappa}_{\alpha\delta} - \frac{\partial}{\partial x_\delta} \check{R}_{\mathrm{Ric},\gamma\kappa} = 0. \tag{2.214}$$

In the second term, one can, based on (2.171), replace $\check{R}^{\gamma\kappa}_{\alpha\delta}$ by $-\check{R}^{\alpha\kappa}_{\gamma\delta}$. If we set $\gamma = \delta$ and sum over γ, we obtain from (2.214) with the "curvature scalar" $\check{R} \stackrel{\mathrm{def}}{=} \sum_{\gamma=0}^{3} \check{R}_{\mathrm{Ric},\gamma\gamma}$, the trace of the Ricci matrix $\check{\boldsymbol{R}}_{\mathrm{Ric}}$:

$$\frac{\partial}{\partial x_\kappa} \check{R} - \sum_{\alpha=0}^{3} \frac{\partial}{\partial x_\alpha} \check{R}_{\mathrm{Ric},\alpha\kappa} - \sum_{\gamma=0}^{3} \frac{\partial}{\partial x_\gamma} \check{R}_{\mathrm{Ric},\gamma\kappa} = 0. \tag{2.215}$$

If in the last sum the summation index γ is replaced by α, we can finally summarize

$$\frac{\partial}{\partial x_\kappa} \check{R} - 2\sum_{\alpha=0}^{3} \frac{\partial}{\partial x_\alpha} \check{R}_{\mathrm{Ric},\alpha\kappa} = 0. \tag{2.216}$$

We would get the same result, when this equation was assumed:

$$\frac{\partial}{\partial x_\kappa} \check{R}^{\alpha\beta}_{\gamma\delta} - 2\frac{\partial}{\partial x_\beta} \check{R}^{\alpha\kappa}_{\gamma\delta} = 0. \tag{2.217}$$

Indeed, setting $\delta = \gamma$ and summing over γ, we receive first:

$$\frac{\partial}{\partial x_\kappa}\check{R}_{\mathrm{Ric},\alpha\beta} - 2\frac{\partial}{\partial x_\beta}\check{R}_{\mathrm{Ric},\alpha\kappa} = 0.$$

Substituting now $\alpha = \beta$ and summing over α, we receive again (2.216).

A different result is obtained when starting at first from (2.217) (with ν instead of α), multiplying this equation by $g_{\alpha\nu}^{[-1]}$,

$$\frac{\partial}{\partial x_\kappa} g_{\alpha\nu}^{[-1]}\check{R}_{\gamma\delta}^{\nu\beta} - 2\frac{\partial}{\partial x_\beta} g_{\alpha\nu}^{[-1]}\check{R}_{\gamma\delta}^{\nu\kappa} = 0,$$

then setting $\gamma = \delta$, summing over γ and ν, and using (2.214):

$$\sum_\gamma\sum_\nu \frac{\partial}{\partial x_\kappa} g_{\alpha\nu}^{[-1]}\check{R}_{\gamma\delta}^{\nu\beta} - 2\sum_\gamma\sum_\nu \frac{\partial}{\partial x_\beta} g_{\alpha\nu}^{[-1]}\check{R}_{\gamma\delta}^{\nu\kappa}$$

$$= \frac{\partial}{\partial x_\kappa}R_{\mathrm{Ric},\alpha\beta} - 2\frac{\partial}{\partial x_\beta}R_{\mathrm{Ric},\alpha\kappa} = 0.$$

If we now set $\alpha = \beta$ and sum over α, we finally obtain the important relationship

$$\frac{\partial}{\partial x_\kappa}R - 2\sum_\alpha \frac{\partial}{\partial x_\alpha}R_{\mathrm{Ric},\alpha\kappa} = 0. \tag{2.218}$$

These are the four equations for the four spacetime coordinates $x_0, \ldots, x_3$. Finally, the overall result can be represented as

$$\boxed{\vec{\nabla}^\mathsf{T}\left(\boldsymbol{R}_{\mathrm{Ric}} - \frac{1}{2}R\boldsymbol{I}_4\right) = \boldsymbol{0}^\mathsf{T}.} \tag{2.219}$$

2.11 General Theory of Gravitation

2.11.1 The Einstein's Matrix $\mathfrak{G}$

With the Einstein's matrix

$$\boxed{\mathfrak{G} \stackrel{\mathrm{def}}{=} \boldsymbol{R}_{\mathrm{Ric}} - \frac{1}{2}R\boldsymbol{I}_4,} \tag{2.220}$$

taking into account that the matrix $\mathfrak{G}$ is symmetric, (2.219) can be restated as

$$\boxed{\vec{\nabla}^\mathsf{T}\mathfrak{G} = \boldsymbol{0}^\mathsf{T}.} \tag{2.221}$$

The following is a very important property of the Einstein's matrix:

The divergence of the Einstein's matrix $\mathfrak{G}$ vanishes!

2.11.2 Newton's Theory of Gravity

According to Newton, for the acceleration the following is valid:

$$\frac{\mathrm{d}^2\boldsymbol{x}}{\mathrm{d}t^2} = -\boldsymbol{\nabla}\phi(\boldsymbol{x}), \tag{2.222}$$

where $\phi(\boldsymbol{x})$ is the gravitational potential and $\boldsymbol{x} \in \mathbb{R}^3$. One can also write

$$\frac{\mathrm{d}^2\boldsymbol{x}}{\mathrm{d}t^2} + \boldsymbol{\nabla}\phi(\boldsymbol{x}) = \boldsymbol{0}. \tag{2.223}$$

The Newtonian universal time is a parameter which has two degrees of freedom, namely the time origin t_0 and a, the unit of time, both can be chosen arbitrarily: $t = t_0 + a\tau$. Thus we obtain

$$\frac{\mathrm{d}^2 t}{\mathrm{d}\tau^2} = 0, \qquad \frac{\mathrm{d}^2\boldsymbol{x}}{\mathrm{d}\tau^2} + \frac{\partial\phi}{\partial\boldsymbol{x}}\left(\frac{\mathrm{d}t}{\mathrm{d}\tau}\right)^2 = \boldsymbol{0}. \tag{2.224}$$

This can also be written with the spacetime vector $\vec{\boldsymbol{x}} = \binom{ct}{\boldsymbol{x}} \in \mathbb{R}^4$:

$$\frac{\mathrm{d}^2\vec{\boldsymbol{x}}}{\mathrm{d}\tau^2} + \left(\boldsymbol{I}_4 \otimes \dot{\vec{\boldsymbol{x}}}^{\mathsf{T}}\right)\boldsymbol{\Gamma}\dot{\vec{\boldsymbol{x}}} = \boldsymbol{0}. \tag{2.225}$$

Here $\boldsymbol{\Gamma} \in \mathbb{R}^{16\times 4}$ has the form

$$\boldsymbol{\Gamma} = \left(\begin{array}{cccc} 0 & 0 & 0 & 0 \\ 0 & 0 & 0 & 0 \\ 0 & 0 & 0 & 0 \\ 0 & 0 & 0 & 0 \\ \hline \frac{\partial\phi}{\partial x_1} & 0 & 0 & 0 \\ 0 & 0 & 0 & 0 \\ 0 & 0 & 0 & 0 \\ 0 & 0 & 0 & 0 \\ \hline \frac{\partial\phi}{\partial x_2} & 0 & 0 & 0 \\ 0 & 0 & 0 & 0 \\ 0 & 0 & 0 & 0 \\ 0 & 0 & 0 & 0 \\ \hline \frac{\partial\phi}{\partial x_3} & 0 & 0 & 0 \\ 0 & 0 & 0 & 0 \\ 0 & 0 & 0 & 0 \\ 0 & 0 & 0 & 0 \end{array}\right).$$

Now, how can one get the expression of the Poisson equation

$$\Delta\phi(\boldsymbol{x}) = \frac{\partial^2\phi}{\partial x_1^2} + \frac{\partial^2\phi}{\partial x_2^2} + \frac{\partial^2\phi}{\partial x_3^2} = 4\pi G\rho(\boldsymbol{x})? \tag{2.226}$$

For the curvature matrix $\boldsymbol{R}$ one needs the matrix

$$
\frac{\partial \boldsymbol{\Gamma}}{\partial \boldsymbol{x}^{\top}} = \left(\begin{array}{cccc|cccc|cccc|cccc}
0 & 0 & 0 & 0 & 0 & 0 & 0 & 0 & 0 & 0 & 0 & 0 & 0 & 0 & 0 & 0 \\
0 & 0 & 0 & 0 & 0 & 0 & 0 & 0 & 0 & 0 & 0 & 0 & 0 & 0 & 0 & 0 \\
0 & 0 & 0 & 0 & 0 & 0 & 0 & 0 & 0 & 0 & 0 & 0 & 0 & 0 & 0 & 0 \\
0 & 0 & 0 & 0 & 0 & 0 & 0 & 0 & 0 & 0 & 0 & 0 & 0 & 0 & 0 & 0 \\
\hline
0 & 0 & 0 & 0 & \frac{\partial^2 \phi}{\partial x_1^2} & 0 & 0 & 0 & \frac{\partial^2 \phi}{\partial x_1 \partial x_2} & 0 & 0 & 0 & \frac{\partial^2 \phi}{\partial x_1 \partial x_3} & 0 & 0 & 0 \\
0 & 0 & 0 & 0 & 0 & 0 & 0 & 0 & 0 & 0 & 0 & 0 & 0 & 0 & 0 & 0 \\
0 & 0 & 0 & 0 & 0 & 0 & 0 & 0 & 0 & 0 & 0 & 0 & 0 & 0 & 0 & 0 \\
0 & 0 & 0 & 0 & 0 & 0 & 0 & 0 & 0 & 0 & 0 & 0 & 0 & 0 & 0 & 0 \\
\hline
0 & 0 & 0 & 0 & \frac{\partial^2 \phi}{\partial x_2 \partial x_1} & 0 & 0 & 0 & \frac{\partial^2 \phi}{\partial x_2^2} & 0 & 0 & 0 & \frac{\partial^2 \phi}{\partial x_2 \partial x_3} & 0 & 0 & 0 \\
0 & 0 & 0 & 0 & 0 & 0 & 0 & 0 & 0 & 0 & 0 & 0 & 0 & 0 & 0 & 0 \\
0 & 0 & 0 & 0 & 0 & 0 & 0 & 0 & 0 & 0 & 0 & 0 & 0 & 0 & 0 & 0 \\
0 & 0 & 0 & 0 & 0 & 0 & 0 & 0 & 0 & 0 & 0 & 0 & 0 & 0 & 0 & 0 \\
\hline
0 & 0 & 0 & 0 & \frac{\partial^2 \phi}{\partial x_3 \partial x_1} & 0 & 0 & 0 & \frac{\partial^2 \phi}{\partial x_3 \partial x_2} & 0 & 0 & 0 & \frac{\partial^2 \phi}{\partial x_3^2} & 0 & 0 & 0 \\
0 & 0 & 0 & 0 & 0 & 0 & 0 & 0 & 0 & 0 & 0 & 0 & 0 & 0 & 0 & 0 \\
0 & 0 & 0 & 0 & 0 & 0 & 0 & 0 & 0 & 0 & 0 & 0 & 0 & 0 & 0 & 0 \\
0 & 0 & 0 & 0 & 0 & 0 & 0 & 0 & 0 & 0 & 0 & 0 & 0 & 0 & 0 & 0
\end{array}\right)
$$

and the matrix

$$
\boldsymbol{\Gamma}^{*} - \boldsymbol{U}_{4\times 4}\boldsymbol{\Gamma} = \left(\begin{array}{cccc}
0 & 0 & 0 & 0 \\
\frac{\partial \phi}{\partial x_1} & 0 & 0 & 0 \\
\frac{\partial \phi}{\partial x_2} & 0 & 0 & 0 \\
\frac{\partial \phi}{\partial x_3} & 0 & 0 & 0 \\
\hline
0 & 0 & 0 & 0 \\
0 & 0 & 0 & 0 \\
0 & 0 & 0 & 0 \\
0 & 0 & 0 & 0 \\
\hline
0 & 0 & 0 & 0 \\
0 & 0 & 0 & 0 \\
0 & 0 & 0 & 0 \\
0 & 0 & 0 & 0 \\
\hline
0 & 0 & 0 & 0 \\
0 & 0 & 0 & 0 \\
0 & 0 & 0 & 0 \\
0 & 0 & 0 & 0
\end{array}\right)
$$

in order to determine the following matrix

$$
\overline{\boldsymbol{\Gamma}} = \left(\begin{array}{cccc|cccc|cccc|cccc}
0 & 0 & 0 & 0 & 0 & 0 & 0 & 0 & 0 & 0 & 0 & 0 & 0 & 0 & 0 & 0 \\
\frac{\partial \phi}{\partial x_1} & 0 & 0 & 0 & 0 & 0 & 0 & 0 & 0 & 0 & 0 & 0 & 0 & 0 & 0 & 0 \\
\frac{\partial \phi}{\partial x_2} & 0 & 0 & 0 & 0 & 0 & 0 & 0 & 0 & 0 & 0 & 0 & 0 & 0 & 0 & 0 \\
\frac{\partial \phi}{\partial x_3} & 0 & 0 & 0 & 0 & 0 & 0 & 0 & 0 & 0 & 0 & 0 & 0 & 0 & 0 & 0
\end{array}\right).
$$

However, for the product $(\overline{\boldsymbol{\Gamma}} \otimes \boldsymbol{I}_4)(\boldsymbol{I}_4 \otimes \boldsymbol{\Gamma})$ we get the zero matrix, so that the curvature matrix is only composed as

$$\boldsymbol{R} = \frac{\partial \boldsymbol{\Gamma}}{\partial \boldsymbol{x}^{\mathsf{T}}} (\boldsymbol{U}_{4\times 4} - \boldsymbol{I}_{16}).$$

It is true that

$$\frac{\partial \boldsymbol{\Gamma}}{\partial \boldsymbol{x}^{\mathsf{T}}} \boldsymbol{U}_{4\times 4} = \left(\begin{array}{cccc|cccc|cccc|cccc}
0 & 0 & 0 & 0 & 0 & 0 & 0 & 0 & 0 & 0 & 0 & 0 & 0 & 0 & 0 & 0 \\
0 & 0 & 0 & 0 & 0 & 0 & 0 & 0 & 0 & 0 & 0 & 0 & 0 & 0 & 0 & 0 \\
0 & 0 & 0 & 0 & 0 & 0 & 0 & 0 & 0 & 0 & 0 & 0 & 0 & 0 & 0 & 0 \\
0 & 0 & 0 & 0 & 0 & 0 & 0 & 0 & 0 & 0 & 0 & 0 & 0 & 0 & 0 & 0 \\
\hline
0 & \frac{\partial^2 \phi}{\partial x_1^2} & \frac{\partial^2 \phi}{\partial x_1 \partial x_2} & \frac{\partial^2 \phi}{\partial x_1 \partial x_3} & 0 & 0 & 0 & 0 & 0 & 0 & 0 & 0 & 0 & 0 & 0 & 0 \\
0 & 0 & 0 & 0 & 0 & 0 & 0 & 0 & 0 & 0 & 0 & 0 & 0 & 0 & 0 & 0 \\
0 & 0 & 0 & 0 & 0 & 0 & 0 & 0 & 0 & 0 & 0 & 0 & 0 & 0 & 0 & 0 \\
0 & 0 & 0 & 0 & 0 & 0 & 0 & 0 & 0 & 0 & 0 & 0 & 0 & 0 & 0 & 0 \\
\hline
0 & \frac{\partial^2 \phi}{\partial x_2 \partial x_1} & \frac{\partial^2 \phi}{\partial x_2^2} & \frac{\partial^2 \phi}{\partial x_2 \partial x_3} & 0 & 0 & 0 & 0 & 0 & 0 & 0 & 0 & 0 & 0 & 0 & 0 \\
0 & 0 & 0 & 0 & 0 & 0 & 0 & 0 & 0 & 0 & 0 & 0 & 0 & 0 & 0 & 0 \\
0 & 0 & 0 & 0 & 0 & 0 & 0 & 0 & 0 & 0 & 0 & 0 & 0 & 0 & 0 & 0 \\
\hline
0 & \frac{\partial^2 \phi}{\partial x_3 \partial x_1} & \frac{\partial^2 \phi}{\partial x_3 \partial x_2} & \frac{\partial^2 \phi}{\partial x_3^2} & 0 & 0 & 0 & 0 & 0 & 0 & 0 & 0 & 0 & 0 & 0 & 0 \\
0 & 0 & 0 & 0 & 0 & 0 & 0 & 0 & 0 & 0 & 0 & 0 & 0 & 0 & 0 & 0 \\
0 & 0 & 0 & 0 & 0 & 0 & 0 & 0 & 0 & 0 & 0 & 0 & 0 & 0 & 0 & 0
\end{array}\right),$$

finally resulting in

$$\boldsymbol{R} = \left(\begin{array}{cccc|cccc|cccc|cccc}
0 & 0 & 0 & 0 & 0 & 0 & 0 & 0 & 0 & 0 & 0 & 0 & 0 & 0 & 0 & 0 \\
0 & 0 & 0 & 0 & 0 & 0 & 0 & 0 & 0 & 0 & 0 & 0 & 0 & 0 & 0 & 0 \\
0 & 0 & 0 & 0 & 0 & 0 & 0 & 0 & 0 & 0 & 0 & 0 & 0 & 0 & 0 & 0 \\
0 & 0 & 0 & 0 & 0 & 0 & 0 & 0 & 0 & 0 & 0 & 0 & 0 & 0 & 0 & 0 \\
\hline
0 & \frac{\partial^2 \phi}{\partial x_1^2} & \frac{\partial^2 \phi}{\partial x_1 \partial x_2} & \frac{\partial^2 \phi}{\partial x_1 \partial x_3} & -\frac{\partial^2 \phi}{\partial x_1^2} & 0 & 0 & 0 & -\frac{\partial^2 \phi}{\partial x_1 \partial x_2} & 0 & 0 & 0 & -\frac{\partial^2 \phi}{\partial x_1 \partial x_3} & 0 & 0 & 0 \\
0 & 0 & 0 & 0 & 0 & 0 & 0 & 0 & 0 & 0 & 0 & 0 & 0 & 0 & 0 & 0 \\
0 & 0 & 0 & 0 & 0 & 0 & 0 & 0 & 0 & 0 & 0 & 0 & 0 & 0 & 0 & 0 \\
0 & 0 & 0 & 0 & 0 & 0 & 0 & 0 & 0 & 0 & 0 & 0 & 0 & 0 & 0 & 0 \\
\hline
0 & \frac{\partial^2 \phi}{\partial x_2 \partial x_1} & \frac{\partial^2 \phi}{\partial x_2^2} & \frac{\partial^2 \phi}{\partial x_2 \partial x_3} & -\frac{\partial^2 \phi}{\partial x_2 \partial x_1} & 0 & 0 & 0 & -\frac{\partial^2 \phi}{\partial x_2^2} & 0 & 0 & 0 & -\frac{\partial^2 \phi}{\partial x_2 \partial x_3} & 0 & 0 & 0 \\
0 & 0 & 0 & 0 & 0 & 0 & 0 & 0 & 0 & 0 & 0 & 0 & 0 & 0 & 0 & 0 \\
0 & 0 & 0 & 0 & 0 & 0 & 0 & 0 & 0 & 0 & 0 & 0 & 0 & 0 & 0 & 0 \\
0 & 0 & 0 & 0 & 0 & 0 & 0 & 0 & 0 & 0 & 0 & 0 & 0 & 0 & 0 & 0 \\
\hline
0 & \frac{\partial^2 \phi}{\partial x_3 \partial x_1} & \frac{\partial^2 \phi}{\partial x_3 \partial x_2} & \frac{\partial^2 \phi}{\partial x_3^2} & -\frac{\partial^2 \phi}{\partial x_3 \partial x_1} & 0 & 0 & 0 & -\frac{\partial^2 \phi}{\partial x_3 \partial x_2} & 0 & 0 & 0 & -\frac{\partial^2 \phi}{\partial x_3^2} & 0 & 0 & 0 \\
0 & 0 & 0 & 0 & 0 & 0 & 0 & 0 & 0 & 0 & 0 & 0 & 0 & 0 & 0 & 0 \\
0 & 0 & 0 & 0 & 0 & 0 & 0 & 0 & 0 & 0 & 0 & 0 & 0 & 0 & 0 & 0 \\
0 & 0 & 0 & 0 & 0 & 0 & 0 & 0 & 0 & 0 & 0 & 0 & 0 & 0 & 0 & 0
\end{array}\right). \tag{2.227}$$

The 4×4-sub-matrices on the main diagonal of the curvature matrix $\boldsymbol{R}$ contain exactly the components of the left-hand side of the Poisson equation. This subsequently provides the motivation for the introduction of the Ricci matrix. Here it is specifically equal to

$$\boldsymbol{R}_{\text{Ric}} = \begin{pmatrix} 0 & 0 & 0 & 0 \\ 0 & -\frac{\partial^2\phi}{\partial x_1^2} & -\frac{\partial^2\phi}{\partial x_1 \partial x_2} & -\frac{\partial^2\phi}{\partial x_1 \partial x_3} \\ 0 & -\frac{\partial^2\phi}{\partial x_2 \partial x_1} & -\frac{\partial^2\phi}{\partial x_2^2} & -\frac{\partial^2\phi}{\partial x_2 \partial x_3} \\ 0 & -\frac{\partial^2\phi}{\partial x_3 \partial x_1} & -\frac{\partial^2\phi}{\partial x_3 \partial x_2} & -\frac{\partial^2\phi}{\partial x_3^2} \end{pmatrix}. \tag{2.228}$$

If we form the trace of the Ricci matrix, then

$$R = \text{trace}(\boldsymbol{R}_{\text{Ric}}) = -\left(\frac{\partial^2\phi}{\partial x_1^2} + \frac{\partial^2\phi}{\partial x_2^2} + \frac{\partial^2\phi}{\partial x_3^2}\right), \tag{2.229}$$

so that finally the Poisson equation can also be succinctly written as

$$\boxed{-R = 4\pi G\rho.} \tag{2.230}$$

The relationship in Newtonian mechanics between the gravitational potential ϕ and matter is the Poisson equation

$$\Delta\phi = 4\pi G\rho,$$

where ρ is the mass density and G the Newtonian Gravitational Constant. In the theory of General Relativity, it is now required to set up general invariant equations of gravitation between the g_{ik}'s and matter. It is useful to characterise the matter by the energy–momentum matrix $\boldsymbol{T} = \boldsymbol{T}_{\text{total}}$.

For the rotating system in Sect. 2.5, we had $\ddot{r} = r(\omega + \dot{\varphi})^2$ so that there the potential was $\phi = -\frac{\omega^2 r^2}{2}$ and $g_{00} = 1 - \frac{r^2\omega^2}{c^2} = 1 + \frac{2\phi}{c^2}$. On the other hand, in Newtonian non-relativistic limit $\frac{v^2}{c^2} \leqslant 1$, we obtained for the energy–momentum matrix (see Sect. 1.9.2) $T_{00} = c^2\rho_0$ and for the remaining $T_{ij} \approx 0$.

With $\phi = \frac{c^2}{2}(g_{00} - 1)$ we obtain $\Delta\phi = \frac{c^2}{2}\Delta g_{00}$. So one can write for the above Poisson equation

$$\Delta g_{00} = \frac{8\pi G}{c^4} T_{00}. \tag{2.231}$$

2.11.3 The Einstein's Equation with $\mathfrak{G}$

If one assumes that in the general case, i.e. in the presence of gravitational fields on the right-hand side of (2.231), the symmetric energy–momentum matrix is $\boldsymbol{T}$, a matrix containing the second partial derivatives of the elements of the metric matrix $\boldsymbol{G}$

must appear on the left-hand side of the equation. If therefore the Einstein's matrix $\mathfrak{G}$ is taken, one obtains the Einstein's Field Equation

$$\mathfrak{G} = -\frac{8\pi G}{c^4}\boldsymbol{T}. \tag{2.232}$$

Because the matrix $\boldsymbol{T} \in \mathbb{R}^{4\times 4}$ is symmetric, also the Einstein's matrix

$$\mathfrak{G} = \boldsymbol{R}_{\mathrm{Ric}} - \frac{1}{2}R\boldsymbol{I}_4 \tag{2.233}$$

must be symmetric. This is indeed the case because both the Ricci matrix $\boldsymbol{R}_{\mathrm{Ric}}$ and the diagonal matrix $R\boldsymbol{I}_4$ are symmetric. $\boldsymbol{R}_{\mathrm{Ric}}$ is extracted, according to (2.200), from the Riemannian curvature matrix

$$\boldsymbol{R} = \left(\frac{\partial \boldsymbol{\Gamma}}{\partial \boldsymbol{x}^{\mathsf{T}}} + (\overline{\boldsymbol{\Gamma}} \otimes \boldsymbol{I}_4)(\boldsymbol{I}_4 \otimes \boldsymbol{\Gamma})\right)(\boldsymbol{U}_{4\times 4} - \boldsymbol{I}_{16}). \tag{2.234}$$

Final Form of the Einstein's Equation The energy–momentum matrix $\boldsymbol{T}$ on the right-hand side of the Einstein's field equation (2.232) has the property that $\boldsymbol{T}\vec{\nabla} = \boldsymbol{0}$, if one considers a closed system, i.e. if there are no external forces acting. The same must apply on the left: $\mathfrak{G}\vec{\nabla} = \boldsymbol{0}$. But that is, due to (2.221), true for the symmetric matrix $\mathfrak{G}$!

Overall, the final Einstein's Field Equation is fixed as an axiom as

$$\boxed{\mathfrak{G} = \boldsymbol{R}_{\mathrm{Ric}} - \frac{R}{2}\boldsymbol{I}_4 = -\frac{8\pi G}{c^4}\boldsymbol{T}.} \tag{2.235}$$

This is a matrix differential equation for determining the metric matrix $\boldsymbol{G}$. This is no longer an "action-at-a-distance-equation" as Newton's, but describes the relationships at a spacetime point $\vec{x}$! $\boldsymbol{R}_{\mathrm{Ric}}$ depends linearly on the second order derivatives of g_{ik}, and depends nonlinearly on the g_{ik} directly. So the Einstein's equation is a coupled system of second-order nonlinear partial differential equations for the determination of the components g_{ik} of the metric matrix $\boldsymbol{G}$ as a function of the given distribution of matter, i.e. the matrix $\boldsymbol{T}$, the source of the gravitational field.

By taking the trace, from (2.235) it follows that

$$R - \frac{R}{2}\cdot 4 = -\frac{8\pi G}{c^4}T,$$

so

$$R = \frac{8\pi G}{c^4}T, \tag{2.236}$$

where

$$T \stackrel{\mathrm{def}}{=} \mathrm{trace}(\boldsymbol{T}).$$

Substituting (2.236) into (2.235), one obtains the following form of the Einstein's Field Equation:

$$\boxed{\boldsymbol{R}_{\text{Ric}} = \frac{8\pi G}{c^4}\left(\frac{T}{2}\boldsymbol{I}_4 - \boldsymbol{T}\right).} \tag{2.237}$$

By the way, the constant factor $\frac{8\pi G}{c^4}$ has the numerical value

$$\underline{\underline{\frac{8\pi G}{c^4} = 1.86 \cdot 10^{-27} \text{cm/g}.}} \tag{2.238}$$

2.12 Summary

2.12.1 Covariance Principle

Einstein postulated the *Equivalence Principle:*

Gravitational forces are equivalent to inertial forces.

Gravitational fields can be eliminated by transitioning to an accelerated coordinate system. In this new local inertial frame, the laws of the theory of special relativity are valid.

Directly from the equivalence principle follows the *Covariance Principle:*

The equations of physics must be invariant under general coordinate transformations.

In particular, this means that even in a local inertial frame the equations must be valid, so the transition from a metric matrix $\boldsymbol{G}$ to a Minkowski matrix $\boldsymbol{M}$ results in the laws of special relativity.

The connection is $\boldsymbol{G} = \boldsymbol{J}^{\mathsf{T}}\boldsymbol{M}\boldsymbol{J}$ and $\boldsymbol{M} = \boldsymbol{J}^{-1\mathsf{T}}\boldsymbol{G}\boldsymbol{J}^{-1}$, i.e. using the special transformation matrix $\boldsymbol{T}(\vec{x}) = \boldsymbol{J}^{-1}(\vec{x})$ leads for a particular event $\vec{x}$ to a local inertial frame in which the laws of special relativity apply. Conversely, one passes from the special local inertial frame with the event $\vec{\xi}$ to the same event in the general coordinate system $\vec{x}$ by the transformation $\vec{x} = \boldsymbol{J}^{-1}\vec{\xi}$.

The physical equations in the theory of general relativity must be formulated so that they are invariant (covariant) with respect to general coordinate transformations.

Above we derived: The formulas in the theory of general relativity are invariant under coordinate transformations if one replaces in the formulas of special relativity the ordinary derivatives $\frac{\partial \boldsymbol{a}}{\partial \boldsymbol{x}^{\mathsf{T}}}$ by the covariant derivatives $\boldsymbol{a}_{\|\boldsymbol{x}^{\mathsf{T}}}$. Then we're done! The law applies generally.

For a particle on which no force acts, for example, in an inertial system with $\dot{\vec{\xi}} \overset{\text{def}}{=} \frac{\mathrm{d}\vec{\xi}}{\mathrm{d}\tau}$, applies

$$\frac{\mathrm{d}\dot{\vec{\xi}}}{\mathrm{d}\tau} = \mathbf{0}. \tag{2.239}$$

Herein the ordinary differential $\mathrm{d}\dot{\vec{\xi}}$ must be replaced by the covariant differential $\mathrm{D}\vec{\boldsymbol{u}}$, with $\vec{\boldsymbol{u}} = \dot{\vec{\boldsymbol{x}}} \overset{\text{def}}{=} \boldsymbol{J}^{-1}\dot{\vec{\xi}}$. According to (2.145), we first obtain

$$\mathrm{D}\boldsymbol{u} = \vec{\boldsymbol{u}}_{\|\vec{x}^{\mathsf{T}}}\mathrm{d}\vec{\boldsymbol{x}} = \frac{\partial \vec{\boldsymbol{u}}}{\partial \vec{\boldsymbol{x}}^{\mathsf{T}}}\mathrm{d}\vec{\boldsymbol{x}} + \left(\boldsymbol{I}_4 \otimes \vec{\boldsymbol{u}}^{\mathsf{T}}\right)\boldsymbol{\Gamma}\mathrm{d}\vec{\boldsymbol{x}}, \tag{2.240}$$

so

$$\frac{\mathrm{D}\vec{\boldsymbol{u}}}{\mathrm{d}\tau} = \vec{\boldsymbol{u}}_{\|\vec{x}^{\mathsf{T}}}\frac{\mathrm{d}\vec{\boldsymbol{x}}}{\mathrm{d}\tau} = \frac{\mathrm{d}\vec{\boldsymbol{u}}}{\mathrm{d}\tau} + \left(\boldsymbol{I}_4 \otimes \vec{\boldsymbol{u}}^{\mathsf{T}}\right)\boldsymbol{\Gamma}\frac{\mathrm{d}\vec{\boldsymbol{x}}}{\mathrm{d}\tau}.$$

This used in (2.239) instead of $\frac{\mathrm{d}\vec{\boldsymbol{u}}}{\mathrm{d}\tau}$ generally yields

$$\boxed{\frac{\mathrm{D}\vec{\boldsymbol{u}}}{\mathrm{d}\tau} = \mathbf{0},} \tag{2.241}$$

or

$$\underline{\underline{\frac{\mathrm{d}\vec{\boldsymbol{u}}}{\mathrm{d}\tau} = -\left(\boldsymbol{I}_4 \otimes \vec{\boldsymbol{u}}^{\mathsf{T}}\right)\boldsymbol{\Gamma}\vec{\boldsymbol{u}}.}} \tag{2.242}$$

Through the Christoffel matrix $\boldsymbol{\Gamma}$, the effect of the gravitational field is expressed. If no gravitational field is present, then $\boldsymbol{\Gamma} = \mathbf{0}$ and (2.239) is again obtained. A comparison of (2.242) with the equation of a geodesic shows that a material particle moves on a geodesic, so to speak, "on the shortest path in curved space".

If, in addition to gravitational forces, other forces appear, e.g. exerted by electric fields, then for an inertial frame one has the equation

$$m\frac{\mathrm{d}\dot{\vec{\xi}}}{\mathrm{d}\tau} = \vec{\boldsymbol{f}}. \tag{2.243}$$

This equation multiplied from the left with $\boldsymbol{J}^{-1}$ and again used with the covariant derivative leads with

$$\vec{\boldsymbol{f}}_x \overset{\text{def}}{=} \boldsymbol{J}^{-1}\vec{\boldsymbol{f}}$$

to

$$\boxed{m\frac{\mathrm{D}\vec{\boldsymbol{u}}}{\mathrm{d}\tau} = \vec{\boldsymbol{f}}_x,} \tag{2.244}$$

or

$$\underline{\underline{m\frac{\mathrm{d}\vec{\boldsymbol{u}}}{\mathrm{d}\tau} = \vec{\boldsymbol{f}}_x - m\left(\boldsymbol{I}_4 \otimes \vec{\boldsymbol{u}}^{\mathsf{T}}\right)\boldsymbol{\Gamma}\vec{\boldsymbol{u}},}} \tag{2.245}$$

where on the right-hand side, next to the other forces $\vec{f}_x$, the gravitational forces are written.

2.12.2 Einstein's Field Equation and Momentum

The Einstein's field equation expresses that each form of matter and energy is a source of the gravitational field. The gravitational field is described by the metric matrix $\boldsymbol{G}$, the components of which must be determined by using Einstein's Field Equation

$$\boldsymbol{R}_{\mathrm{Ric}} - \frac{R}{2}\boldsymbol{I}_4 = -\frac{8\pi G}{c^4}\boldsymbol{T}. \tag{2.246}$$

So one basically has to perform the same procedure as in Newton's dynamics:

1. Solve the Poisson equation $\Delta\phi(\boldsymbol{x}) = 4\pi G\rho(\boldsymbol{x})$, to determine the potential function ϕ.
2. Establish the solution of the equation $\frac{\mathrm{d}^2\boldsymbol{x}}{\mathrm{d}t^2} = -\boldsymbol{\nabla}\phi(\boldsymbol{x})$.

In the theory of general relativity, we now arrive at the following procedure:

1. Solve Einstein's Field Equation (2.246) $\boldsymbol{R}_{\mathrm{Ric}} - \frac{R}{2}\boldsymbol{I}_4 = -\frac{8\pi G}{c^4}\boldsymbol{T}$, to determine the metric matrix $\boldsymbol{G}$.
2. Establish the solution of (2.245)

$$m\frac{\mathrm{d}\vec{u}}{\mathrm{d}\tau} = \vec{f}_x - m\left(\boldsymbol{I}_4 \otimes \vec{u}^{\top}\right)\boldsymbol{\Gamma}\vec{u}.$$

2.13 Hilbert's Action Functional

Above, the Einstein's equation was postulated as an axiom. Einstein has found this equation through years of hard work. As a supplement, we will now shown that one can obtain the Einstein's equation with the help of the calculus of variations. Following Hilbert, this is derived from a variational principle, but initially only for the free gravitational field, i.e. when $\boldsymbol{T} = \boldsymbol{0}$.

The gravitational field is determined primarily by the metric matrix $\boldsymbol{G}$, i.e. the curvature of space caused by this field. All curvature parameters are concentrated in the curvature scalar R which is formed through the trace of the Ricci matrix $\boldsymbol{R}_{\mathrm{Ric}}$. Now a variational functional, the Hilbert's action functional, is formed so that the space curvature is minimal:

$$W_{\mathrm{Grav}} = \int R\,\mathrm{d}V. \tag{2.247}$$

However, this integral is not invariant under coordinate transformations $\boldsymbol{\Theta}$. Therefore, we need the following considerations: Let

$$\boldsymbol{\Theta}^{\mathsf{T}}(\vec{x})\boldsymbol{G}(\vec{x})\boldsymbol{\Theta}(\vec{x}) = \boldsymbol{M}, \tag{2.248}$$

i.e. $\boldsymbol{\Theta}(\vec{x})$ is the matrix which at the point $\vec{x}$ transforms the metric matrix $\boldsymbol{G}$ into the Minkowski matrix $\boldsymbol{M}$. Computing the determinants on both sides of (2.248), we obtain

$$\underbrace{\det(\boldsymbol{G})}_{g}\,\underbrace{\det(\boldsymbol{\Theta})^2}_{\Theta^2} = \det(\boldsymbol{M}) = -1, \tag{2.249}$$

i.e.

$$\sqrt{-g} = \frac{1}{\Theta}.$$

In a Cartesian coordinate system, the integral of a scalar with respect to the scalar $\mathrm{d}V = \mathrm{d}x_0 \cdot \mathrm{d}x_1 \cdot \mathrm{d}x_2 \cdot \mathrm{d}x_3$ is also a scalar. When passing to curvilinear coordinates $\vec{x}'$, the integration element $\mathrm{d}V$ is

$$\frac{1}{\Theta}\mathrm{d}V' = \sqrt{-g'}\,\mathrm{d}V'.$$

In curvilinear coordinates, $\sqrt{-g}\,\mathrm{d}V$ behaves as an invariant when integrated over any area of the four-dimensional space. If f is a scalar, then $f\sqrt{-g}$, which is invariant when integrated over $\mathrm{d}V$, is named the *scalar density*. This quantity provides a scalar for its multiplication with the four-dimensional volume element $\mathrm{d}V$.

For this reason, we now consider only the effect

$$W_{\text{Grav}} = \int R\left(\boldsymbol{\Gamma}(\boldsymbol{x}), \frac{\partial \boldsymbol{\Gamma}}{\partial \boldsymbol{x}^{\mathsf{T}}}\right)\sqrt{-g}\,\mathrm{d}^4\vec{x}. \tag{2.250}$$

Under $\mathrm{d}^4\vec{x}$ the four-dimensional volume element $\mathrm{d}x_0 \cdot \mathrm{d}x_1 \cdot \mathrm{d}x_2 \cdot \mathrm{d}x_3$ is understood. The Einstein's equation will follow from (2.250) and the condition $\delta W_{\text{Grav}} = 0$ for any variation δg_{ik}. $R\sqrt{-g}$ is a Lagrange density which is integrated over the volume. The elements of $\Gamma^k_{\alpha\beta}$ of the Christoffel matrix $\boldsymbol{\Gamma}$ are (2.62)

$$\Gamma^k_{\alpha\beta} = \sum_{i=0}^{3} \frac{g^{[-1]}_{ki}}{2}\left(\frac{\partial g_{\beta i}}{\partial x_\alpha} + \frac{\partial g_{\alpha i}}{\partial x_\beta} - \frac{\partial g_{\alpha\beta}}{\partial x_i}\right) \tag{2.251}$$

and the elements of the Riemannian curvature matrix $\boldsymbol{R}$ are, according to (2.165),

$$R^{\gamma\delta}_{\alpha\beta} = \frac{\partial}{\partial x_\beta}\Gamma^\gamma_{\alpha\delta} - \frac{\partial}{\partial x_\delta}\Gamma^\gamma_{\alpha\beta} + \sum_\nu \Gamma^\gamma_{\beta\nu}\Gamma^\nu_{\delta\alpha} - \sum_\nu \Gamma^\gamma_{\delta\nu}\Gamma^\nu_{\alpha\beta}. \tag{2.252}$$

The curvature scalar R, due to (2.207), is obtained from the Ricci matrix by

$$R = \operatorname{trace}(\boldsymbol{R}_{\mathrm{Ric}}) = \sum_{\alpha}\sum_{\nu} R^{\alpha\alpha}_{\nu\nu}$$
$$= \sum_{\alpha}\sum_{\nu}\sum_{\mu} g^{[-1]}_{\alpha\mu} \check{R}^{\mu\alpha}_{\nu\nu} = \sum_{\alpha}\sum_{\mu} g^{[-1]}_{\alpha\mu} \check{R}_{\mathrm{Ric},\mu\alpha}, \tag{2.253}$$

which, in accordance with (2.210), is

$$\check{R}_{\mathrm{Ric},\gamma\delta} = \sum_{\nu=0}^{3}\left(\frac{\partial}{\partial x_\delta}\check{\Gamma}^{\nu}_{\gamma\nu} - \frac{\partial}{\partial x_\nu}\check{\Gamma}^{\nu}_{\gamma\delta} + \sum_{\mu=0}^{3}\Gamma^{\mu}_{\gamma\delta}\check{\Gamma}^{\mu}_{\nu\nu} - \sum_{\mu=0}^{3}\Gamma^{\mu}_{\gamma\nu}\check{\Gamma}^{\mu}_{\nu\delta}\right). \tag{2.254}$$

The Lagrange/Hamilton theory applied to the action integral (2.250) provides the Euler/Lagrange equations for the variational problem associated to (2.250). We consider the elements of $g^{[-1]}_{ki}$ and $\Gamma^{k}_{\alpha\beta}$ as independent functions $f_i(\vec{x})$. The integral therefore is a functional of the form

$$\int L\left(f_i(\vec{x}), \frac{\partial f_i}{\partial x_k}(\vec{x})\right) \mathrm{d}^4\vec{x}. \tag{2.255}$$

This yields the Euler/Lagrange equations

$$\frac{\partial L}{\partial f_i} = \sum_{k=0}^{3} \frac{\partial}{\partial x_k} \frac{\partial L}{\partial(\frac{\partial f_i}{\partial x_k})}. \tag{2.256}$$

The Euler/Lagrange equations for (2.250) are

$$\frac{\partial}{\partial g^{[-1]}_{\alpha\beta}}\left(\sqrt{-g}\sum_{\delta}\sum_{\mu} g^{[-1]}_{\mu\delta}\check{R}_{\mathrm{Ric},\mu\delta}\right) = 0, \tag{2.257}$$

$$\frac{\partial}{\partial \Gamma^{\gamma}_{\alpha\beta}}\left(\sqrt{-g}\sum_{\delta}\sum_{\mu} g^{[-1]}_{\mu\delta}\check{R}_{\mathrm{Ric},\mu\delta}\right) = \sum_{\delta=0}^{3}\frac{\partial}{\partial x_\delta}\left(\sqrt{-g}\sum_{\delta}\sum_{\mu} g^{[-1]}_{\mu\delta}\frac{\partial \check{R}_{\mathrm{Ric},\mu\delta}}{\partial(\frac{\partial \Gamma^{\gamma}_{\alpha\beta}}{\partial x_\delta})}\right). \tag{2.258}$$

The first equation (2.257) provides the Einstein's field equation. Indeed, one can first write this compactly:

$$\frac{\partial L}{\partial \boldsymbol{G}^{-1}} = \frac{\partial(\sqrt{-g}R)}{\partial \boldsymbol{G}^{-1}} = \frac{\partial\sqrt{-g}}{\partial \boldsymbol{G}^{-1}}R + \sqrt{-g}\frac{\partial R}{\partial \boldsymbol{G}^{-1}} = 0. \tag{2.259}$$

For $\frac{\partial\sqrt{-g}}{\partial \boldsymbol{G}^{-1}}$, one then obtains

$$\frac{\partial\sqrt{-g}}{\partial \boldsymbol{G}^{-1}} = \frac{-1}{2\sqrt{-g}} \cdot \frac{\partial g}{\partial \boldsymbol{G}^{-1}}. \tag{2.260}$$

In addition,

$$\frac{\partial}{\partial \boldsymbol{G}^{-1}}\left(\frac{1}{g}\cdot g\right)=\boldsymbol{0}=\frac{1}{g}\frac{\partial g}{\partial \boldsymbol{G}^{-1}}+g\frac{\partial(1/g)}{\partial \boldsymbol{G}^{-1}}. \tag{2.261}$$

According to the Laplace expansion theorem for determinants ("The sum of the products of all elements of a row with their cofactors is equal to the value of the determinant"), we obtain by developing the determinant of $\boldsymbol{G}^{-1}$ along the γth row:

$$\det\left(\boldsymbol{G}^{-1}\right)=\frac{1}{g}=g_{\gamma 0}^{[-1]}A_{\gamma 0}^{[-1]}+\cdots+g_{\gamma\beta}^{[-1]}A_{\gamma\beta}^{[-1]}+\cdots+g_{\gamma 3}^{[-1]}A_{\gamma 3}^{[-1]}, \tag{2.262}$$

where $A_{\gamma\beta}^{[-1]}$ is the element in the γth row and βth column of the adjoint of $\boldsymbol{G}^{-1}$; $g_{\gamma\beta}^{[-1]}$ is the $(\gamma\beta)$-element of $\boldsymbol{G}^{-1}$. It is, of course, true that

$$\boldsymbol{G}=\frac{\operatorname{adj}(\boldsymbol{G}^{-1})}{\det(\boldsymbol{G}^{-1})}$$

so element-wise $g_{\beta\gamma}=g\cdot A_{\gamma\beta}^{[-1]}$, or $A_{\gamma\beta}^{[-1]}=\frac{1}{g}g_{\beta\gamma}$. Thus, we obtain by partial differentiation of (2.262) with respect to $g_{\gamma\beta}^{[-1]}$

$$\frac{\partial(1/g)}{\partial g_{\gamma\beta}^{[-1]}}=A_{\gamma\beta}^{[-1]}=\frac{1}{g}g_{\beta\gamma},$$

so

$$\frac{\partial(1/g)}{\partial \boldsymbol{G}^{-1}}=\frac{1}{g}\boldsymbol{G}. \tag{2.263}$$

Using this in (2.261) provides

$$\frac{\partial g}{\partial \boldsymbol{G}^{-1}}=-g\boldsymbol{G}, \tag{2.264}$$

and, together with (2.260), finally yields

$$\underline{\underline{\frac{\partial\sqrt{-g}}{\partial \boldsymbol{G}^{-1}}=\frac{-1}{2}\sqrt{-g}\cdot\boldsymbol{G}.}} \tag{2.265}$$

All what is missing in (2.259) is $\frac{\partial R}{\partial \boldsymbol{G}^{-1}}$. Using $R=\sum_\alpha\sum_\mu g_{\alpha\mu}^{[-1]}\check{R}_{\text{Ric},\mu\alpha}$,

$$\underline{\underline{\frac{\partial R}{\partial \boldsymbol{G}^{-1}}}}=\begin{pmatrix}\frac{\partial R}{\partial g_{00}^{[-1]}} & \cdots & \frac{\partial R}{\partial g_{03}^{[-1]}}\\ \vdots & \ddots & \vdots\\ \frac{\partial R}{\partial g_{30}^{[-1]}} & \cdots & \frac{\partial R}{\partial g_{33}^{[-1]}}\end{pmatrix}=\begin{pmatrix}\check{R}_{\text{Ric},00} & \cdots & \check{R}_{\text{Ric},03}\\ \vdots & \ddots & \vdots\\ \check{R}_{\text{Ric},30} & \cdots & \check{R}_{\text{Ric},33}\end{pmatrix}=\underline{\underline{\check{\boldsymbol{R}}_{\text{Ric}}}}. \tag{2.266}$$

Multiplying this matrix with G^{-1}, by the way, yields

$$\boldsymbol{R}_{\text{Ric}} = \boldsymbol{G}^{-1}\check{\boldsymbol{R}}_{\text{Ric}}. \tag{2.267}$$

Inserting (2.265) and (2.266) into (2.259) yields first

$$\sqrt{-g}\left(\frac{-1}{2}\cdot \boldsymbol{G}R + \check{\boldsymbol{R}}_{\text{Ric}}\right) = \boldsymbol{0}, \tag{2.268}$$

i.e. the following particular form of the Einstein's field equation

$$\check{\boldsymbol{R}}_{\text{Ric}} - \frac{R}{2}\cdot \boldsymbol{G} = \boldsymbol{0}. \tag{2.269}$$

Multiplying this equation from the left with the inverse metric matrix $\boldsymbol{G}^{-1}$, we finally obtain the Einstein's field equation for a source-free gravitational field ($\boldsymbol{T} = \boldsymbol{0}$) as in (2.235)

$$\boxed{\boldsymbol{R}_{\text{Ric}} - \frac{R}{2}\boldsymbol{I}_4 = \boldsymbol{0}.} \tag{2.270}$$

2.13.1 *Effects of Matter*

So far, only the gravitational field in vacuum has been treated. To account for the sources of the gravitational field, e.g. matter, the action functional must include an additive term W_M that describes the source:

$$W \stackrel{\text{def}}{=} W_{\text{Grav}} + W_M = \int (kR + \mathcal{L}_M)\sqrt{-g}\,\mathrm{d}^4\vec{\boldsymbol{x}}. \tag{2.271}$$

The Lagrange equation with respect to $\boldsymbol{G}^{-1}$ is

$$\begin{aligned}\frac{\partial(\sqrt{-g}(kR + \mathcal{L}_M))}{\partial \boldsymbol{G}^{-1}} &= \boldsymbol{0}\\ &= k\left(\frac{\partial\sqrt{-g}}{\partial \boldsymbol{G}^{-1}}R + \sqrt{-g}\frac{\partial R}{\partial \boldsymbol{G}^{-1}}\right) + \frac{\partial(\sqrt{-g}\mathcal{L}_M)}{\partial \boldsymbol{G}^{-1}}. \end{aligned} \tag{2.272}$$

The term inside the large parentheses gives the left-hand side of (2.268), i.e. together we first get for (2.272)

$$\boldsymbol{0} = k\sqrt{-g}\left(\frac{-1}{2}\cdot \boldsymbol{G}R + \check{\boldsymbol{R}}_{\text{Ric}}\right) + \sqrt{-g}\left(\frac{\partial \mathcal{L}_M}{\partial \boldsymbol{G}^{-1}} + \frac{\mathcal{L}_M}{\sqrt{-g}}\underbrace{\frac{\partial\sqrt{-g}}{\partial \boldsymbol{G}^{-1}}}_{-\frac{1}{2}\sqrt{-g}\boldsymbol{G}}\right).$$

Now, if we define the energy–momentum matrix as

$$-\frac{1}{2}\check{\boldsymbol{T}} \stackrel{\text{def}}{=} \frac{\partial \mathcal{L}_M}{\partial \boldsymbol{G}^{-1}} - \frac{1}{2}\mathcal{L}_M \boldsymbol{G}, \tag{2.273}$$

then we finally obtain with $k = c^4/(16\pi G)$ by left multiplication of (2.273) by $\boldsymbol{G}^{-1}$ and with $\boldsymbol{T} \stackrel{\text{def}}{=} \boldsymbol{G}^{-1}\check{\boldsymbol{T}}$ the Einstein's field equation

$$\boxed{\boldsymbol{R}_{\text{Ric}} - \frac{R}{2}\boldsymbol{I}_4 = \frac{8\pi G}{c^4}\boldsymbol{T}.} \tag{2.274}$$

2.14 Most Important Definitions and Formulas

Christoffel Matrix (2.68):

$$\boxed{\boldsymbol{\Gamma} \stackrel{\text{def}}{=} \begin{pmatrix} \boldsymbol{\Gamma}_0 \\ \vdots \\ \boldsymbol{\Gamma}_3 \end{pmatrix} = \frac{1}{2}(\boldsymbol{G}^{-1} \otimes \boldsymbol{I}_4)\left[\begin{pmatrix} \frac{\partial \boldsymbol{g}_0^{\mathsf{T}}}{\partial \vec{\boldsymbol{x}}} \\ \vdots \\ \frac{\partial \boldsymbol{g}_3^{\mathsf{T}}}{\partial \vec{\boldsymbol{x}}} \end{pmatrix} + \begin{pmatrix} \frac{\partial \boldsymbol{g}_0}{\partial \vec{\boldsymbol{x}}^{\mathsf{T}}} \\ \vdots \\ \frac{\partial \boldsymbol{g}_3}{\partial \vec{\boldsymbol{x}}^{\mathsf{T}}} \end{pmatrix} - \frac{\partial \boldsymbol{G}}{\partial \vec{\boldsymbol{x}}}\right] \in \mathbb{R}^{16\times 4}}$$

with the components given, in accordance with (2.63) and (2.64), as

$$\Gamma_{\alpha\beta}^{k} = \sum_{i=0}^{3} g_{ki}^{[-1]} \frac{1}{2}\left(\frac{\partial g_{\beta i}}{\partial x_\alpha} + \frac{\partial g_{\alpha i}}{\partial x_\beta} - \frac{\partial g_{\alpha\beta}}{\partial x_i}\right).$$

Motion in a Gravitational Field (2.67):

$$\boxed{\ddot{\vec{\boldsymbol{x}}} = -(\boldsymbol{I}_4 \otimes \dot{\vec{\boldsymbol{x}}}^{\mathsf{T}})\boldsymbol{\Gamma}\dot{\vec{\boldsymbol{x}}}.}$$

Riemannian Curvature Matrix (2.160):

$$\boxed{\boldsymbol{R} \stackrel{\text{def}}{=} \left[\frac{\partial \boldsymbol{\Gamma}}{\partial \boldsymbol{x}^{\mathsf{T}}} + (\overline{\boldsymbol{\Gamma}} \otimes \boldsymbol{I}_4)(\boldsymbol{I}_4 \otimes \boldsymbol{\Gamma})\right](\boldsymbol{U}_{4\times 4} - \boldsymbol{I}_{16}) \in \mathbb{R}^{16\times 16},}$$

with the components (2.165)

$$R_{\alpha\beta}^{\gamma\delta} = \frac{\partial}{\partial x_\beta}\Gamma_{\alpha\delta}^{\gamma} - \frac{\partial}{\partial x_\delta}\Gamma_{\alpha\beta}^{\gamma} + \sum_\nu \Gamma_{\beta\nu}^{\gamma}\Gamma_{\delta\alpha}^{\nu} - \sum_\nu \Gamma_{\delta\nu}^{\gamma}\Gamma_{\alpha\beta}^{\nu}.$$

Ricci Matrix $R_{\text{Ric}} \in \mathbb{R}^{4\times 4}$ (2.200):

$$R_{\text{Ric},\gamma\delta} \stackrel{\text{def}}{=} \text{trace}\big(\boldsymbol{R}^{\gamma\delta}\big) = \sum_{\nu=0}^{3} R_{\nu\nu}^{\gamma\delta},$$

consisting of the traces of the sub-matrices $\boldsymbol{R}^{\gamma\delta}$ of $\boldsymbol{R}$ with the components in accordance with (2.209):

$$R_{\text{Ric},\gamma\delta} = \sum_{\nu=0}^{3}\left(\frac{\partial}{\partial x_\delta}\Gamma_{\gamma\nu}^{\nu} - \frac{\partial}{\partial x_\nu}\Gamma_{\gamma\delta}^{\nu} + \sum_{\mu=0}^{3}\Gamma_{\delta\mu}^{\nu}\Gamma_{\nu\gamma}^{\mu} - \sum_{\mu=0}^{3}\Gamma_{\nu\mu}^{\mu}\Gamma_{\gamma\delta}^{\mu}\right).$$

Curvature Scalar R (2.200):

$$R \stackrel{\text{def}}{=} \text{trace}(\boldsymbol{R}_{\text{Ric}}) = \sum_{\alpha}\sum_{\nu} R_{\nu\nu}^{\alpha\alpha},$$

which is obtained by taking the trace of the Ricci matrix.

Einstein's Field Equation (2.235) (fixed as an axiom):

$$\boldsymbol{R}_{\text{Ric}} - \frac{R}{2}\boldsymbol{I}_4 = -\frac{8\pi G}{c^4}\boldsymbol{T},$$

or with $T \stackrel{\text{def}}{=} \text{trace}\,\boldsymbol{T}$, see (2.237):

$$\boldsymbol{R}_{\text{Ric}} = \frac{8\pi G}{c^4}\left(\frac{T}{2}\boldsymbol{I}_4 - \boldsymbol{T}\right).$$

Chapter 3
Gravitation of a Spherical Mass

The chapter begins with the first solution of Einstein's equations obtained in 1916 by Schwarzschild for a spherical mass. From these results, the influence of a mass in time and space and the redshift of the spectral lines are given. This gives also a first indication of the existence of "black holes". Finally, a brief look at the effect of rotating masses and on the Lense–Thirring effect is taken.

3.1 Schwarzschild's Solution

Because of its simplicity, the solution of Einstein's field equation will now be determined for the *outside* of a spherically symmetric, uniform, time-invariant mass distribution. This first exact solution of Einstein's field equation was given in 1916 by Schwarzschild.

As $r \to \infty$, the desired metric should be the Minkowski metric, i.e.

$$\mathrm{d}s^2 = c^2\,\mathrm{d}t^2 - \mathrm{d}r^2 - r^2\big(\mathrm{d}\theta^2 + \sin^2\theta\,\mathrm{d}\varphi^2\big). \tag{3.1}$$

where r, θ and φ are the spherical coordinates. For a gravitational field, we now start with

$$\mathrm{d}s^2 = A(r)\,\mathrm{d}t^2 - B(r)\,\mathrm{d}r^2 - r^2\big(\mathrm{d}\theta^2 + \sin^2\theta\,\mathrm{d}\varphi^2\big). \tag{3.2}$$

Since the field must be spherically symmetric, the factors A and B may depend only on r and not on θ or φ. Due to (3.1), the factor $A(r)$ must tend to c^2 as $r \to \infty$, and the factor $B(r) \to 1$. The metric matrix thus has the form

$$\boldsymbol{G} = \begin{pmatrix} A(r) & 0 & 0 & 0 \\ 0 & -B(r) & 0 & 0 \\ 0 & 0 & -r^2 & 0 \\ 0 & 0 & 0 & -r^2\sin^2\theta \end{pmatrix}. \tag{3.3}$$

G. Ludyk, *Einstein in Matrix Form*, Graduate Texts in Physics,
DOI 10.1007/978-3-642-35798-5_3, © Springer-Verlag Berlin Heidelberg 2013

3.1.1 Christoffel Matrix $\boldsymbol{\Gamma}$

Due to (2.57)

$$\hat{\boldsymbol{\Gamma}}_k \stackrel{\text{def}}{=} \left(\boldsymbol{I}_4 \otimes \boldsymbol{g}_k^{-T}\right)\left[\boldsymbol{I}_{16} - \frac{1}{2}\boldsymbol{U}_{4\times 4}\right]\frac{\partial \boldsymbol{G}}{\partial \vec{\boldsymbol{x}}}$$

and

$$\boldsymbol{G}^{-1} = \begin{pmatrix} \frac{1}{A(r)} & 0 & 0 & 0 \\ 0 & \frac{-1}{B(r)} & 0 & 0 \\ 0 & 0 & \frac{-1}{r^2} & 0 \\ 0 & 0 & 0 & \frac{-1}{r^2 \sin^2\theta} \end{pmatrix}, \tag{3.4}$$

with

$$\frac{\partial \boldsymbol{G}}{\partial \vec{\boldsymbol{x}}} = \left(\begin{array}{cccc} & \boldsymbol{0}_4 & & \\ \hline A'(r) & 0 & 0 & 0 \\ 0 & -B'(r) & 0 & 0 \\ 0 & 0 & -2r & 0 \\ 0 & 0 & 0 & -2r\sin^2\theta \\ \hline 0 & 0 & 0 & 0 \\ 0 & 0 & 0 & 0 \\ 0 & 0 & 0 & 0 \\ 0 & 0 & 0 & -2r^2\sin\theta\cos\theta \\ \hline & \boldsymbol{0}_4 & & \end{array}\right),$$

in which $A' \stackrel{\text{def}}{=} \frac{\partial A}{\partial r}$ and $\boldsymbol{0}_4$ is a 4×4-zero-matrix, and

$$\left[\boldsymbol{I}_{16} - \frac{1}{2}\boldsymbol{U}_{4\times 4}\right]\frac{\partial \boldsymbol{G}}{\partial \vec{\boldsymbol{x}}} = \left(\begin{array}{cccc} 0 & 0 & 0 & 0 \\ -\frac{A'}{2} & 0 & 0 & 0 \\ 0 & 0 & 0 & 0 \\ 0 & 0 & 0 & 0 \\ \hline A' & 0 & 0 & 0 \\ 0 & -\frac{B'}{2} & 0 & 0 \\ 0 & 0 & -2r & 0 \\ 0 & 0 & 0 & -2r\sin^2\theta \\ \hline 0 & 0 & 0 & 0 \\ 0 & 0 & r & 0 \\ 0 & 0 & 0 & 0 \\ 0 & 0 & 0 & -2r^2\sin\theta\cos\theta \\ \hline 0 & 0 & 0 & 0 \\ 0 & 0 & 0 & r\sin^2\theta \\ 0 & 0 & 0 & r^2\sin\theta\cos\theta \\ 0 & 0 & 0 & 0 \end{array}\right),$$

one obtains

$$\hat{\boldsymbol{\Gamma}}_0 = (\boldsymbol{I}_4 \otimes [1/A(r)|0|0|0])\left[\boldsymbol{I}_{16} - \frac{1}{2}\boldsymbol{U}_{4\times 4}\right]\begin{pmatrix} \frac{\partial \boldsymbol{G}}{\partial t} \\ \frac{\partial \boldsymbol{G}}{\partial r} \\ \frac{\partial \boldsymbol{G}}{\partial \theta} \\ \frac{\partial \boldsymbol{G}}{\partial \varphi} \end{pmatrix}$$

$$= \begin{pmatrix} 0 & 0 & 0 & 0 \\ \frac{A'}{A} & 0 & 0 & 0 \\ 0 & 0 & 0 & 0 \\ 0 & 0 & 0 & 0 \end{pmatrix}. \tag{3.5}$$

Accordingly, we obtain

$$\hat{\boldsymbol{\Gamma}}_1 = \begin{pmatrix} \frac{A'}{2B} & 0 & 0 & 0 \\ 0 & \frac{B'}{2B} & 0 & 0 \\ 0 & 0 & -\frac{r}{B} & 0 \\ 0 & 0 & 0 & -\frac{r}{B}\sin^2\theta \end{pmatrix},$$

$$\hat{\boldsymbol{\Gamma}}_2 = \begin{pmatrix} 0 & 0 & 0 & 0 \\ 0 & 0 & \frac{2}{r} & 0 \\ 0 & 0 & 0 & 0 \\ 0 & 0 & 0 & -\sin\theta\cos\theta \end{pmatrix}$$

and

$$\hat{\boldsymbol{\Gamma}}_3 = \begin{pmatrix} 0 & 0 & 0 & 0 \\ 0 & 0 & 0 & \frac{2}{r} \\ 0 & 0 & 0 & 2\cot\theta \\ 0 & 0 & 0 & 0 \end{pmatrix}.$$

Symmetrizing according to

$$\boldsymbol{\Gamma}_k = \frac{1}{2}(\hat{\boldsymbol{\Gamma}}_k + \hat{\boldsymbol{\Gamma}}_k^{\mathsf{T}})$$

yields the symmetric matrices

$$\boldsymbol{\Gamma}_0 = \begin{pmatrix} 0 & \frac{A'}{(2A)} & 0 & 0 \\ \frac{A'}{(2A)} & 0 & 0 & 0 \\ 0 & 0 & 0 & 0 \\ 0 & 0 & 0 & 0 \end{pmatrix},$$

$$\boldsymbol{\Gamma}_1 = \hat{\boldsymbol{\Gamma}}_1,$$

$$\boldsymbol{\Gamma}_2 = \begin{pmatrix} 0 & 0 & 0 & 0 \\ 0 & 0 & \frac{1}{r} & 0 \\ 0 & \frac{1}{r} & 0 & 0 \\ 0 & 0 & 0 & -\sin\theta\cos\theta \end{pmatrix}$$

and

$$\boldsymbol{\Gamma}_3 = \begin{pmatrix} 0 & 0 & 0 & 0 \\ 0 & 0 & 0 & \frac{1}{r} \\ 0 & 0 & 0 & \cot\theta \\ 0 & \frac{1}{r} & \cot\theta & 0 \end{pmatrix},$$

which put altogether give

$$\boldsymbol{\Gamma} = \left(\begin{array}{cccc} 0 & \frac{A'}{2A} & 0 & 0 \\ \frac{A'}{2A} & 0 & 0 & 0 \\ 0 & 0 & 0 & 0 \\ 0 & 0 & 0 & 0 \\ \hline \frac{A'}{2B} & 0 & 0 & 0 \\ 0 & \frac{B'}{2B} & 0 & 0 \\ 0 & 0 & -\frac{r}{B} & 0 \\ 0 & 0 & 0 & -\frac{r}{B}\sin^2\theta \\ \hline 0 & 0 & 0 & 0 \\ 0 & 0 & \frac{1}{r} & 0 \\ 0 & \frac{1}{r} & 0 & 0 \\ 0 & 0 & 0 & -\sin\theta\cos\theta \\ \hline 0 & 0 & 0 & 0 \\ 0 & 0 & 0 & \frac{1}{r} \\ 0 & 0 & 0 & \cot\theta \\ 0 & \frac{1}{r} & \cot\theta & 0 \end{array}\right). \tag{3.6}$$

The nonzero Christoffel elements can be read from the matrices as:

$$\Gamma^0_{10} = \Gamma^0_{01} = \frac{A'}{2A}, \qquad \Gamma^1_{00} = \frac{A'}{2B}, \qquad \Gamma^1_{11} = \frac{B'}{2B}, \qquad \Gamma^1_{22} = -\frac{r}{B},$$

$$\Gamma^1_{33} = -\frac{r}{B}\sin^2\theta, \qquad \Gamma^2_{12} = \Gamma^2_{21} = \frac{1}{r}, \qquad \Gamma^2_{33} = -\sin\theta\cos\theta,$$

$$\Gamma^3_{13} = \Gamma^3_{31} = \frac{1}{r} \quad \text{and} \quad \Gamma^3_{23} = \Gamma^3_{32} = \cot\theta.$$

3.1.2 Ricci Matrix $\boldsymbol{R}_{\text{Ric}}$

The Ricci matrix $\boldsymbol{R}_{\text{Ric}}$ is obtained from the Riemannian curvature matrix

$$\boldsymbol{R} = \left[\frac{\partial \boldsymbol{\Gamma}}{\partial \boldsymbol{x}^{\mathsf{T}}} + (\overline{\boldsymbol{\Gamma}} \otimes \boldsymbol{I}_4)(\boldsymbol{I}_4 \otimes \boldsymbol{\Gamma})\right](\boldsymbol{U}_{4\times 4} - \boldsymbol{I}_{16})$$

by summing the 4×4-sub-matrices on the main diagonal. For the calculation of $\boldsymbol{R}$, in addition to $\boldsymbol{\Gamma}$, we need

$$
\frac{\partial \boldsymbol{\Gamma}}{\partial \vec{x}^{\mathsf{T}}}
= \left(\begin{array}{c|cccc|cccc|c}
 & 0 & \frac{A''A-A'^2}{2A^2} & 0 & 0 & & & & & \\
\mathbf{0}_4 & \frac{A''A-A'^2}{2A^2} & 0 & 0 & 0 & & \mathbf{0}_4 & & & \mathbf{0}_4 \\
 & 0 & 0 & 0 & 0 & & & & & \\
 & 0 & 0 & 0 & 0 & & & & & \\
\hline
 & \frac{A''B-A'B'}{2B^2} & 0 & 0 & 0 & & & & & \\
\mathbf{0}_4 & 0 & \frac{B''B-B'^2}{2B^2} & 0 & 0 & & \mathbf{0}_{3\times 4} & & & \mathbf{0}_4 \\
 & 0 & 0 & -\frac{B-rB'}{B^2} & 0 & & & & & \\
 & 0 & 0 & 0 & -\sin^2\theta\frac{B-rB'}{B^2} & 0 & 0 & 0 & -\frac{2r}{b}\sin\theta\cos\theta & \\
\hline
 & 0 & 0 & 0 & 0 & & & & & \\
\mathbf{0}_4 & 0 & 0 & -\frac{1}{r^2} & 0 & & \mathbf{0}_{3\times 4} & & & \mathbf{0}_4 \\
 & 0 & -\frac{1}{r^2} & 0 & 0 & & & & & \\
 & 0 & 0 & 0 & 0 & 0 & 0 & 0 & -\cos^2\theta+\sin^2\theta & \\
\hline
 & 0 & 0 & 0 & 0 & 0 & 0 & 0 & 0 & \\
\mathbf{0}_4 & 0 & 0 & 0 & \frac{-1}{r^2} & 0 & 0 & 0 & 0 & \mathbf{0}_4 \\
 & 0 & 0 & 0 & 0 & 0 & 0 & 0 & \frac{-1}{\sin^2\theta} & \\
 & 0 & \frac{-1}{r^2} & 0 & 0 & 0 & 0 & \frac{-1}{\sin^2\theta} & 0 &
\end{array}\right)
$$

and

$$
\overline{\boldsymbol{\Gamma}} = \left(\begin{array}{cccc|cccc|cccc|cccc}
0 & \frac{A'}{2A} & 0 & 0 & \frac{A'}{2A} & 0 & 0 & 0 & 0 & 0 & 0 & 0 & 0 & 0 & 0 & 0 \\
\frac{A'}{2B} & 0 & 0 & 0 & 0 & \frac{B'}{2B} & 0 & 0 & 0 & 0 & -\frac{r}{B} & 0 & 0 & 0 & 0 & -\frac{r}{B}\sin^2\theta \\
0 & 0 & 0 & 0 & 0 & 0 & \frac{1}{r} & 0 & 0 & \frac{1}{r} & 0 & 0 & 0 & 0 & 0 & -\sin\theta\cos\theta \\
0 & 0 & 0 & 0 & 0 & 0 & 0 & \frac{1}{r} & 0 & 0 & 0 & \cot\theta & 0 & \frac{1}{r} & \cot\theta & 0
\end{array}\right)
$$

For the Ricci matrix $\boldsymbol{R}_{\mathrm{Ric}}$ the 16×16-matrix $\frac{\partial \boldsymbol{\Gamma}}{\partial \vec{x}^{\mathsf{T}}}(\boldsymbol{U}_{4\times 4} - \boldsymbol{I}_{16})$ contributes the term

$$
\begin{pmatrix}
\frac{A'B'-A''B}{2B^2} & 0 & 0 & 0 \\
0 & \frac{A''A-A'^2}{2A^2} - \frac{2}{r^2} & 0 & 0 \\
0 & 0 & \frac{B-rB'}{B^2} - \frac{1}{\sin^2\theta} & 0 \\
0 & 0 & 0 & \sin^2\theta\frac{B-rB'}{B^2} + \cos^2\theta - \sin^2\theta
\end{pmatrix}
$$

and the matrix $(\overline{\boldsymbol{\Gamma}} \otimes \boldsymbol{I}_4)(\boldsymbol{I}_4 \otimes \boldsymbol{\Gamma})(\boldsymbol{U}_{4\times 4} - \boldsymbol{I}_{16})$ the term

$$
\begin{pmatrix}
\frac{A'^2}{4AB} - \frac{A'B'}{4B^2} - \frac{A'}{rB} & 0 & 0 & 0 \\
0 & \frac{A'^2}{4A^2} - \frac{A'B'}{4AB} - \frac{B'}{rB} + \frac{2}{r^2} & 0 & 0 \\
0 & 0 & \frac{rB'}{B^2} + \frac{rA'}{2AB} + \cot^2\theta & 0 \\
0 & 0 & 0 & \frac{rB'}{B^2}\sin^2\theta + \frac{rA'}{2AB}\sin^2\theta - \cos^2\theta
\end{pmatrix}.
$$

Finally, we obtain for the four elements on the main diagonal of the Ricci matrix $\boldsymbol{R}_{\text{Ric}}$:

$$R_{\text{Ric},00} = -\frac{A''}{2B} + \frac{A'}{4B}\left(\frac{A'}{A} + \frac{B'}{B}\right) - \frac{A'}{rB}, \tag{3.7}$$

$$R_{\text{Ric},11} = \frac{A''}{2A} - \frac{A'}{4A}\left(\frac{A'}{A} + \frac{B'}{B}\right) - \frac{B'}{rB}, \tag{3.8}$$

$$R_{\text{Ric},22} = \frac{1}{B} + \frac{r}{2B}\left(\frac{A'}{A} - \frac{B'}{B}\right) - 1 \tag{3.9}$$

and

$$R_{\text{Ric},33} = \sin^2\theta\, R_{\text{Ric},22}. \tag{3.10}$$

The other matrix elements are equal to zero: $R_{\text{Ric},\nu\mu} = 0$ for $\nu \neq \mu$.

3.1.3 The Factors $A(r)$ and $B(r)$

To finally write down the metric matrix $\boldsymbol{G}$, the two factors $A(r)$ and $B(r)$ are required. Since we search for the solution of Einstein's field equation outside the sphere containing the mass and having the energy–momentum matrix $\boldsymbol{T}$ equal to the zero matrix, the Ansatz $\boldsymbol{R}_{\text{Ric}} = \boldsymbol{0}$ thus yields

$$-\frac{A''}{2B} + \frac{A'}{4B}\left(\frac{A'}{A} + \frac{B'}{B}\right) - \frac{A'}{rB} = 0, \tag{3.11}$$

$$\frac{A''}{2A} - \frac{A'}{4A}\left(\frac{A'}{A} + \frac{B'}{B}\right) - \frac{B'}{rB} = 0, \tag{3.12}$$

and

$$\frac{1}{B} + \frac{r}{2B}\left(\frac{A'}{A} - \frac{B'}{B}\right) - 1 = 0. \tag{3.13}$$

Adding to (3.11) divided by A (3.12) divided by B, we obtain the condition

$$A'B + AB' = 0, \tag{3.14}$$

which means nothing else than that

$$AB = \text{constant}. \tag{3.15}$$

However, since $A(\infty) = c^2$ and $B(\infty) = 1$ as $r \to \infty$, the constant must be equal to c^2, so

$$A(r)B(r) = c^2 \quad \text{and} \quad B(r) = \frac{c^2}{A(r)}.$$

Substituting this into (3.13), we obtain $A + rA' = c^2$, which we also can write

$$\frac{\mathrm{d}(rA)}{\mathrm{d}r} = c^2. \tag{3.16}$$

Integrating this equation yields

$$rA = c^2(r + K), \tag{3.17}$$

so

$$A(r) = c^2\left(1 + \frac{K}{r}\right) \quad \text{and} \quad B(r) = \left(1 + \frac{K}{r}\right)^{-1}. \tag{3.18}$$

K still remains to be determined. In Sect. 2.8, a uniform rotating system was treated. In (2.85), we obtained the element g_{00} of the metric matrix $\boldsymbol{G}$ as

$$g_{00} = 1 - \frac{r^2\omega^2}{c^2} \tag{3.19}$$

and in (2.88) the centrifugal force $mr\omega^2$. According to Sect. 2.8, the centrifugal force $\boldsymbol{f}$ depends, on the other hand, on the centrifugal potential φ, so together

$$\boldsymbol{f} = -m\nabla\varphi. \tag{3.20}$$

Taking

$$\varphi = -\frac{r^2\omega^2}{2} \tag{3.21}$$

(3.20) provides just the above centrifugal force. Equation (3.21) inserted into (3.19) provides the general connection

$$g_{00} = 1 + \frac{2\varphi}{c^2}. \tag{3.22}$$

Outside of our spherical mass M, as Newton's approximation the gravitational potential

$$\varphi = -\frac{GM}{r} \tag{3.23}$$

is obtained. This used in (3.22) and compared with (3.18) yields

$$K = -\frac{2GM}{c^2}, \tag{3.24}$$

so that we finally obtain the Schwarzschild's metric in matrix form:

$$\boxed{ds^2 = d\vec{x}^{\mathsf{T}}\, \boldsymbol{G}\, d\vec{x} = d\vec{x}^{\mathsf{T}} \begin{pmatrix} 1-\frac{2GM}{c^2 r} & 0 & 0 & 0 \\ 0 & -(1-\frac{2GM}{c^2 r})^{-1} & 0 & 0 \\ 0 & 0 & -r^2 & 0 \\ 0 & 0 & 0 & -r^2\sin^2\theta \end{pmatrix} d\vec{x},} \tag{3.25}$$

with

$$d\vec{x} \stackrel{\text{def}}{=} \begin{pmatrix} c\,dt \\ dr \\ d\theta \\ d\varphi \end{pmatrix},$$

and the calculated Schwarzschild's metric is then

$$\boxed{ds^2 = \left(1-\frac{2GM}{c^2 r}\right)c^2\,dt^2 - \left(1-\frac{2GM}{c^2 r}\right)^{-1} dr^2 - r^2\left(d\theta^2 + \sin^2\theta\, d\varphi^2\right).} \tag{3.26}$$

For the so-called Schwarzschild's radius

$$\boxed{r_S \stackrel{\text{def}}{=} \frac{2GM}{c^2},} \tag{3.27}$$

the Schwarzschild's metric shows a singularity as for $r = 0$. Of these, however, only the singularity at $r = 0$ is a real singularity. This is suggested even by the fact that, when $r = r_S$, the determinant of $\boldsymbol{G}$, namely $g = -r^4\sin^2\theta$, has no singularity. The singularity at $r = r_S$ is not a *physical singularity* but a *coordinate singularity*, which depends entirely on the choice of the coordinate system, i.e. there would exist no singularity at this place if a different coordinate system were chosen! More about this in the next section on black holes.

3.2 Influence of a Massive Body on the Environment

3.2.1 Introduction

The Schwarzschild's solution (3.26) is valid only outside the solid sphere of radius r_M generating the gravity, so only for $r_M < r < \infty$. Because for the Schwarzschild's radius r_S the element $g_{11} = (1-\frac{r_S}{r})^{-1}$ goes to infinity, r_S is also a limit. In general, $r_S \ll r_M$, but for the so-called *Black Holes* one has $r_M < r_S$, and in this case the solution is limited to $r_S < r < \infty$.

3.2.2 Changes to Length and Time

How does a length change in the environment of a massive body? For a constant time, i.e. for $\mathrm{d}t = 0$ one receives, due to (3.26),

$$\mathrm{d}\ell^2 = \left(1 - \frac{2GM}{c^2 r}\right)^{-1} \mathrm{d}r^2 + r^2\left(\mathrm{d}\theta^2 + \sin^2\theta\,\mathrm{d}\varphi^2\right). \tag{3.28}$$

On the surface of a sphere of radius $r > r_M$ with the centre being the centre of mass, as also for $\mathrm{d}r = 0$, the tangential infinitely small distances are given by

$$\mathrm{d}L = r\left(\mathrm{d}\theta^2 + \sin^2\theta\,\mathrm{d}\varphi^2\right)^{1/2}. \tag{3.29}$$

This is a result that is valid on every sphere, whether with or without gravity. But what happens with a distance in the radial direction? In this case, $\mathrm{d}\theta$ and $\mathrm{d}\varphi$ are equal to zero, thus for infinitely small distances in the radial direction, according to (3.28), one obtains

$$\boxed{\mathrm{d}R = \left(1 - \frac{2GM}{c^2 r}\right)^{-1/2} \mathrm{d}r,} \tag{3.30}$$

so $\mathrm{d}R > \mathrm{d}r$, and so the length is much longer, the larger the mass M and the smaller the distance r from the mass! $\mathrm{d}r$ is through the mass 'elongated', caused by a curvature of space.

Let us now turn to the time. For a clock at the point, where r, θ, and φ are constant, we obtain from the Schwarzschild's metric (3.26)

$$\mathrm{d}s^2 = c^2\,\mathrm{d}\tau^2 = c^2\left(1 - \frac{2GM}{c^2 r}\right)\mathrm{d}t^2,$$

so

$$\boxed{\mathrm{d}\tau = \left(1 - \frac{2GM}{c^2 r}\right)^{1/2} \mathrm{d}t.} \tag{3.31}$$

If $\mathrm{d}\tau < \mathrm{d}t$, the closer one is located to the mass, the shorter are the time intervals, i.e. the slower the time passes! For an observer the time goes slower the closer he is to the mass. Particularly large is the time dilation near a *Black Hole* in which the mass is so concentrated that the body radius is smaller than its own Schwarzschild's radius. For a black hole with a mass of ten solar masses, the Schwarzschild's radius is $r_S = 30$ km. At a distance of 1 cm from the so-called horizon, which is the spherical shell around the centre of mass with radius r_S, $\gamma = (1 - \frac{2GM}{c^2 r})^{1/2} = 1.826 \cdot 10^{-5}$, so the time goes by about 55000 times slower than far away: after a year near the horizon of the black hole, in a place far away, e.g. on the home planet of an astronaut, 55000 years would have gone by! All is summed up into a popular statement:

If you climb a mountain, you are smaller and age faster!

The relations (3.30) and (3.31) are very similar to the relationships of space and time contractions for the relatively to each other moving reference frames in special relativity; see Chap. 1. There with

$$\gamma \stackrel{\text{def}}{=} \left(1 - \frac{v^2}{c^2}\right)^{-1/2}$$

for the length contraction we had

$$\mathrm{d}\ell = \frac{1}{\gamma}\,\mathrm{d}\ell_0$$

and for the time dilation

$$\mathrm{d}t = \gamma\,\mathrm{d}\tau.$$

If we introduce the pseudo-velocity

$$v_G^2(r) \stackrel{\text{def}}{=} \frac{2GM}{r}$$

then with

$$\gamma_G(r) \stackrel{\text{def}}{=} \left(1 - \frac{v_G^2(r)}{c^2}\right)^{-1/2}$$

the above gravitational relationships can be written as follows:

$$\mathrm{d}r = \frac{1}{\gamma_G(r)}\,\mathrm{d}R \quad \text{and} \quad \mathrm{d}t = \gamma_G(r)\,\mathrm{d}\tau.$$

By the way, $v_G(r)$ is the escape velocity of a planet with the diameter $2r$ and mass M.

3.2.3 *Redshift of Spectral Lines*

Suppose in the gravitational field of a mass a light signal is sent from a transmitter at a fixed point $\boldsymbol{x}_T = [r_T, \theta_T, \varphi_T]^\mathsf{T}$ and it drifts along a geodesic line to a fixed receiver at the point $\boldsymbol{x}_R = [r_R, \theta_R, \varphi_R]^\mathsf{T}$. Let, furthermore, λ be any parameter along the geodesic line, with $\lambda = \lambda_T$ for the sent event and $\lambda = \lambda_R$ for the received event. Then for a photon, by (2.11), the following is valid:

$$\frac{\mathrm{d}\vec{x}^\mathsf{T}}{\mathrm{d}\lambda}\boldsymbol{G}\frac{\mathrm{d}\vec{x}}{\mathrm{d}\lambda} = 0,$$

so here

$$c^2\left(1-\frac{2GM}{c^2r}\right)\left(\frac{\mathrm{d}t}{\mathrm{d}\lambda}\right)^2-\left(1-\frac{2GM}{c^2r}\right)^{-1}\left(\frac{\mathrm{d}r}{\mathrm{d}\lambda}\right)^2-r^2\left(\left(\frac{\mathrm{d}\theta}{\mathrm{d}\lambda}\right)^2+\sin^2\left(\frac{\mathrm{d}\varphi}{\mathrm{d}\lambda}\right)^2\right)=0,$$

i.e.

$$\frac{\mathrm{d}t}{\mathrm{d}\lambda}=\frac{1}{c}\left[\left(1-\frac{2GM}{c^2r}\right)^{-2}\left(\frac{\mathrm{d}r}{\mathrm{d}\lambda}\right)^2+\left(1-\frac{2GM}{c^2r}\right)^{-1}r^2\left(\left(\frac{\mathrm{d}\theta}{\mathrm{d}\lambda}\right)^2+\sin^2\left(\frac{\mathrm{d}\varphi}{\mathrm{d}\lambda}\right)^2\right)\right]^{1/2}.$$

From this we obtain the signal transmission time

$$t_R-t_T=\frac{1}{c}\int_{\lambda_T}^{\lambda_R}\left[\left(1-\frac{2GM}{c^2r}\right)^{-2}\left(\frac{\mathrm{d}r}{\mathrm{d}\lambda}\right)^2\right.$$
$$\left.+\left(1-\frac{2GM}{c^2r}\right)^{-1}r^2\left(\left(\frac{\mathrm{d}\theta}{\mathrm{d}\lambda}\right)^2+\sin^2\left(\frac{\mathrm{d}\varphi}{\mathrm{d}\lambda}\right)^2\right)\right]^{1/2}\mathrm{d}\lambda.$$

This time depends only on the path that the light takes between the spatially fixed transmitter and the spatially fixed receiver. So for two consecutive transmitted signals 1 and 2 the duration is equal:

$$t_{E,1}-t_{S,1}=t_{E,2}-t_{S,2},$$

and also

$$\Delta t_R\stackrel{\mathrm{def}}{=}t_{E,2}-t_{E,1}=t_{S,2}-t_{S,1}\stackrel{\mathrm{def}}{=}\Delta t_T,$$

so the Schwarzschild's time difference at the transmitter is equal to the Schwarzschild's time difference at the receiver, though the clock of an observer at the transmitter location would show the proper time difference to

$$\Delta\tau_T=\left(1-\frac{2GM}{c^2r_T}\right)^{1/2}\Delta t_T$$

and accordingly at the receiving location the proper time difference would be

$$\Delta\tau_R=\left(1-\frac{2GM}{c^2r_R}\right)^{1/2}\Delta t_R.$$

Since $\Delta t_R=\Delta t_T$, one obtains the ratio

$$\frac{\Delta\tau_R}{\Delta\tau_T}=\left(\frac{1-\frac{2GM}{c^2r_R}}{1-\frac{2GM}{c^2r_T}}\right)^{1/2}.$$

If the transmitter is a pulsating atom, emitting N pulses in the proper time interval $\Delta\tau_T$, an observer at the transmitter would assign to the signal a proper frequency

$\nu_T \stackrel{\text{def}}{=} \frac{N}{\Delta\tau_T}$. An observer at the receiver will see the N pulses arriving in the proper time $\Delta\tau_R$, thus at the frequency $\nu_R = \frac{N}{\Delta\tau_R}$. Then the frequency ratio is obtained as

$$\boxed{\frac{\nu_R}{\nu_T} = \left(\frac{1-\frac{2GM}{c^2 r_T}}{1-\frac{2GM}{c^2 r_R}}\right)^{1/2}.} \tag{3.32}$$

If the transmitter is closer to the mass compared to the receiver ($r_T < r_R$), then a shift to the red in the "colour" of the signal takes place. Conversely, if the transmitter is further away than the receiver, one gets a blueshift. For $r_S, r_E \gg 2GM$ one obtains the approximation

$$\frac{\nu_R}{\nu_T} \approx 1 + \frac{GM}{c^2}\left(\frac{1}{r_R} - \frac{1}{r_T}\right)$$

and the relative frequency change

$$\frac{\Delta\nu}{\nu_T} \stackrel{\text{def}}{=} \frac{\nu_R - \nu_T}{\nu_T} \approx \frac{GM}{c^2}\left(\frac{1}{r_R} - \frac{1}{r_T}\right).$$

If the transmitter (e.g. a radiating atom) is on the solar surface and the observing receiver on the surface of the Earth, then

$$\frac{\Delta\nu}{\nu_T} \approx 2 \cdot 10^{-6}.$$

This effect is indeed poorly measurable due to a variety of disturbances by the atmosphere. However, this redshift is measurable using the Mössbauer's effect (see the relevant Physics literature).

3.2.4 Deflection of Light

According to (2.11), for light one has

$$\dot{\vec{x}}^{\mathsf{T}} \boldsymbol{G} \dot{\vec{x}} = 0. \tag{3.33}$$

Differentiating (3.33) with respect to the orbital parameter λ, one obtains

$$\frac{\partial \vec{x}^{\mathsf{T}}}{\partial\lambda} \boldsymbol{G} \frac{\partial \vec{x}}{\partial\lambda} = 0. \tag{3.34}$$

Setting for $\boldsymbol{G}$ the Schwarzschild's metric, we obtain

$$c^2\left(1-\frac{r_S}{r}\right)\left(\frac{\partial t}{\partial\lambda}\right)^2 - \left(1-\frac{r_S}{r}\right)^{-1}\left(\frac{\partial r}{\partial\lambda}\right)^2 - r^2\left(\frac{\partial\theta}{\partial\lambda}^2 + \sin^2\theta\left(\frac{\partial\varphi}{\partial\lambda}\right)^2\right) = 0. \tag{3.35}$$

Without loss of generality, especially in the present centrally symmetric solution, we can assume that $\theta = \pi/2$ and $\frac{\partial\theta}{\partial\lambda} = 0$, and that the solution is in a plane through the centre of mass, then (3.35) simplifies to

$$c^2\left(1-\frac{r_S}{r}\right)\left(\frac{\partial t}{\partial\lambda}\right)^2 - \left(1-\frac{r_S}{r}\right)^{-1}\left(\frac{\partial r}{\partial\lambda}\right)^2 - r^2\left(\frac{\partial\varphi}{\partial\lambda}\right)^2 = 0. \tag{3.36}$$

For a light ray in the gravitational field the following holds:

$$\frac{\partial^2\vec{x}}{\partial\lambda^2} = -\left(\boldsymbol{I}_4 \otimes \frac{\partial\vec{x}}{\partial\lambda}^{\mathsf{T}}\right)\boldsymbol{\Gamma}\frac{\partial\vec{x}}{\partial\lambda}. \tag{3.37}$$

With the Christoffel elements from Sect. 3.1.1 and $\vec{x}^{\mathsf{T}} = [x_0|r|\theta|\varphi]$ expanding (3.37) yields

$$\frac{d^2x_0}{d\lambda^2} = -\frac{A'}{A}\frac{dx_0}{d\lambda}\frac{dr}{d\lambda}, \tag{3.38}$$

$$\frac{d^2r}{d\lambda^2} = -\frac{A'}{B}\left(\frac{dx_0}{d\lambda}\right)^2 - \frac{B'}{2B}\left(\frac{dr}{d\lambda}\right)^2 + \frac{r}{B}\left(\frac{d\theta}{d\lambda}\right)^2 + \frac{r\sin^2\theta}{B}\left(\frac{d\varphi}{d\lambda}\right)^2, \tag{3.39}$$

$$\frac{d^2\theta}{d\lambda^2} = -\frac{2}{r}\frac{d\theta}{d\lambda}\frac{dr}{d\lambda} + \sin\theta\cos\theta\left(\frac{d\varphi}{d\lambda}\right)^2, \tag{3.40}$$

$$\frac{d^2\varphi}{d\lambda^2} = -\frac{2}{r}\frac{d\varphi}{d\lambda}\frac{dr}{d\lambda} - \cot\theta\frac{d\theta}{d\lambda}\frac{d\varphi}{d\lambda}. \tag{3.41}$$

If the coordinate system is selected so that for the start time λ_0 one has

$$\theta = \pi/2 \quad \text{and} \quad \frac{\partial\theta}{\partial\lambda} = 0, \tag{3.42}$$

then from (3.40) $\frac{d^2\theta}{d\lambda^2} = 0$, so $\theta(\lambda) \equiv \pi/2$, i.e. the entire path remains in the plane through the centre of mass. With this θ value from (3.41) one obtains

$$\frac{d^2\varphi}{d\lambda^2} + \frac{2}{r}\frac{d\varphi}{d\lambda}\frac{dr}{d\lambda} = 0,$$

which can be summarized as follows:

$$\frac{1}{r^2}\frac{d}{d\lambda}\left(r^2\frac{d\varphi}{d\lambda}\right) = 0. \tag{3.43}$$

It is therefore true that

$$\underline{\underline{r^2\frac{d\varphi}{d\lambda} = \text{const.} = h.}} \tag{3.44}$$

Equation (3.38) can be converted to

$$\frac{\mathrm{d}}{\mathrm{d}\lambda}\left(\ln\frac{\mathrm{d}x_0}{\mathrm{d}\lambda}+\ln A\right)=0 \tag{3.45}$$

and integrated to yield

$$A\frac{\mathrm{d}x_0}{\mathrm{d}\lambda}=\text{const.}=k. \tag{3.46}$$

Equations (3.42), (3.44) and (3.46) used in (3.39) provide

$$\frac{\mathrm{d}^2r}{\mathrm{d}\lambda^2}+\frac{k^2A'}{2A^2B}+\frac{B'}{2B}\left(\frac{\mathrm{d}r}{\mathrm{d}\lambda}\right)^2-\frac{h^2}{Br^3}=0. \tag{3.47}$$

Multiplying this equation by $2B\frac{\mathrm{d}r}{\mathrm{d}\lambda}$, we obtain first

$$\frac{\mathrm{d}}{\mathrm{d}\lambda}\left(B\frac{\mathrm{d}r}{\mathrm{d}\lambda}+\frac{h^2}{r^2}-\frac{k^2}{A}\right)=0 \tag{3.48}$$

and after integration

$$B\left(\frac{\mathrm{d}r}{\mathrm{d}\lambda}\right)^2+\frac{h^2}{r^2}-\frac{k^2}{A}=\text{const.}=0, \tag{3.49}$$

or

$$\frac{\mathrm{d}r}{\mathrm{d}\lambda}=\sqrt{\frac{\frac{k^2}{A}-\frac{h^2}{r^2}}{B}}. \tag{3.50}$$

But we don't look for $r(\lambda)$, we want $\varphi(r)$. Since

$$\frac{\mathrm{d}\varphi}{\mathrm{d}r}=\frac{\mathrm{d}\varphi}{\mathrm{d}\lambda}\frac{\mathrm{d}\lambda}{\mathrm{d}r},$$

it follows by (3.44) and (3.50) that

$$\frac{\mathrm{d}\varphi}{\mathrm{d}r}=\frac{h}{r^2}\sqrt{\frac{B}{\frac{k^2}{A}-\frac{h^2}{r^2}}} \tag{3.51}$$

and integrated

$$\varphi(r)=\varphi(r_0)+\int_{r_0}^{r}\frac{\sqrt{B(\psi)}}{\psi^2\sqrt{\frac{k^2}{A(\psi)h^2}-\frac{1}{\psi^2}}}\,\mathrm{d}\psi. \tag{3.52}$$

Let r_0 be the smallest distance from the centre of mass, having a passing flying photon. It moves in the plane through the centre of mass with $\theta=\pi/2$. The coordinate system is still placed such that $\phi(r_0)=0$. If gravity is not present, the photon will continue to fly straight and as $r \to \infty$ we will have $\varphi(\infty) = \pi/2$. It comes from $\varphi(-\infty) = -\pi/2$. Overall, therefore, the photon has covered an angle of $\Delta\varphi = \varphi(\infty) - \varphi(-\infty) = \pi$. Considering the gravity, the photon is "bent" to the mass. It passes from $(r_0, \varphi(r_0))$ to $\varphi(\infty) = \pi/2 + \alpha/2$ as $r \to \infty$. For reasons of symmetry, it is then $\varphi(-\infty) = \pi/2 - \alpha/2$ as $r \to -\infty$, so that the photon has covered an overall angle of $\Delta\varphi = \varphi(\infty) - \varphi(-\infty) = \pi + \alpha$. This angle α is now calculated for the solar mass $M_\odot = 1.9891 \times 10^{33}$ g.

Since r_0 should be the minimum distance, it is true that

$$\left.\frac{\mathrm{d}r}{\mathrm{d}\varphi}\right|_{r_0} = 0. \tag{3.53}$$

On the other hand, $\frac{\mathrm{d}\varphi}{\mathrm{d}r}$ is precisely the integrand of (3.52), whence it follows

$$\frac{k^2}{h^2} = \frac{A(r_0)}{r_0^2}. \tag{3.54}$$

The integral used in (3.52) yields

$$\varphi(\infty) = \int_{r_0}^{\infty} \frac{\sqrt{B(r)}}{r\sqrt{\frac{r^2 A(r_0)}{r_0^2 A(r)} - 1}}\,\mathrm{d}r. \tag{3.55}$$

For this integral no simple antiderivative exists so that it should be calculated by approximations. One has

$$A(r) = 1 - \frac{2GM_\odot}{c^2 r} \tag{3.56}$$

and

$$B(r) = \left(1 - \frac{2GM_\odot}{c^2 r}\right)^{-1} \approx 1 + \frac{2GM_\odot}{c^2 r}. \tag{3.57}$$

Furthermore,

$$\begin{aligned}\frac{A(r_0)}{A(r)} &= \left(1 - \frac{2GM_\odot}{c^2 r_0}\right)\left(1 - \frac{2GM_\odot}{c^2 r}\right)^{-1} \\ &\approx \left(1 - \frac{2GM_\odot}{c^2 r_0}\right)\left(1 + \frac{2GM_\odot}{c^2 r}\right) = 1 + \frac{2GM_\odot}{c^2}\left(\frac{1}{r} - \frac{1}{r_0}\right)\end{aligned}$$

and

$$\frac{r^2 A(r_0)}{r_0^2 A(r)} - 1 \approx \left(\frac{r^2}{r_0^2} - 1\right)\left(1 - \frac{2GM_\odot r}{c^2 r_0 (r + r_0)}\right). \tag{3.58}$$

Therefore, the integral (3.55) becomes

$$\varphi(\infty) = \int_{r_0}^{\infty} \frac{r_0}{\sqrt{r^2 - r_0^2}} \left(\frac{1}{r} + \frac{GM_\odot}{c^2 r^2} + \frac{GM_\odot}{c^2 (r + r_0)} \right) \mathrm{d}r$$

$$= \left[\arccos \frac{r_0}{r} + \frac{GM_\odot}{c^2 r_0} \frac{\sqrt{r^2 - r_0^2}}{r} + \frac{GM_\odot}{c^2 r_0} \sqrt{\frac{r - r_0}{r + r_0}} \right]_{r_0}^{\infty}, \tag{3.59}$$

so

$$\underline{\underline{\varphi(\infty) = \frac{\pi}{2} + \frac{2GM_\odot}{c^2 r_0}}}, \tag{3.60}$$

from which one can read off

$$\underline{\underline{\alpha = \frac{4GM_\odot}{c^2 r_0}}}. \tag{3.61}$$

If r_0 is precisely the solar radius $R_\odot = 6.957 \times 10^8$ m, we obtain

$$\underline{\underline{\alpha = 1.75''}}. \tag{3.62}$$

This prediction was one of the first confirmations of the Theory of General Relativity in 1919 at a solar eclipse measured by Eddington.

3.3 Schwarzschild's Inner Solution

For the determination of the metric inside a symmetric sphere, the right-hand side of Einstein's field equation with the energy–momentum matrix is needed. The hydromechanical energy–momentum matrix was deduced in Chap. 1 in (1.183) as

$$\boldsymbol{T}_{\text{mech}} = \left(\rho_0 + \frac{p}{c^2} \right) \vec{u}\vec{u}^{\mathsf{T}} - p\boldsymbol{M}.$$

This was done in the Lorentz basis of an inertial frame. It can thus also be the local inertial system of a general coordinate system. If for the relationship between the local coordinates $\mathrm{d}\vec{x}$ and the global coordinates $\mathrm{d}\vec{x}'$ the transformation equation is $\mathrm{d}\vec{x} = \boldsymbol{J}\,\mathrm{d}\vec{x}'$, then for the velocities one has $\vec{u} = \boldsymbol{J}\vec{u}'$. If used above this yields

$$\boldsymbol{T}_{\text{mech}} = \left(\rho_0 + \frac{p}{c^2} \right) \boldsymbol{J}\vec{u}'\vec{u}'^{\mathsf{T}} \boldsymbol{J}^{\mathsf{T}} - p\boldsymbol{M}. \tag{3.63}$$

Now multiplying this equation from the left with the matrix product $\boldsymbol{J}^{\mathsf{T}}\boldsymbol{M}$ and from the right with the matrix product $\boldsymbol{M}\boldsymbol{J}$, we finally obtain with $\boldsymbol{M}\boldsymbol{M} = \boldsymbol{I}$ and $\boldsymbol{J}^{\mathsf{T}}\boldsymbol{M}\boldsymbol{J} = \boldsymbol{G}$

$$\boldsymbol{T}_{\text{mech,Riemann}} = \left(\varrho_0 + \frac{p}{c^2} \right) \boldsymbol{G}\vec{u}'\vec{u}'^{\mathsf{T}}\boldsymbol{G} - p\boldsymbol{G}. \tag{3.64}$$

Now the metric matrix $\boldsymbol{G}$ appears globally.

The Einstein's field equation is used in the form (2.237)

$$\boldsymbol{R}_{\text{Ric}} = \frac{8\pi G}{c^4}\left(\frac{T}{2}\boldsymbol{I}_4 - \boldsymbol{T}\right). \tag{3.65}$$

It is assumed that the masses inside the sphere do not move, the velocity components are zero: $u'_i = 0$, for $i = 1, 2, 3$. Therefore, from the condition

$$c^2 = \vec{u}'^{\mathsf{T}}\boldsymbol{G}\vec{u}'$$

there remains

$$c^2 = g_{00}u_0^2.$$

If the metric matrix is chosen again as in (3.2), namely

$$\mathrm{d}s^2 = A(r)\,\mathrm{d}t^2 - B(r)\,\mathrm{d}r^2 - r^2\left(\mathrm{d}\theta^2 + \sin^2\theta\,\mathrm{d}\varphi^2\right), \tag{3.66}$$

then

$$u_0 = \frac{c}{\sqrt{A(r)}}.$$

Thus we get for $\boldsymbol{T}$ the diagonal matrix

$$\boldsymbol{T} = \operatorname{diag}\left(\varrho c^2 A(r), pB(r), pr^2, pr^2\sin^2\theta\right). \tag{3.67}$$

With the trace $T = c^2\varrho - 3p$ of the local energy–momentum matrix, we obtain for the right-hand side of the Einstein's field equation

$$\frac{8\pi G}{c^4}\left(\frac{T}{2}\boldsymbol{I}_4 - \boldsymbol{T}\right)$$
$$= \frac{4\pi G}{c^4}\operatorname{diag}\left((\varrho c^2 + 3p)A, (\varrho c^2 - p)B, (\varrho c^2 - p)r^2, (\varrho c^2 - p)r^2\sin^2\theta\right). \tag{3.68}$$

This with the elements (3.7), (3.8), (3.9) and (3.10) on the main diagonal of the Ricci matrix $\boldsymbol{R}_{\text{Ric}}$ gives

$$R_{\text{Ric},00} = -\frac{A''}{2B} + \frac{A'}{4B}\left(\frac{A'}{A} + \frac{B'}{B}\right) - \frac{A'}{rB},$$

$$R_{\text{Ric},11} = \frac{A''}{2A} - \frac{A'}{4A}\left(\frac{A'}{A} + \frac{B'}{B}\right) - \frac{B'}{rB},$$

$$R_{\text{Ric},22} = \frac{1}{B} + \frac{r}{2B}\left(\frac{A'}{A} - \frac{B'}{B}\right) - 1$$

and

$$R_{\mathrm{Ric},33} = \sin^2\theta \, R_{\mathrm{Ric},22}$$

(the remaining matrix elements are equal to zero: $R_{\mathrm{Ric},\nu\mu} = 0$ for $\nu \neq \mu$), combined with Einstein's field equation yielding the three determining equations for the factors $A(r)$ and $B(r)$:

$$\frac{A''}{2B} - \frac{A'}{4B}\left(\frac{A'}{A} + \frac{B'}{B}\right) + \frac{A'}{rB} = \frac{4\pi G}{c^4}(\varrho c^2 + 3p)A, \tag{3.69}$$

$$-\frac{A''}{2A} + \frac{A'}{4A}\left(\frac{A'}{A} + \frac{B'}{B}\right) + \frac{B'}{rB} = \frac{4\pi G}{c^4}(\varrho c^2 - p)B \tag{3.70}$$

and

$$-\frac{1}{B} - \frac{r}{2B}\left(\frac{A'}{A} - \frac{B'}{B}\right) + 1 = \frac{4\pi G}{c^4}(\varrho c^2 - p)r^2. \tag{3.71}$$

Adding together (3.69) multiplied by $r^2/(2A)$, (3.70) multiplied by $r^2/(2B)$ and (3.71), we obtain

$$\frac{B'r}{B^2} + 1 - \frac{1}{B} = \frac{8\pi G}{c^2}\varrho r^2. \tag{3.72}$$

This can also be rewritten as

$$\frac{\mathrm{d}}{\mathrm{d}r}\frac{r}{B(r)} = 1 - \frac{8\pi G}{c^2}\varrho r^2. \tag{3.73}$$

Integrating this equation provides

$$\frac{r}{B(r)} = \int_0^r \left(1 - \frac{8\pi G}{c^2}\varrho(\alpha)\alpha^2\right) \mathrm{d}\alpha = r - \frac{2G}{c^2}\mathcal{M}(r), \tag{3.74}$$

with

$$\mathcal{M}(r) \overset{\text{def}}{=} 4\pi \int_0^r \varrho(\alpha)\alpha^2 \, \mathrm{d}\alpha. \tag{3.75}$$

Solving (3.74) for $B(r)$ finally yields

$$\underline{\underline{B(r) = \left(1 - \frac{2G\mathcal{M}(r)}{c^2 r}\right)^{-1}}}. \tag{3.76}$$

The density function $\varrho(r)$ is zero for $r > R$, i.e. outside the spherical mass, so $\mathcal{M}(r) = \mathcal{M}(\mathcal{R}) = M$ for $r > R$. We thus obtain the same coefficient $B(r)$ for $r > R$ as for the Schwarzschild's outer solution.

Now $A(r)$ is still to be determined. For that, $B(r)$ and

$$B'(r) = B^2(r)\left(\frac{8\pi G}{c^2}\varrho(r)r - \frac{2G\mathcal{M}(r)}{c^2 r^2}\right)$$

is used in (3.71), which gives

$$-\frac{A'(r)r}{2A(r)}\left(1-\frac{2G\mathcal{M}(r)}{c^2r}\right)+\frac{4\pi G}{c^2}\varrho(r)r^2+\frac{G\mathcal{M}(r)}{c^2r}=\frac{4\pi G}{c^4}(\varrho c^2-p)r^2,$$

i.e.

$$\frac{A'(r)}{A(r)}=\frac{\mathrm{d}}{\mathrm{d}r}(\ln A(r)+\text{const.})=\left(\frac{8\pi G}{c^4}p(r)r+\frac{2G\mathcal{M}(r)}{c^2r^2}\right)\left(1-\frac{2G\mathcal{M}(r)}{c^2r}\right)^{-1}$$
$$\stackrel{\text{def}}{=} f(r).$$

The sum $\ln A(r)+\text{const.}$ is thus the antiderivative for the right-hand side $f(r)$ of this equation. Therefore, $\ln A(r)$ equals the integral of $f(r)$. If we use as integration limits r and ∞ and $A(\infty)=1$, we finally obtain

$$A(r)=\exp\left[-\frac{2G}{c^2}\int_r^\infty\frac{1}{\alpha^2}\left(\mathcal{M}(\alpha)+4\pi\alpha^3p(\alpha)/c^2\right)\left(1-\frac{2G\mathcal{M}}{c^2\alpha}\right)^{-1}\mathrm{d}\alpha\right]. \quad (3.77)$$

For $r>R$, this solution is equal to the Schwarzschild's outer solution because for $r>R$ one has $\varrho(r)=p(r)=0$ and $\mathcal{M}(r)=\mathcal{M}(R)=M$, so

$$A(r)=\exp\left[-\frac{2G}{c^2}\int_r^\infty\frac{M}{\alpha^2}\left(1-\frac{r_S}{\alpha}\right)^{-1}\mathrm{d}\alpha\right]. \quad (3.78)$$

With the new integration variable $\xi=1-\frac{r_S}{r}$ we obtain for $r>R$

$$A(r)=\exp\left[\int_1^{1-\xi}\frac{1}{\xi'}\mathrm{d}\xi'\right]=1-\frac{r_S}{r}, \quad (3.79)$$

i.e. the same solution as the Schwarzschild's outer solution. The calculated factors $A(r)$ and $B(r)$ therefore apply both inside and outside the spherical mass of radius R.

3.4 Black Holes

3.4.1 Astrophysics

The three most interesting celestial objects are *white dwarfs, neutron stars* and *black holes*. Stars like the Sun are built up by compression of interstellar clouds, caused by gravity. The contraction process comes to a standstill when the temperature inside becomes so great that nuclear fusion starts. By the nuclear fusion, hydrogen is converted into helium. The further fate of a star depends on its initial mass: The initial mass equal to

- $\approx$0.05 solar mass ends up as a *brown dwarf*;
- $\approx$1 solar mass and of volume comparable to that of the Earth grows into a red giant and ends after an explosion (supernova) as a *white dwarf*;
- $\approx$3 solar masses, with a corresponding radius of about 12 km, grows into a supergiant and ends after explosion (supernova) as a *neutron star*;
- $\approx$30 solar masses grows into a supergiant and ends after explosion (supernova) as a *black hole*.

A white dwarf is composed mostly of carbon and oxygen, produced by nuclear fusion, that are left over when the nuclear solar fuel is consumed. A white dwarf has a mass about that of our Sun and the diameter of the Earth, so a fairly compact structure. However, a neutron star is much more compact, i.e. it has a factor of 10^9 higher density. It has roughly the mass of a white dwarf but a diameter of only 24 km! A black hole, on the other hand, has different masses, sizes, and densities, as is further explained below.

The Schwarzschild's solution (3.26)

$$\mathrm{d}s^2 = \left(1 - \frac{r_S}{r}\right)c^2\,\mathrm{d}t^2 - \left(1 - \frac{r_S}{r}\right)^{-1}\mathrm{d}r^2 - r^2\left(\mathrm{d}\theta^2 + \sin^2\theta\,\mathrm{d}\varphi^2\right)$$

applies only outside the considered spherical mass. This formula gives the tiny spacetime interval $\mathrm{d}s$ which mass elements cover, moving from an event A to a closely adjacent event B. The Schwarzschild's radius is therefore only interesting for very large masses or masses with sufficient density, where $r_S > R$ is the radius of the spherical mass.

However, for the solar mass $M_\odot \approx 2 \cdot 10^{30}$ kg the Schwarzschild's radius is

$$r_{S\odot} \approx 3\ \text{km},$$

i.e. it is much smaller than the solar radius $r_\odot \approx 7 \cdot 10^5$ km, so our solution is valid only outside the mass of the Sun! For the Earth with a mass of $M_\oplus \approx 6 \cdot 10^{24}$ kg we even obtain a Schwarzschild's radius of

$$r_{S\oplus} \approx 9\ \text{mm!}$$

It seems that the Schwarzschild's radius only plays a role for highly concentrated matter. But that is not the case. Indeed, consider just the radius of a spherical mass $R = r_S$, that is, $R = \frac{2GM}{c^2}$. The so-called Schwarzschild's *density* ρ_S would in this case be

$$\underline{\underline{\rho_S}} \stackrel{\text{def}}{=} \frac{M}{\frac{4}{3}\pi r_S^3} = \frac{M}{\frac{4}{3}\pi\left(\frac{2GM}{c^2}\right)^3} = \frac{3c^6}{32\pi G^3 M^2} = \underline{\underline{2.33 \cdot 10^{71} \cdot M_{[\text{kg}]}^{-2}\left[\frac{\text{kg}}{\text{cm}^3}\right]}}.$$

The required density decreases inversely proportionally with the square of the mass! If one wants to do a black hole from Earth's mass, i.e. shrink the Earth to a radius of 9 mm, then the resulting body would have a density of $8.58 \cdot 10^{22}$ kg/cm^3. For the Sun a density of approximately $8.6 \cdot 10^{10}$ kg/cm^3 would be necessary; that is, 86 million tons per cubic centimetre! So still the enormous density of the neutron liquid[1] is in a neutron star. In the centre of the Milky Way, there is a black hole with a mass of $5.2 \cdot 10^{36}$ kg, or 2.6 million solar masses, and in the centre of the Virgo galaxy cluster, a black hole with the mass of $6 \cdot 10^{39}$ kg (3 billion solar masses).

3.4.2 Further Details about "Black Holes"

A spherical mass of radius less than the Schwarzschild's radius, $R < r_S$, is called a *black hole*. The name is explained as follows. The Schwarzschild's solution (3.26) for constant angles θ and φ, i.e. for $d\theta = d\varphi = 0$, with $r_S = \frac{2GM}{c^2}$, is

$$ds^2 = \left(1 - \frac{r_S}{r}\right)c^2\,dt^2 - \left(1 - \frac{r_S}{r}\right)^{-1} dr^2. \tag{3.80}$$

For light, i.e. photons, one has $ds^2 = 0$, so from (3.80) follows

$$c^2\,dt^2 = \left(1 - \frac{r_S}{r}\right)^{-2} dr^2,$$

giving

$$c\,dt = \pm\left(1 - \frac{r_S}{r}\right)^{-1} dr = \pm\frac{r}{r - r_S}\,dr,$$

which integrated yields

$$c\int_{t_0}^{t} dt = ct - ct_0 = \int_{r_0}^{r}\frac{r}{r - r_S}\,dr = \pm\left[r + r_S\ln\frac{r - r_S}{r_0 - r_S} - r_0\right]$$

and with $c_0 \overset{\text{def}}{=} ct_0 - r_0 - r_S\ln(r_0 - r_S)$ finally

$$\underline{\underline{ct = \pm\big(r + r_S\ln(r - r_S) + c_0\big)}}, \tag{3.81}$$

where the constant c_0 depends on the start time t_0 and on the starting location R_0, and ct has the dimension of a length. The substance of this formula can be pictured in the $(ct - r)$-half-plane, as shown in Fig. 3.1.

The light cones are to the right of the dividing line at $r = r_S$ upward open towards the time axis, i.e. particles (including photons) in the picture fly only upwards. In particular, even photons on the minus-sign trajectories seemingly not exceed the

[1] Neutron liquid consists predominantly of neutrons, and their average density is approximately equal to that of atomic nuclei.

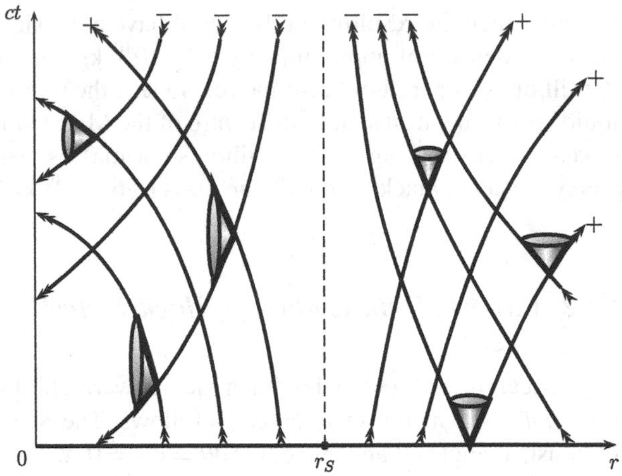

Fig. 3.1 Schwarzschild's solution

dividing line $r = r_S$, so can never arrive at the field with $r < r_S$. This is particularly recognised at the speed of the particles or photons, resulting from the above equations in

$$\frac{\mathrm{d}r}{\mathrm{d}t} = \pm\left(1 - \frac{r_S}{r}\right)c.$$

This velocity goes to zero as r goes to r_S. For $r < r_S$ the light cone shows the future in the direction $r = 0$. No particle or photon can escape from the field $r < r_S$! No light can overstep the limit of r_S; the "black hole" is, in fact, "black"! For a particle at r_S as is in Fig. 3.1, the time t decreases, so runs backwards! It is obvious that for this region, the time t is not very suitable. On the other hand, it is a fallacy that the boundary $r = r_S$ cannot be exceeded from the outside because the formulas report only the behaviour which would be seen by an observer. An observer, flying with the particle, would normally exceed the limit $r = r_S$ because for him the speed in the vicinity $r = r_S$ would not tend to zero, he would fly continuously through this sphere with finite speed. This can be shown as follows:

Due to Newton's theory, we arrive at the following context. Integrating the equation

$$m_0 a(t) = m_0 \frac{\mathrm{d}v}{\mathrm{d}t} = G\frac{Mm_0}{r^2}$$

with respect to

$$a = \frac{\mathrm{d}v}{\mathrm{d}t} = \frac{\mathrm{d}v}{\mathrm{d}r}\frac{\mathrm{d}r}{\mathrm{d}t} = v\frac{\mathrm{d}v}{\mathrm{d}r},$$

one obtains

$$\int_0^v \tilde{v}\,\mathrm{d}\tilde{v} = \frac{v^2}{2} = \int_\infty^r a\,\mathrm{d}\tilde{r} = \int_\infty^r \frac{GM}{\tilde{r}^2}\,\mathrm{d}\tilde{r} = \frac{GM}{r},$$

or solving for the velocity

$$v = -\sqrt{\frac{2GM}{r}} = -c\sqrt{\frac{r_S}{r}}. \tag{3.82}$$

The mass m_0 would pass the Schwarzschild's spherical shell for $r = r_S$ and according to (3.82) with the speed of light c, which is not possible according to the special theory of relativity! For $r < r_S$ the velocity would even be greater than light's! The velocity equations have to be modified. What happens then? If one passes through the Schwarzschild's spherical shell in the direction $r \to 0$, we will never be able to learn from him what he experienced because neither he nor a signal from him to us outside the Schwarzschild's spherical shell will ever reach us. We only can analyse theoretically what happens.

Within the Schwarzschild's Radius From (3.82) for the proper time we have

$$\mathrm{d}\tau = -\frac{1}{c}\left(\frac{r}{r_S}\right)^{1/2} \mathrm{d}r. \tag{3.83}$$

From this we get the time τ needed when flying through the Schwarzschild's spherical shell to reach the singularity $r = 0$. It is

$$\tau_S \overset{\text{def}}{=} -\frac{1}{c}\int_{r_S}^{0}\left(\frac{r}{r_S}\right)^{1/2} \mathrm{d}r = \frac{2}{3}\frac{r_S}{c}, \tag{3.84}$$

therefore, a finite time.

If a particle is located to the left of the parting line $r = r_S$, then the particle flies left in the direction $r \to 0$. Photons, starting left of the parting line cannot also exceed this border. Therefore, no light (photon) is coming outward from inside the sphere of radius r_S, a "black hole" is given there. Further, it is interesting that an ensemble of boundary trajectories (each characterized with a minus sign) proceeds in the negative direction of time, thus decreasing time. If one starts, for instance, at the point $(t_0 = 0, r = r_0 < r_S)$ on a 'minus-trajectory", then one travels to the past! The result, however, contains no logical contradiction because one cannot act from the inside of the sphere with the radius r_S on the outside: one only passively receives signals, i.e. one can "look into the past".

3.4.3 Singularities

The solution (3.26) has two singularities: a real singularity at $r = 0$ and an apparent one at $r = r_S$. The apparent singularity is a so-called *coordinate singularity* which could have been avoided by a better coordinate choice. Such a favourable coordinate system would have been obtained, for example, if we had introduced a different radial coordinate r^* as follows:

$$r = \left(1 + \frac{r_S}{4r^*}\right)^2 r^*.$$

Then

$$\frac{\mathrm{d}r}{\mathrm{d}r^*} = \left(1 - \frac{r_S}{4r^*}\right)\left(1 + \frac{r_S}{4r^*}\right),$$

so

$$\mathrm{d}r = \left(1 - \frac{r_S}{4r^*}\right)\left(1 + \frac{r_S}{4r^*}\right)\mathrm{d}r^*.$$

Used in (3.26) this yields

$$\mathrm{d}s^2 = \left(\frac{1 - \frac{r_S}{4r^*}}{1 + \frac{r_S}{4r^*}}\right)^2 c^2\,\mathrm{d}t^2 - \left(1 + \frac{r_S}{4r^*}\right)^4 \left(\mathrm{d}r^{*2} + r^{*2}\left(\mathrm{d}\theta^2 + \sin^2\theta\,\mathrm{d}\varphi^2\right)\right).$$

In these coordinates, in fact, only a singularity at $r^* = 0$ exists!

The authenticity of the singularity at $r = 0$ can be seen by examining the invariants. A function of coordinates is invariant under a transformation if it remains unchanged applying the transformation to the coordinates. This is precisely the mark of a true singularity, which does not depend on the randomly selected coordinate system. Such invariants were already encountered in the investigation of electromagnetic fields and the Lorentz transformation, e.g.

$$-\frac{1}{2}\mathrm{trace}\left(\boldsymbol{F}^*_{B,e}\boldsymbol{F}_{B,e}\right) = b^2 - e^2 = b'^2 - e'^2$$

and

$$-\frac{1}{4}\mathrm{trace}\left(\boldsymbol{F}^*_{B,e}\boldsymbol{F}_{E,b}\right) = \boldsymbol{e}^\intercal\boldsymbol{b} = \boldsymbol{e}'^\intercal\boldsymbol{b}'.$$

Such an invariant for our problem now is the so-called Kretschmann's invariant, which is defined with the aid of the modified symmetric Riemannian curvature matrix

$$\boldsymbol{R}^* \stackrel{\text{def}}{=} \left(\boldsymbol{I}_4 \otimes \boldsymbol{G}^{-1}\right)\boldsymbol{R}$$

as follows:

$$I_K \stackrel{\text{def}}{=} \mathrm{trace}\left(\boldsymbol{R}^*\boldsymbol{R}^*\right). \tag{3.85}$$

Because the Ricci matrix $\boldsymbol{R}_{\mathrm{Ric}}$ equals the zero matrix, it is unsuitable for an invariant. The newly defined matrix $\boldsymbol{R}^*$ will now be derived step by step for the

Schwarzschild's metric. First, we have

$$
\boldsymbol{\Gamma} = \left(\begin{array}{cccc}
0 & \frac{m}{r^2 h} & 0 & 0 \\
\frac{m}{r^2 h} & 0 & 0 & 0 \\
0 & 0 & 0 & 0 \\
0 & 0 & 0 & 0 \\
\hline
\frac{mh}{r^2} & 0 & 0 & 0 \\
0 & -\frac{m}{r^2 h} & 0 & 0 \\
0 & 0 & -rh & 0 \\
0 & 0 & 0 & -rh\sin^2\theta \\
\hline
0 & 0 & 0 & 0 \\
0 & 0 & \frac{1}{r} & 0 \\
0 & \frac{1}{r} & 0 & 0 \\
0 & 0 & 0 & -\sin\theta\cos\theta \\
\hline
0 & 0 & 0 & 0 \\
0 & 0 & 0 & \frac{1}{r} \\
0 & 0 & 0 & \cot\theta \\
0 & \frac{1}{r} & \cot\theta & 0
\end{array}\right),
$$

where

$$
m \stackrel{\text{def}}{=} \frac{GM}{c^2} \quad \text{and} \quad h \stackrel{\text{def}}{=} 1 - \frac{2m}{r}.
$$

With the help of (2.165) $\boldsymbol{R}$ can be calculated. The result is

$$
\boldsymbol{R} = \left(\begin{array}{cccc|cccc}
0 & 0 & 0 & 0 & 0 & 0 & 0 & 0 \\
0 & -\frac{2m}{r^3 h} & 0 & 0 & \frac{2m}{r^3 h} & 0 & 0 & 0 \\
0 & 0 & \frac{m}{r} & 0 & 0 & 0 & 0 & 0 \\
0 & 0 & 0 & \frac{m\sin^2\theta}{r} & 0 & 0 & 0 & 0 \\
\hline
0 & -\frac{2mh}{r^3} & 0 & 0 & \frac{2mh}{r^3} & 0 & 0 & 0 \\
0 & 0 & 0 & 0 & 0 & 0 & 0 & 0 \\
0 & 0 & 0 & 0 & 0 & 0 & \frac{m}{r} & 0 \\
0 & 0 & 0 & 0 & 0 & 0 & 0 & \frac{m\sin^2\theta}{r} \\
\hline
0 & 0 & \frac{mh}{r^3} & 0 & 0 & 0 & 0 & 0 \\
0 & 0 & 0 & 0 & 0 & 0 & -\frac{m}{r^3 h} & 0 \\
0 & 0 & 0 & 0 & 0 & 0 & 0 & 0 \\
0 & 0 & 0 & 0 & 0 & 0 & 0 & 0 \\
\hline
0 & 0 & 0 & \frac{mh}{r^3} & 0 & 0 & 0 & 0 \\
0 & 0 & 0 & 0 & 0 & 0 & 0 & -\frac{m}{r^3 h} \\
0 & 0 & 0 & 0 & 0 & 0 & 0 & 0 \\
0 & 0 & 0 & 0 & 0 & 0 & 0 & 0
\end{array}\right|
$$

$$
\left.\begin{array}{cccc|cccc}
0 & 0 & 0 & 0 & 0 & 0 & 0 & 0 \\
0 & 0 & 0 & 0 & 0 & 0 & 0 & 0 \\
-\frac{m}{r} & 0 & 0 & 0 & 0 & 0 & 0 & 0 \\
0 & 0 & 0 & 0 & -\frac{m\sin^2\theta}{r} & 0 & 0 & 0 \\
\hline
0 & 0 & 0 & 0 & 0 & 0 & 0 & 0 \\
0 & 0 & 0 & 0 & 0 & 0 & 0 & 0 \\
0 & -\frac{m}{r} & 0 & 0 & 0 & 0 & 0 & 0 \\
0 & 0 & 0 & 0 & 0 & -\frac{m\sin^2\theta}{r} & 0 & 0 \\
\hline
-\frac{mh}{r^3} & 0 & 0 & 0 & 0 & 0 & 0 & 0 \\
0 & \frac{m}{r^3 h} & 0 & 0 & 0 & 0 & 0 & 0 \\
0 & 0 & 0 & 0 & 0 & 0 & 0 & 0 \\
0 & 0 & 0 & -\frac{2m\sin^2\theta}{r} & 0 & 0 & \frac{2m\sin^2\theta}{r} & 0 \\
\hline
0 & 0 & 0 & 0 & -\frac{mh}{r^3} & 0 & 0 & 0 \\
0 & 0 & 0 & 0 & 0 & \frac{m}{r^3 h} & 0 & 0 \\
0 & 0 & 0 & \frac{2m}{r} & 0 & 0 & -\frac{2m}{r} & 0 \\
0 & 0 & 0 & 0 & 0 & 0 & 0 & 0
\end{array}\right).
$$

The first, (4 + 2)th, (8 + 3)th and 16th row/column are zero rows/columns, as it must always be in the matrix $\boldsymbol{R}$. With $\boldsymbol{R}$ the following symmetric matrix is obtained

$$
\left(\boldsymbol{I}_4 \otimes \boldsymbol{G}^{-1}\right)\boldsymbol{R}
= \left(\begin{array}{cccc|cccc}
0 & 0 & 0 & 0 & 0 & 0 & 0 & 0 \\
0 & \frac{2m}{r^3} & 0 & 0 & -\frac{2m}{r^3} & 0 & 0 & 0 \\
0 & 0 & -\frac{m}{r^3} & 0 & 0 & 0 & 0 & 0 \\
0 & 0 & 0 & -\frac{m}{r^3} & 0 & 0 & 0 & 0 \\
\hline
0 & -\frac{2m}{r^3} & 0 & 0 & \frac{2m}{r^3} & 0 & 0 & 0 \\
0 & 0 & 0 & 0 & 0 & 0 & 0 & 0 \\
0 & 0 & 0 & 0 & 0 & 0 & -\frac{m}{r^3} & 0 \\
0 & 0 & 0 & 0 & 0 & 0 & 0 & -\frac{m}{r^3} \\
\hline
0 & 0 & \frac{m}{r^3} & 0 & 0 & 0 & 0 & 0 \\
0 & 0 & 0 & 0 & 0 & 0 & \frac{m}{r^3} & 0 \\
0 & 0 & 0 & 0 & 0 & 0 & 0 & 0 \\
0 & 0 & 0 & 0 & 0 & 0 & 0 & 0 \\
\hline
0 & 0 & 0 & \frac{m}{r^3} & 0 & 0 & 0 & 0 \\
0 & 0 & 0 & 0 & 0 & 0 & 0 & \frac{m}{r^3} \\
0 & 0 & 0 & 0 & 0 & 0 & 0 & 0 \\
0 & 0 & 0 & 0 & 0 & 0 & 0 & 0
\end{array}\right|
$$

$$\left.\begin{array}{cccc|cccc}
0 & 0 & 0 & 0 & 0 & 0 & 0 & 0 \\
0 & 0 & 0 & 0 & 0 & 0 & 0 & 0 \\
\frac{m}{r^3} & 0 & 0 & 0 & 0 & 0 & 0 & 0 \\
0 & 0 & 0 & 0 & \frac{m}{r^3} & 0 & 0 & 0 \\
\hline
0 & 0 & 0 & 0 & 0 & 0 & 0 & 0 \\
0 & 0 & 0 & 0 & 0 & 0 & 0 & 0 \\
0 & \frac{m}{r^3} & 0 & 0 & 0 & 0 & 0 & 0 \\
0 & 0 & 0 & 0 & 0 & \frac{m}{r^3} & 0 & 0 \\
\hline
-\frac{m}{r^3} & 0 & 0 & 0 & 0 & 0 & 0 & 0 \\
0 & -\frac{m}{r^3} & 0 & 0 & 0 & 0 & 0 & 0 \\
0 & 0 & 0 & 0 & 0 & 0 & 0 & 0 \\
0 & 0 & 0 & \frac{2m}{r^3} & 0 & 0 & -\frac{2m}{r^3} & 0 \\
\hline
0 & 0 & 0 & 0 & -\frac{m}{r^3} & 0 & 0 & 0 \\
0 & 0 & 0 & 0 & 0 & -\frac{m}{r^3} & 0 & 0 \\
0 & 0 & 0 & -\frac{2m}{r^3} & 0 & 0 & \frac{2m}{r^3} & 0 \\
0 & 0 & 0 & 0 & 0 & 0 & 0 & 0
\end{array}\right),$$

and finally the Kretschmann's invariant is

$$\underline{\underline{I_K = \mathrm{trace}\left(\boldsymbol{R}^* \boldsymbol{R}^*\right) = 48\frac{m^2}{r^6} = 12\frac{r_S^2}{r^6}.}}$$

Hence, it is now quite apparent that the only real singularity occurs at $r = 0$!

Event Horizon Detector In [13], a possibility is given, how to determine if one approaches the event horizon (the spherical shell of radius r_S) of a black hole or even exceeds it. To this end, the authors give the invariant

$$I_1 = -\frac{720M^2(2M - r)}{r^9},$$

which for $r = 2M = r_S$ is zero and outside the horizon is positive with a maximum at $r = 9M/4$. An observer falling to the black hole can detect the presence of a horizon by observation of I_1. If the event horizon is crossed, it is too late for the observer. But he may, if the maximum is exceeded, use it as a warning and quickly initiate the trajectory reversal.

3.4.4 *Eddington's Coordinates*

The coordinate singularity at $r = r_S$ of the Schwarzschild's metric can be eliminated, for example, by using the following coordinate transformation.

In (3.81), from the Schwarzschild's solution we derived

$$ct = \pm\left(r + r_S \ln|r - r_S| + c_0\right).$$

If we define

$$r^* \stackrel{\text{def}}{=} r_S \ln(r - r_S), \tag{3.86}$$

we get from

$$\frac{dr^*}{dr} = \frac{r_S}{r - r_S}$$

the differential

$$dr^* = \frac{r_S}{r}\frac{r}{r - r_S}\,dr = \frac{r_S}{r}\left(\frac{1}{1 - \frac{r_S}{r}}\right)^{-1} dr. \tag{3.87}$$

It looks like the term

$$-\left(\frac{1}{1 - \frac{r_S}{r}}\right)^{-1} dr^2$$

in the Schwarzschild's metric, which was the reason for the coordinate singularity! If (3.87) were squared, indeed dr^2 would appear, but the parenthesis would have too much negative power. If one installs, however, the term r^* into a new time coordinate t^*, then within the Schwarzschild's metric it would again be multiplied by $\frac{r}{r - r_S}$, i.e. have the "right" power! Now the new time coordinate is set as

$$ct^* \stackrel{\text{def}}{=} ct + r^* = ct + r_S \ln|r - r_S|. \tag{3.88}$$

Differentiation with respect to r gives

$$c\frac{dt^*}{dr} = c\frac{dt}{dr} + \frac{dr^*}{dr}, \tag{3.89}$$

so

$$c\,dt^* = c\,dt + dr^* = c\,dt + \frac{r_S}{r - r_S}\,dr,$$

i.e.

$$c\,dt = c\,dt^* - \frac{r_S}{r - r_S}\,dr,$$

and squared

$$c^2\,dt^2 = c^2\,dt^{*2} - 2\frac{r_S}{r - r_S}c\,dt^*\,dr + \left(\frac{r_S}{r - r_S}\right)^2 dr^2.$$

Used in the Schwarzschild's metric this yields

$$ds^2 = \left(1 - \frac{r_S}{r}\right)c^2\,dt^{*2} - 2\frac{r_S}{r}c\,dt^*\,dr + \frac{r - r_S}{r}\left(\frac{r_S}{r - r_S}\right)^2 dr^2 - \left(\frac{r}{r - r_S}\right)dr^2,$$

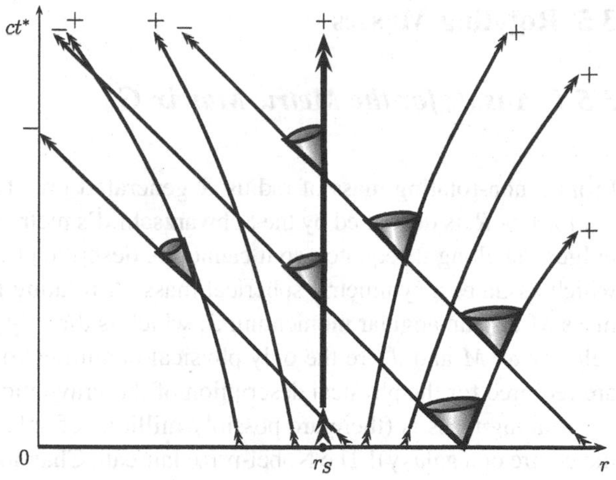

Fig. 3.2 Eddington's solution

so finally,

$$ds^2 = \frac{r - r_S}{r} c^2 \, dt^{*2} - 2\frac{r_S}{r} c \, dt^* \, dr - \frac{r + r_S}{r} \, dr^2. \tag{3.90}$$

When we remove the restriction $d\theta = d\varphi = 0$, we finally get the full Schwarzschild's metric in Eddington's coordinates

$$ds^2 = \frac{r - r_S}{r} c^2 \, dt^{*2} - 2\frac{r_S}{r} c \, dt^* \, dr - \frac{r + r_S}{r} \, dr^2 - r^2 \left(d\theta^2 + \sin^2\theta \, d\varphi^2\right). \tag{3.91}$$

This metric has, in fact, a singularity only at $r = 0$!

How will we see movements in the new (ct^*, r)-plane? For photons one has $ds^2 = 0$; so, if we divide (3.90) by $c^2 \, dt^{*2}$, we receive

$$0 = \frac{r - r_S}{r} - 2\frac{r_S}{r} \frac{dr}{c \, dt^*} - \frac{r + r_S}{r} \left(\frac{dr}{c \, dt^*}\right)^2.$$

This quadratic equation has the solutions

$$\frac{dr}{c \, dt^*} = \frac{-r_S \pm r}{r + r_S} = \frac{r - r_S}{r + r_S} \text{ and } -1, \tag{3.92}$$

or

$$\frac{dr}{dt^*} = \frac{-r_S \pm r}{r + r_S} c = \frac{r - r_S}{r + r_S} c \text{ and } -c. \tag{3.93}$$

The first solution provides for $r = r_S$ the zero slope, i.e. a vertical line; for $r < r_S$ the slope is negative, and it is positive for $r > r_S$. The second solution in the (ct^*, r)-plane is an inclined at 45° straight line. It shows that the velocity of an ingoing photon is constant and equal to c. Overall, one obtains Fig. 3.2.

3.5 Rotating Masses

3.5.1 *Ansatz for the Metric Matrix G*

From a non-rotating mass of radius R generated gravitation effect on the outside, i.e. for $r > R$, is described by the Schwarzschild's metric. A non-rigid rotating mass bulges out along its equator, so it cannot be described by a Schwarzschild's metric, which assumes a symmetric spherical mass. A rotating mass has in addition to the mass M also an angular momentum J, which is directly proportional to the angular velocity ω. M and J are the only physical quantities for a rotating black hole that are required for the physical description of the gravitational field and how powerful the rotating mass is (there are possibly millions of solar masses in a black hole in the centre of a galaxy)! The Nobel-prize laureate Chandrasekhar has expressed it as:

> Rotating black holes are the most perfect macroscopic objects in the universe. And as the theory of general relativity provides a single, unique, two-parameter solution for their description, they are also the simplest objects.

For the mathematical description, spatial polar coordinates r, θ and φ are selected. The polar axis is the axis of rotation around which the body rotates with constant angular velocity ω. The elements of the metric matrix $\mathbf{G}$ are allowed to depend neither on the angle θ nor on time t. One can also use a priori, for reasons of symmetry, some matrix elements equal to zero. Because if one takes a time reflection $t \to -t$, this also changes the direction of rotation $\omega = \frac{\mathrm{d}\varphi}{\mathrm{d}t} \to -\omega = -\frac{\mathrm{d}\varphi}{\mathrm{d}t} = \frac{\mathrm{d}(-\varphi)}{\mathrm{d}t}$. But if we introduce two transformations $t \to -t$ and $\varphi \to -\varphi$ simultaneously, then the gravitational field is not allowed to change at all, also the metric matrix $\mathbf{G}$ is not changed. So that being the case, the matrix elements $g_{t\theta}$ and g_{tr} must be zero because in the calculation of the metric here—with the indicated transformations—$\mathrm{d}t\,\mathrm{d}\theta \to -\mathrm{d}t\,\mathrm{d}\theta$ and $\mathrm{d}t\,\mathrm{d}r \to -\mathrm{d}t\,\mathrm{d}r$. And so one can start from the following the metric matrix which must always be symmetric:

$$\mathbf{G} = \begin{pmatrix} g_{tt} & 0 & 0 & g_{t\varphi} \\ 0 & g_{rr} & g_{r\theta} & 0 \\ 0 & g_{r\theta} & g_{\theta\theta} & 0 \\ g_{t\varphi} & 0 & 0 & g_{\varphi\varphi} \end{pmatrix}.$$

The symmetric 2×2-matrix in the centre of the matrix $\mathbf{G}$ can be brought to diagonal form with the help of a similarity transformation so that, without loss of generality, one may start with the symmetric matrix:

$$\mathbf{G} = \begin{pmatrix} g_{tt} & 0 & 0 & g_{t\varphi} \\ 0 & g_{rr} & 0 & 0 \\ 0 & 0 & g_{\theta\theta} & 0 \\ g_{t\varphi} & 0 & 0 & g_{\varphi\varphi} \end{pmatrix}.$$

3.5.2 Kerr's Solution in Boyer–Lindquist Coordinates

Without derivation we here directly give as solution the so-called Kerr's metric in Boyer–Lindquist coordinates:

$$\underline{\underline{\mathrm{d}s^2 = \left(1 - \frac{2mr}{\rho^2}\right)c^2\,\mathrm{d}t^2 + 4ma\frac{r\sin^2\theta}{\rho^2}c\,\mathrm{d}t\,\mathrm{d}\varphi - \frac{\rho^2}{\Delta}\,\mathrm{d}r^2}}$$

$$\underline{\underline{-\rho^2\,\mathrm{d}\theta^2 - \left(r^2 + a^2 + \frac{2mr}{r^2}a^2\sin^2\theta\right)\sin^2\theta\,\mathrm{d}\varphi^2,}} \tag{3.94}$$

with

$$\Delta \overset{\text{def}}{=} r^2 - 2mr + a^2,$$

$$\rho^2 \overset{\text{def}}{=} r^2 + a^2\cos^2\theta,$$

$$m \overset{\text{def}}{=} \frac{MG}{c^2}$$

and the angular momentum (spin) J per unit mass

$$a \overset{\text{def}}{=} \frac{J}{m}.$$

As system parameters, in fact, only the two physical parameters occur: the modified *mass*, m, and the *angular momentum per unit mass*, a! For $J = 0$, i.e. for a non-rotating mass, one obtains, of course, the Schwarzschild's solution.

A simpler form of the solution is obtained for $M/r \ll 1$ and $a/r \ll 1$, i.e. for weak fields and slow rotation, namely

$$\mathrm{d}s^2 \cong \left(1 - \frac{2m}{r}\right)c^2\,\mathrm{d}t^2 + \frac{4J}{r}\sin^2\theta\,c\,\mathrm{d}t\,\mathrm{d}\varphi - \left(1 + \frac{2m}{r}\right)\mathrm{d}r^2$$
$$- r^2\left(\mathrm{d}\theta^2 + \sin^2\theta\,\mathrm{d}\varphi^2\right).$$

3.5.3 The Lense–Thirring Effect

Two electric charges q_1 and q_2 of opposite polarity attract according to the Coulomb's law with the force

$$\boldsymbol{f}_{\text{electr}} = \frac{1}{4\pi\epsilon_0}\frac{q_1q_2}{r_{12}^2}\frac{\boldsymbol{r}_{12}}{r_{12}}.$$

Here the vector $\boldsymbol{r}_{12} \in \mathbb{R}^3$ shows from q_1 to q_2. Newton's law of attraction between two masses m_1 and m_2 has almost the same shape, namely

$$\boldsymbol{f}_{\text{mech}} = G\frac{m_1 m_2}{r_{12}^2}\frac{\boldsymbol{r}_{12}}{r_{12}}.$$

Going over in both laws from the action at a distance theory to the theory of contiguous action, i.e. the field theory, with the electric field strength vector

$$\boldsymbol{e}_e \stackrel{\text{def}}{=} \frac{1}{4\pi\epsilon_0}\frac{q_1}{r_{12}^2}\frac{\boldsymbol{r}_{12}}{r_{12}}$$

one can also write the electric force as

$$\boldsymbol{f}_{\text{electr}} = \boldsymbol{e}_e q_2.$$

With the appropriate nomenclature for the mechanical parameters and the field strength

$$\boldsymbol{e}_m \stackrel{\text{def}}{=} G\frac{m_1}{r_{12}^2}\frac{\boldsymbol{r}_{12}}{r_{12}},$$

one finally obtains

$$\boldsymbol{f}_{\text{mech}} = \boldsymbol{e}_m m_2.$$

If an electric charge is moving, then there is in addition a magnetic field $\boldsymbol{b}$, which depends on the speed and the load of the charge. The magnetic field $\boldsymbol{b}$ acts on the moving charge q having speed $\boldsymbol{v}$, generating, together with the electric field $\boldsymbol{e}$, the force:

$$\boldsymbol{f} = q\left(\boldsymbol{e} + \frac{1}{c}\boldsymbol{v} \times \boldsymbol{b}\right).$$

The question now is: Does a moving *mass* have a similar additional effect to another moving mass? This is indeed the case, for example, the precession of gyroscopes in the vicinity of large rotating masses like the Earth. This first was treated by Föppl [11]. The Austrian physicists Lense and Thirring calculated exactly this effect from the gravitational equations of Einstein in 1918.

One speaks in this context of a so-called *gravitomagnetic* field in analogy with classical electromagnetism. To this end, we first define with the help of the angular momentum $\boldsymbol{J}$ of a spherical rotating mass the field

$$\boldsymbol{h}(\boldsymbol{r}) \stackrel{\text{def}}{=} -2\frac{\boldsymbol{J} \times \boldsymbol{r}}{r^3} = -\frac{4GMR^2}{5c^3}\frac{\boldsymbol{\omega} \times \boldsymbol{r}}{r^3}, \tag{3.95}$$

where $\boldsymbol{\omega}$ is the angular velocity of the rotating mass M with the radius R, and $r > R$ is the distance from the centre of mass. Next we define the *gravitomagnetic field*:

$$\bar{\boldsymbol{h}} \stackrel{\text{def}}{=} \nabla \times \boldsymbol{h} = 2\frac{\boldsymbol{J} - 3(\boldsymbol{r}^\top \boldsymbol{J})\boldsymbol{r}}{r^3} = \frac{2GMR^2}{5c^2}\frac{3(\boldsymbol{\omega}^\top \boldsymbol{r})\boldsymbol{r} - \boldsymbol{\omega} r^2}{r^5}. \tag{3.96}$$

The angular momentum here thus plays the same role as the magnetic dipole moment in electrodynamics, and the vector $\boldsymbol{h}$ plays the same role as the vector potential. In summary, we obtain an analogue of the Lorentz force, namely

$$\underline{\underline{m\frac{\mathrm{d}^2\boldsymbol{x}}{\mathrm{d}t^2} = m\left(\boldsymbol{e}_{\text{mech}} + \frac{\mathrm{d}\boldsymbol{x}}{\mathrm{d}t} \times \bar{\boldsymbol{h}}\right).}} \tag{3.97}$$

In the case of stars orbiting close to a spinning, supermassive black hole, the Lense–Thirring effect should cause the star's orbital plane to precess about the black hole's spin axis. This effect should be detectable within the next few years via astrometric monitoring of stars at the centre of the Milky Way galaxy.

3.6 Summary of Results for the Gravitation of a Spherical Mass

The solution of Einstein's field equation for the *outside* of a spherically symmetric, uniform, time-invariant mass distribution is this Schwarzschild's metric (3.26):

$$\boxed{\mathrm{d}s^2 = \left(1 - \frac{2GM}{c^2 r}\right)c^2\,\mathrm{d}t^2 - \left(1 - \frac{2GM}{c^2 r}\right)^{-1}\mathrm{d}r^2 - r^2\left(\mathrm{d}\theta^2 + \sin^2\theta\,\mathrm{d}\varphi^2\right),}$$

and the so-called Schwarzschild's radius is (3.28):

$$\boxed{r_S \stackrel{\text{def}}{=} \frac{2GM}{c^2}.}$$

The Schwarzschild's metric in matrix form is (3.25):

$$\boxed{\mathrm{d}s^2 = \mathrm{d}\vec{x}^\mathsf{T}\,\boldsymbol{G}\,\mathrm{d}\vec{x} = \mathrm{d}\vec{x}^\mathsf{T}\begin{pmatrix} 1-\frac{r_S}{r} & 0 & 0 & 0 \\ 0 & -(1-\frac{r_S}{r})^{-1} & 0 & 0 \\ 0 & 0 & -r^2 & 0 \\ 0 & 0 & 0 & -r^2\sin^2\theta \end{pmatrix}\mathrm{d}\vec{x},}$$

with

$$\boxed{\mathrm{d}\vec{x} \stackrel{\text{def}}{=} \begin{pmatrix} c\,\mathrm{d}t \\ \mathrm{d}r \\ \mathrm{d}\theta \\ \mathrm{d}\varphi \end{pmatrix}.}$$

This changes the length (3.30) and time (3.31) according to:

$$\boxed{\mathrm{d}R = \left(1 - \frac{2GM}{c^2 r}\right)^{-1/2}\mathrm{d}r,}$$

and

$$d\tau = \left(1 - \frac{2GM}{c^2 r}\right)^{1/2} dt.$$

For the redshift of spectral lines the frequency ratio is obtained (3.32):

$$\frac{\nu_R}{\nu_T} = \left(\frac{1 - \frac{2GM}{c^2 r_T}}{1 - \frac{2GM}{c^2 r_R}}\right)^{1/2}.$$

The necessary Schwarzschild's density ρ_S for the existence of a *black hole* is

$$\rho_S \stackrel{\text{def}}{=} \frac{M}{\frac{4}{3}\pi r_S^3} = \frac{M}{\frac{4}{3}\pi (\frac{2GM}{c^2})^3}$$
$$= \frac{3c^6}{32\pi G^3 M^2} = 2.33 \cdot 10^{71} \cdot M_{[\mathrm{kg}]}^{-2} \left[\frac{\mathrm{kg}}{\mathrm{cm}^3}\right].$$

The Eddington's coordinates for a black hole are given with the new time coordinate (3.88)

$$ct^* \stackrel{\text{def}}{=} ct + r_S \ln |r - r_S|,$$

so

$$c\, dt^* = c\, dt + \frac{r_S}{r - r_S}\, dr.$$

The Schwarzschild's metric in Eddington's coordinates is (3.91):

$$ds^2 = \frac{r - r_S}{r} c^2\, dt^{*2} - 2\frac{r_S}{r} c\, dt^*\, dr - \frac{r + r_S}{r}\, dr^2 - r^2\left(d\theta^2 + \sin^2\theta\, d\varphi^2\right).$$

The Lense–Thirring effect for a rotating mass generates a *gravitomagnetic* field in analogy with classical electromagnetism. With the help of the angular momentum $\boldsymbol{J}$ of a spherical rotating mass, the field is (3.95):

$$\boldsymbol{h}(\boldsymbol{r}) \stackrel{\text{def}}{=} -2\frac{\boldsymbol{J} \times \boldsymbol{r}}{r^3} = -\frac{4GMR^2}{5c^3}\frac{\boldsymbol{\omega} \times \boldsymbol{r}}{r^3},$$

where $\boldsymbol{\omega}$ is the angular velocity of the rotating mass M with the radius R, and $r > R$ is the distance from the centre of mass. Then the *gravitomagnetic field* is defined as (3.96):

$$\bar{\boldsymbol{h}} \stackrel{\text{def}}{=} \nabla \times \boldsymbol{h} = 2\frac{\boldsymbol{J} - 3(\boldsymbol{r}^\top \boldsymbol{J})\boldsymbol{r}}{r^3} = \frac{2GMR^2}{5c^2}\frac{3(\boldsymbol{\omega}^\top \boldsymbol{r})\boldsymbol{r} - \boldsymbol{\omega} r^2}{r^5}.$$

An analogue of the Lorentz force is obtained (3.97):

$$m\frac{\mathrm{d}^2\boldsymbol{x}}{\mathrm{d}t^2} = m\left(\boldsymbol{e}_{\mathrm{mech}} + \frac{\mathrm{d}\boldsymbol{x}}{\mathrm{d}t} \times \bar{\boldsymbol{h}}\right).$$

3.7 Concluding Remark

In the theory of *Special Relativity*, the effects are most clearly visible when the masses move quickly; however, in the theory of *General Relativity*, the effects are the greatest when the mass densities are very large and thus the spatial curvature is very pronounced.

Appendix A
Vectors and Matrices

A.1 Vectors and Matrices

If the speed of a body is given, then its size and the direction need to be identified. For the description of such a directional quantity, *vectors* are used. These vectors in the three dimensional space require three components which, e.g. in a *column vector*, are summarized as follows:

$$\boldsymbol{v} = \begin{pmatrix} v_1 \\ v_2 \\ v_3 \end{pmatrix}. \tag{A.1}$$

A second possibility is to present a *transposed* column vector

$$\boldsymbol{v}^\top = \begin{pmatrix} v_1 & v_2 & v_3 \end{pmatrix}, \tag{A.2}$$

a *row vector*.

In another way, we get the concept of the vector when the following purely mathematical problem is considered: Find the solutions of the three coupled equations with four unknowns x_1, x_2, x_3 and x_4:

$$a_{11}x_1 + a_{12}x_2 + a_{13}x_3 + a_{14}x_4 = y_1, \tag{A.3}$$

$$a_{21}x_1 + a_{22}x_2 + a_{23}x_3 + a_{24}x_4 = y_2, \tag{A.4}$$

$$a_{31}x_1 + a_{32}x_2 + a_{33}x_3 + a_{34}x_4 = y_3. \tag{A.5}$$

The four unknowns are summarized in the vector

$$\boldsymbol{x} \stackrel{\text{def}}{=} \begin{pmatrix} x_1 \\ x_2 \\ x_3 \\ x_4 \end{pmatrix}, \tag{A.6}$$

G. Ludyk, *Einstein in Matrix Form*, Graduate Texts in Physics,

DOI 10.1007/978-3-642-35798-5,

the variables y_1, y_2 and y_3 form the vector

$$\boldsymbol{y} \stackrel{\text{def}}{=} \begin{pmatrix} y_1 \\ y_2 \\ y_3 \end{pmatrix}, \tag{A.7}$$

and the coefficients a_{ij} are included into the *matrix*

$$\boldsymbol{A} \stackrel{\text{def}}{=} \begin{pmatrix} a_{11} & a_{12} & a_{13} & a_{14} \\ a_{21} & a_{22} & a_{23} & a_{24} \\ a_{31} & a_{32} & a_{33} & a_{34} \end{pmatrix}. \tag{A.8}$$

With the two vectors $\boldsymbol{x}$ and $\boldsymbol{y}$ and the matrix $\boldsymbol{A}$, the system of equations can compactly be written as

$$\boldsymbol{A}\boldsymbol{x} = \boldsymbol{y}. \tag{A.9}$$

If the two systems of equations

$$a_{11}x_1 + a_{12}x_2 = y_1, \tag{A.10}$$

$$a_{21}x_1 + a_{22}x_2 = y_2 \tag{A.11}$$

and

$$a_{11}z_1 + a_{12}z_2 = v_1, \tag{A.12}$$

$$a_{21}z_1 + a_{22}z_2 = v_2 \tag{A.13}$$

are added, one obtains

$$a_{11}(x_1 + z_1) + a_{12}(x_2 + z_2) = (y_1 + v_1), \tag{A.14}$$

$$a_{21}(x_1 + z_1) + a_{22}(x_2 + z_2) = (y_2 + v_2). \tag{A.15}$$

With the aid of vectors and matrices, the two systems of equations can be written as

$$\boldsymbol{A}\boldsymbol{x} = \boldsymbol{y} \quad \text{and} \quad \boldsymbol{A}\boldsymbol{z} = \boldsymbol{v}. \tag{A.16}$$

Adding the two equations in (A.16) is formally accomplished as

$$\boldsymbol{A}\boldsymbol{x} + \boldsymbol{A}\boldsymbol{z} = \boldsymbol{A}(\boldsymbol{x} + \boldsymbol{z}) = \boldsymbol{y} + \boldsymbol{v}. \tag{A.17}$$

A comparison of (A.17) with (A.14) and (A.15) suggests the following definition of the *addition of vectors*:

Definition:

$$\boldsymbol{y}+\boldsymbol{v}=\begin{pmatrix} y_1 \\ y_2 \\ \vdots \\ y_n \end{pmatrix}+\begin{pmatrix} v_1 \\ v_2 \\ \vdots \\ v_n \end{pmatrix}\stackrel{\text{def}}{=}\begin{pmatrix} y_1+v_1 \\ y_2+v_2 \\ \vdots \\ y_n+v_n \end{pmatrix}. \tag{A.18}$$

Accordingly, the product of a vector and a real or complex number c is defined by

Definition:

$$c\cdot\boldsymbol{x}\stackrel{\text{def}}{=}\begin{pmatrix} c\cdot x_1 \\ \vdots \\ c\cdot x_n \end{pmatrix}. \tag{A.19}$$

A.2 Matrices

A.2.1 *Types of Matrices*

In the first section, the concept of the matrix has been introduced.

Definition: If a matrix $\boldsymbol{A}$ has n rows and m columns, it is called an $n\times m$ matrix and denoted $\boldsymbol{A}\in\mathbb{R}^{n\times m}$.

Definition: If $\boldsymbol{A}$ (with the elements a_{ij}) is an $n\times m$ matrix, then the *transpose* of $\boldsymbol{A}$, denoted by $\boldsymbol{A}^{\mathsf{T}}$, is the $m\times n$ matrix with the elements $a^{\mathsf{T}}_{ij}=a_{ji}$.

So the matrix (A.8) has the matrix transpose

$$\boldsymbol{A}^{\mathsf{T}}=\begin{pmatrix} a_{11} & a_{21} & a_{31} \\ a_{12} & a_{22} & a_{32} \\ a_{13} & a_{23} & a_{33} \\ a_{14} & a_{24} & a_{34} \end{pmatrix}. \tag{A.20}$$

In a *square matrix*, one has $n=m$; and in an $n\times n$ *diagonal matrix*, all the elements a_{ij}, $i\neq j$, outside the main diagonal are equal to zero. An *identity matrix* $\boldsymbol{I}$ is a diagonal matrix where all elements on the main diagonal are equal to one. An $r\times r$ identity matrix is denoted by $\boldsymbol{I}_r$. If a transposed matrix $\boldsymbol{A}^{\mathsf{T}}$ is equal to the original matrix $\boldsymbol{A}$, such a matrix is called *symmetric*. In this case, $a_{ij}=a_{ji}$.

A.2.2 Matrix Operations

If the two systems of equations

$$\begin{aligned}
a_{11}x_1 + a_{12}x_2 + \cdots + a_{1m}x_m &= y_1, \\
a_{21}x_1 + a_{22}x_2 + \cdots + a_{2m}x_m &= y_2, \\
&\vdots \\
a_{n1}x_1 + a_{n2}x_2 + \cdots + a_{nm}x_m &= y_n
\end{aligned}$$

and

$$\begin{aligned}
b_{11}x_1 + b_{12}x_2 + \cdots + b_{1m}x_m &= z_1, \\
b_{21}x_1 + b_{22}x_2 + \cdots + b_{2m}x_m &= z_2, \\
&\vdots \\
b_{n1}x_1 + b_{n2}x_2 + \cdots + b_{nm}x_m &= z_n
\end{aligned}$$

are added, one obtains

$$\begin{aligned}
(a_{11}+b_{11})x_1 + (a_{12}+b_{12})x_2 + \cdots + (a_{1m}+b_{1m})x_m &= (y_1+z_1), \\
(a_{21}+b_{21})x_1 + (a_{22}+b_{22})x_2 + \cdots + (a_{2m}+b_{2m})x_m &= (y_2+z_2), \\
&\vdots \\
(a_{n1}+b_{n1})x_1 + (a_{n2}+b_{n2})x_2 + \cdots + (a_{nm}+b_{nm})x_m &= (y_n+z_n),
\end{aligned}$$

or, in vector–matrix notation, with

$$\boldsymbol{A}\boldsymbol{x} = \boldsymbol{y} \quad \text{and} \quad \boldsymbol{B}\boldsymbol{x} = \boldsymbol{z},$$

the same can also be written symbolically as

$$(\boldsymbol{A}+\boldsymbol{B})\boldsymbol{x} = \boldsymbol{y}+\boldsymbol{z}. \tag{A.21}$$

A comparison of the last equations suggests the following definition:

Definition: The *sum* of two $n \times m$ matrices $\boldsymbol{A}$ and $\boldsymbol{B}$ is defined by

$$\boldsymbol{A}+\boldsymbol{B} = \begin{pmatrix} a_{11} & \cdots & a_{1m} \\ \vdots & & \vdots \\ a_{n1} & \cdots & a_{nm} \end{pmatrix} + \begin{pmatrix} b_{11} & \cdots & b_{1m} \\ \vdots & & \vdots \\ b_{n1} & \cdots & b_{nm} \end{pmatrix}$$

$$\stackrel{\text{def}}{=} \begin{pmatrix} (a_{11}+b_{11}) & \cdots & (a_{1m}+b_{1m}) \\ \vdots & & \vdots \\ (a_{n1}+b_{n1}) & \cdots & (a_{nm}+b_{nm}) \end{pmatrix}. \tag{A.22}$$

The sum of two matrices can only be formed when both matrices have the same number of rows and the same number of columns.

If the relations

$$y = Ax \quad \text{and} \quad x = Bz \tag{A.23}$$

are given, what is the connection between the two vectors y and z? One may write

$$\begin{aligned} a_{11}x_1 + a_{12}x_2 + \cdots + a_{1m}x_m &= y_1, \\ a_{21}x_1 + a_{22}x_2 + \cdots + a_{2m}x_m &= y_2, \\ &\vdots \\ a_{n1}x_1 + a_{n2}x_2 + \cdots + a_{nm}x_m &= y_n \end{aligned}$$

and

$$\begin{aligned} b_{11}z_1 + b_{12}z_2 + \cdots + b_{1\ell}z_\ell &= x_1, \\ b_{21}z_1 + b_{22}z_2 + \cdots + b_{2\ell}z_\ell &= x_2, \\ &\vdots \\ b_{m1}z_1 + b_{m2}z_2 + \cdots + b_{m\ell}z_\ell &= x_m, \end{aligned}$$

then one obtains, by inserting the x_i's from the latter system of equations into the former system,

$$\begin{aligned} a_{11}(b_{11}z_1 + \cdots + b_{1\ell}z_\ell) + \cdots + a_{1m}(b_{m1}z_1 + \cdots + b_{m\ell}z_\ell) &= y_1, \\ a_{21}(b_{11}z_1 + \cdots + b_{1\ell}z_\ell) + \cdots + a_{2m}(b_{m1}z_1 + \cdots + b_{m\ell}z_\ell) &= y_2, \\ &\vdots \\ a_{n1}(b_{11}z_1 + \cdots + b_{1\ell}z_\ell) + \cdots + a_{nm}(b_{m1}z_1 + \cdots + b_{m\ell}z_\ell) &= y_n. \end{aligned}$$

Combining the terms with z_i, we obtain

$$\begin{aligned} (a_{11}b_{11} + \cdots + a_{1m}b_{m1})z_1 + \cdots + (a_{11}b_{1\ell} + \cdots + a_{1m}b_{m\ell})z_\ell &= y_1, \\ (a_{21}b_{11} + \cdots + a_{2m}b_{m1})z_1 + \cdots + (a_{21}b_{1\ell} + \cdots + a_{2m}b_{m\ell})z_\ell &= y_2, \\ &\vdots \\ (a_{n1}b_{11} + \cdots + a_{nm}b_{m1})z_1 + \cdots + (a_{n1}b_{1\ell} + \cdots + a_{nm}b_{m\ell})z_\ell &= y_n. \end{aligned}$$

If, on the other hand, we formally insert the right-hand side of (A.23) into the left equation, we obtain

$$y = ABz \stackrel{\text{def}}{=} Cz. \tag{A.24}$$

Definition: The *product* of an $n \times m$ matrix $\boldsymbol{A}$ with an $m \times \ell$ matrix $\boldsymbol{B}$ is the $n \times \ell$ matrix $\boldsymbol{C}$ with the matrix elements

$$c_{ij} = \sum_{k=1}^{m} a_{ik} b_{kj}, \tag{A.25}$$

for $i = 1, 2, \ldots, n$ and $j = 1, 2, \ldots, \ell$.

The element c_{ij} of the product matrix $\boldsymbol{C}$ is obtained by multiplying the elements of the ith row of the first matrix $\boldsymbol{A}$ with the elements of the jth column of the second matrix $\boldsymbol{B}$ and adding these products. It follows that the number of columns of the first matrix must be equal to the number of rows of the second matrix, so that the matrix multiplication can be executed at all. The product matrix has as many rows as the first matrix and as many columns as the second matrix. It follows that, in general, $\boldsymbol{AB} \neq \boldsymbol{BA}$.

We get another matrix operation through the following problem. In

$$\boldsymbol{A}\boldsymbol{x} = \boldsymbol{b}, \tag{A.26}$$

the 3×3 matrix $\boldsymbol{A}$ and the 3×1 vector $\boldsymbol{b}$ shall be given. Wanted is the 3×1 vector $\boldsymbol{x}$ that satisfies the system of equations (A.26). Written out this is the linear system of equations:

$$a_{11}x_1 + a_{12}x_2 + a_{13}x_3 = b_1,$$

$$a_{21}x_1 + a_{22}x_2 + a_{23}x_3 = b_2,$$

$$a_{31}x_1 + a_{32}x_2 + a_{33}x_3 = b_3.$$

Denoting the determinant of the square matrix $\boldsymbol{A}$ by $\det(\boldsymbol{A})$, the solutions are obtained by using Cramer's rule

$$x_1 = \frac{1}{\det(\boldsymbol{A})} \det \begin{pmatrix} b_1 & a_{12} & a_{13} \\ b_2 & a_{22} & a_{23} \\ b_3 & a_{32} & a_{33} \end{pmatrix}, \tag{A.27}$$

$$x_2 = \frac{1}{\det(\boldsymbol{A})} \det \begin{pmatrix} a_{11} & b_1 & a_{13} \\ a_{21} & b_2 & a_{23} \\ a_{31} & b_3 & a_{33} \end{pmatrix}, \tag{A.28}$$

$$x_3 = \frac{1}{\det(\boldsymbol{A})} \det \begin{pmatrix} a_{11} & a_{12} & b_1 \\ a_{21} & a_{22} & b_2 \\ a_{31} & a_{32} & b_3 \end{pmatrix}. \tag{A.29}$$

If we develop in (A.27) the determinant in the numerator with respect to the first column, we obtain

$$\begin{aligned} x_1 &= \frac{1}{\det(\boldsymbol{A})}\left(b_1 \det\begin{pmatrix} a_{22} & a_{23} \\ a_{32} & a_{33}\end{pmatrix} - b_2 \det\begin{pmatrix} a_{12} & a_{13} \\ a_{32} & a_{33}\end{pmatrix} + b_3 \det\begin{pmatrix} a_{12} & a_{13} \\ a_{22} & a_{23}\end{pmatrix}\right) \\ &= \frac{1}{\det(\boldsymbol{A})}(b_1 A_{11} + b_2 A_{21} + b_3 A_{31}) \\ &= \frac{1}{\det(\boldsymbol{A})}\begin{pmatrix} A_{11} & A_{21} & A_{31}\end{pmatrix}\boldsymbol{b}. \end{aligned} \tag{A.30}$$

Accordingly, we obtain from (A.28) and (A.29)

$$x_2 = \frac{1}{\det(\boldsymbol{A})}\begin{pmatrix} A_{12} & A_{22} & A_{32}\end{pmatrix}\boldsymbol{b} \tag{A.31}$$

and

$$x_3 = \frac{1}{\det(\boldsymbol{A})}\begin{pmatrix} A_{13} & A_{23} & A_{33}\end{pmatrix}\boldsymbol{b}. \tag{A.32}$$

Here the *adjuncts* A_{ij} are the determinants which are obtained when the ith row and the jth column of the matrix $\boldsymbol{A}$ are removed, and from the remaining matrix the determinant is computed and this is multiplied by the factor $(-1)^{i+j}$.

Definition: The adjuncts are summarized in the *adjoint matrix*

$$\operatorname{adj}(\boldsymbol{A}) \stackrel{\text{def}}{=} \begin{pmatrix} A_{11} & A_{21} & A_{31} \\ A_{12} & A_{22} & A_{32} \\ A_{13} & A_{23} & A_{33}\end{pmatrix}. \tag{A.33}$$

With this matrix, the three equations (A.30) to (A.32) can be written as one equation

$$\boldsymbol{x} = \frac{\operatorname{adj}(\boldsymbol{A})}{\det(\boldsymbol{A})}\boldsymbol{b}. \tag{A.34}$$

Definition: The $n \times n$ matrix (whenever $\det(\boldsymbol{A}) \neq 0$)

$$\boldsymbol{A}^{-1} \stackrel{\text{def}}{=} \frac{\operatorname{adj}(\boldsymbol{A})}{\det(\boldsymbol{A})} \tag{A.35}$$

is called the *inverse matrix* of the square $n \times n$ matrix $\boldsymbol{A}$.

For a matrix product, the inverse matrix is obtained as

$$(\boldsymbol{A}\boldsymbol{B})^{-1} = \boldsymbol{B}^{-1}\boldsymbol{A}^{-1} \tag{A.36}$$

because

$$(\boldsymbol{A}\boldsymbol{B})(\boldsymbol{B}^{-1}\boldsymbol{A}^{-1}) = \boldsymbol{A}(\boldsymbol{B}\boldsymbol{B}^{-1})\boldsymbol{A}^{-1} = \boldsymbol{A}\boldsymbol{A}^{-1} = \boldsymbol{I}.$$

A.2.3 Block Matrices

Often large matrices have a certain structure, e.g. when one or more sub-arrays are zero matrices. On the other hand, one can make a *block matrix* from each matrix by drawing vertical and horizontal lines. For a system of equations, one then, for example, obtains

$$\left(\begin{array}{c|c|c|c} \boldsymbol{A}_{11} & \boldsymbol{A}_{12} & \cdots & \boldsymbol{A}_{1n} \\ \hline \boldsymbol{A}_{21} & \boldsymbol{A}_{22} & \cdots & \boldsymbol{A}_{2n} \\ \hline \vdots & \vdots & \ddots & \vdots \\ \hline \boldsymbol{A}_{m1} & \boldsymbol{A}_{m2} & \cdots & \boldsymbol{A}_{mn} \end{array}\right) \left(\begin{array}{c} \boldsymbol{x}_1 \\ \hline \boldsymbol{x}_2 \\ \hline \vdots \\ \hline \boldsymbol{x}_n \end{array}\right) = \left(\begin{array}{c} \boldsymbol{y}_1 \\ \hline \boldsymbol{y}_2 \\ \hline \vdots \\ \hline \boldsymbol{y}_m \end{array}\right). \tag{A.37}$$

The $\boldsymbol{A}_{ij}$'s are called *sub-matrices* and the vectors $\boldsymbol{x}_i$ and $\boldsymbol{y}_i$ *sub-vectors*. For two appropriately partitioned block matrices, the product may be obtained by simply carrying out the multiplication as if the sub-matrices were themselves elements:

$$\left(\begin{array}{c|c} \boldsymbol{A}_{11} & \boldsymbol{A}_{12} \\ \hline \boldsymbol{A}_{21} & \boldsymbol{A}_{22} \end{array}\right) \left(\begin{array}{c|c} \boldsymbol{B}_{11} & \boldsymbol{B}_{12} \\ \hline \boldsymbol{B}_{21} & \boldsymbol{B}_{22} \end{array}\right) = \left(\begin{array}{c|c} \boldsymbol{A}_{11}\boldsymbol{B}_{11} + \boldsymbol{A}_{12}\boldsymbol{B}_{21} & \boldsymbol{A}_{11}\boldsymbol{B}_{12} + \boldsymbol{A}_{12}\boldsymbol{B}_{22} \\ \hline \boldsymbol{A}_{21}\boldsymbol{B}_{11} + \boldsymbol{A}_{22}\boldsymbol{B}_{21} & \boldsymbol{A}_{21}\boldsymbol{B}_{12} + \boldsymbol{A}_{22}\boldsymbol{B}_{22} \end{array}\right).$$

In particular, the subdivision into blocks of matrices is of benefit for the calculation of the inverse matrix. Looking at the system of equations

$$\boldsymbol{A}\boldsymbol{x}_1 + \boldsymbol{B}\boldsymbol{x}_2 = \boldsymbol{y}_1, \tag{A.38}$$

$$\boldsymbol{C}\boldsymbol{x}_1 + \boldsymbol{D}\boldsymbol{x}_2 = \boldsymbol{y}_2, \tag{A.39}$$

or combined into the form

$$\left(\begin{array}{c|c} \boldsymbol{A} & \boldsymbol{B} \\ \hline \boldsymbol{C} & \boldsymbol{D} \end{array}\right) \left(\begin{array}{c} \boldsymbol{x}_1 \\ \hline \boldsymbol{x}_2 \end{array}\right) = \left(\begin{array}{c} \boldsymbol{y}_1 \\ \hline \boldsymbol{y}_2 \end{array}\right), \tag{A.40}$$

the inverse of the matrix

$$\boldsymbol{M} = \left(\begin{array}{c|c} \boldsymbol{A} & \boldsymbol{B} \\ \hline \boldsymbol{C} & \boldsymbol{D} \end{array}\right) \tag{A.41}$$

can be expressed by more easily calculated inverse sub-matrices. When the matrix $\boldsymbol{A}$ is invertible, one gets from (A.38)

$$\boldsymbol{x}_1 = \boldsymbol{A}^{-1}\boldsymbol{y}_1 - \boldsymbol{A}^{-1}\boldsymbol{B}\boldsymbol{x}_2. \tag{A.42}$$

This, used in (A.39), yields

$$\boldsymbol{y}_2 = \boldsymbol{C}\boldsymbol{A}^{-1}\boldsymbol{y}_1 - \left(\boldsymbol{C}\boldsymbol{A}^{-1}\boldsymbol{B} - \boldsymbol{D}\right)\boldsymbol{x}_2 \tag{A.43}$$

and, solving for $\boldsymbol{x}_2$,

$$\boldsymbol{x}_2 = \left(\boldsymbol{C}\boldsymbol{A}^{-1}\boldsymbol{B} - \boldsymbol{D}\right)^{-1}\left(\boldsymbol{C}\boldsymbol{A}^{-1}\boldsymbol{y}_1 - \boldsymbol{y}_2\right). \tag{A.44}$$

Equation (A.44) used in (A.42) yields

$$x_1 = \left[A^{-1} - A^{-1}B\left(CA^{-1}B - D\right)^{-1}CA^{-1}\right]y_1 + A^{-1}B\left(CA^{-1}B - D\right)^{-1}y_2. \tag{A.45}$$

Thus, the solution of the system of equations (A.40) is obtained, namely

$$\begin{pmatrix} x_1 \\ \hline x_2 \end{pmatrix} = M^{-1}\begin{pmatrix} y_1 \\ \hline y_2 \end{pmatrix} \tag{A.46}$$

with

$$M^{-1} = \left(\begin{array}{c|c} A^{-1} - A^{-1}B(CA^{-1}B - D)^{-1}CA^{-1} & A^{-1}B(CA^{-1}B - D)^{-1} \\ \hline (CA^{-1}B - D)^{-1}CA^{-1} & -(CA^{-1}B - D)^{-1} \end{array}\right), \tag{A.47}$$

and the inverse matrix of M can be calculated by using the inverse matrices of the smaller sub-matrices A and $(CA^{-1}B - D)$. If the sub-matrix D is invertible, (A.39) can be solved for x_2, and then, in a similar way, also the inverse matrix of M can be calculated. One then obtains a different form of the inverted matrix, namely

$$M^{-1} = \left(\begin{array}{c|c} -(BD^{-1}C - A)^{-1} & (BD^{-1}C - A)^{-1}BD^{-1} \\ \hline D^{-1}C(BD^{-1}C - A)^{-1} & D^{-1} - D^{-1}C(BD^{-1}C - A)^{-1}BD^{-1} \end{array}\right). \tag{A.48}$$

So there are two different results available for the same matrix. It follows that the corresponding sub-matrices must be equal. From the comparison of the *northwestern* blocks, if D is replaced by $-D$, the known matrix-inversion lemma follows:

$$\boxed{\left(A + BD^{-1}C\right)^{-1} = A^{-1} - A^{-1}B\left(CA^{-1}B + D\right)^{-1}CA^{-1}.} \tag{A.49}$$

A special case occurs when we have a block-triangular matrix, for example, if $C = O$. Then we obtain

$$\left(\begin{array}{c|c} A & B \\ \hline O & D \end{array}\right)^{-1} = \left(\begin{array}{c|c} A^{-1} & -A^{-1}BD^{-1} \\ \hline O & D^{-1} \end{array}\right). \tag{A.50}$$

A.3 The Kronecker-Product

A.3.1 Definitions

Definition: The *Kronecker-product* of two matrices $\boldsymbol{A} \in \mathbb{R}^{n\times m}$ and $\boldsymbol{B} \in \mathbb{R}^{p\times q}$ is a matrix $\boldsymbol{C} \in \mathbb{R}^{np\times mq}$, denoted as

$$\boldsymbol{A} \otimes \boldsymbol{B} = \boldsymbol{C}.$$

Here, the sub-matrix $\boldsymbol{C}_{ij} \in \mathbb{R}^{p\times q}$, for $i = 1$ to n and $j = 1$ to m, is defined by

$$\boldsymbol{C}_{ij} \stackrel{\text{def}}{=} a_{ij}\boldsymbol{B},$$

so that, overall, the matrix $\boldsymbol{C}$ has the form

$$\boldsymbol{C} = \begin{pmatrix} a_{11}\boldsymbol{B} & a_{12}\boldsymbol{B} & \dots & a_{1m}\boldsymbol{B} \\ a_{21}\boldsymbol{B} & a_{22}\boldsymbol{B} & \dots & a_{2m}\boldsymbol{B} \\ \vdots & \vdots & \ddots & \vdots \\ a_{n1}\boldsymbol{B} & a_{n2}\boldsymbol{B} & \dots & a_{nm}\boldsymbol{B} \end{pmatrix}.$$

The matrix elements of the product matrix $\boldsymbol{C}$ can be directly calculated using the following formula

$$c_{i,j} = a_{\lfloor \frac{i-1}{p} \rfloor +1, \lfloor \frac{j-1}{q} \rfloor +1} \cdot b_{i-\lfloor \frac{i-1}{p} \rfloor p, j-\lfloor \frac{j-1}{q} \rfloor q},$$

where $\lfloor x \rfloor$ is the integer part of x.

Definition: If a matrix $\boldsymbol{A}$ is composed of the m columns $\boldsymbol{a}_i \in \mathbb{C}^n$,

$$\boldsymbol{A} = \begin{pmatrix} \boldsymbol{a}_1 & \boldsymbol{a}_2 & \dots & \boldsymbol{a}_m \end{pmatrix},$$

the $\boldsymbol{vec}$-operator is defined as follows:

$$\boldsymbol{vec}(\boldsymbol{A}) \stackrel{\text{def}}{=} \begin{pmatrix} \boldsymbol{a}_1 \\ \boldsymbol{a}_2 \\ \vdots \\ \boldsymbol{a}_m \end{pmatrix} \in \mathbb{C}^{nm}.$$

A.3.2 Some Theorems

The following is very interesting:

Lemma: $\boldsymbol{vec}(\boldsymbol{AXB}) = \left(\boldsymbol{B}^{\mathsf{T}} \otimes \boldsymbol{A}\right)\boldsymbol{vec}(\boldsymbol{X}).$ (A.51)

Proof Let $\boldsymbol{B} \in \mathbb{C}^{n \times m}$, then one has

$$\begin{aligned}
\boldsymbol{AXB} &= \begin{pmatrix} \boldsymbol{Ax}_1 & \boldsymbol{Ax}_2 & \dots & \boldsymbol{Ax}_n \end{pmatrix} \boldsymbol{B} \\
&= \begin{pmatrix} \boldsymbol{Ax}_1 & \boldsymbol{Ax}_2 & \dots & \boldsymbol{Ax}_n \end{pmatrix} \begin{pmatrix} \boldsymbol{b}_1 & \boldsymbol{b}_2 & \dots & \boldsymbol{b}_m \end{pmatrix} \\
&= \begin{pmatrix} (b_{11}\boldsymbol{Ax}_1 + b_{21}\boldsymbol{Ax}_2 + \dots + b_{n1}\boldsymbol{Ax}_n) & \dots \\ (b_{1m}\boldsymbol{Ax}_1 + b_{2m}\boldsymbol{Ax}_2 + \dots + b_{nm}\boldsymbol{Ax}_n) & \end{pmatrix}.
\end{aligned}$$

Applying the $\boldsymbol{vec}$-operator to the last equation, we obtain

$$\begin{aligned}
\boldsymbol{vec}(\boldsymbol{AXB}) &= \begin{pmatrix} (b_{11}\boldsymbol{Ax}_1 + b_{21}\boldsymbol{Ax}_2 + \dots + b_{n1}\boldsymbol{Ax}_n) \\ \vdots \\ (b_{1m}\boldsymbol{Ax}_1 + b_{2m}\boldsymbol{Ax}_2 + \dots + b_{nm}\boldsymbol{Ax}_n) \end{pmatrix} \\
&= \begin{pmatrix} b_{11}\boldsymbol{A} & \dots & b_{n1}\boldsymbol{A} \\ \vdots & \ddots & \vdots \\ b_{1m}\boldsymbol{A} & \dots & b_{nm}\boldsymbol{A} \end{pmatrix} \boldsymbol{vec}(\boldsymbol{X}) \\
&= \left(\boldsymbol{B}^{\mathsf{T}} \otimes \boldsymbol{A}\right)\boldsymbol{vec}(\boldsymbol{X}).
\end{aligned}$$

□

As corollaries of this lemma, we get the following results:

$$\boxed{\boldsymbol{vec}(\boldsymbol{AX}) = (\boldsymbol{I} \otimes \boldsymbol{A})\boldsymbol{vec}(\boldsymbol{X}).} \tag{A.52}$$

Proof Set $\boldsymbol{B} = \boldsymbol{I}$ in the lemma. □

$$\boxed{\boldsymbol{vec}(\boldsymbol{XB}) = \left(\boldsymbol{B}^{\mathsf{T}} \otimes \boldsymbol{I}\right)\boldsymbol{vec}(\boldsymbol{X}).} \tag{A.53}$$

Proof Set $\boldsymbol{A} = \boldsymbol{I}$ in the lemma. □

$$\boxed{\boldsymbol{vec}\left(\boldsymbol{ba}^{\mathsf{T}}\right) = (\boldsymbol{a} \otimes \boldsymbol{b}).} \tag{A.54}$$

Proof Simply write $\boldsymbol{vec}(\boldsymbol{ba}^{\mathsf{T}}) = \boldsymbol{vec}(\boldsymbol{b}1\boldsymbol{a}^{\mathsf{T}}) = (\boldsymbol{a} \otimes \boldsymbol{b})\boldsymbol{vec}(1) = \boldsymbol{a} \otimes \boldsymbol{b}$. □

A.3.3 The Permutation Matrix $U_{p\times q}$

Definition: The *permutation matrix*

$$U_{p\times q} \overset{\text{def}}{=} \sum_{i}^{p}\sum_{k}^{q} E_{ik}^{p\times q} \otimes E_{ki}^{q\times p} \in \mathbb{R}^{pq\times qp} \tag{A.55}$$

has just one 1 in each column and each row. In the formation of matrix

$$E_{ik}^{p\times q} \overset{\text{def}}{=} e_i e_k^{\mathsf{T}}, \tag{A.56}$$

e_i is the ith column of I_p and e_k is the kth column of I_q. However, only the matrix element $E_{ik} = 1$; the other matrix elements are zeros.

For example, the permutation matrix $U_{4\times 4} \in \mathbb{R}^{16\times 16}$—often used in this book—has the form

$$U_{4\times 4} = \left(\begin{array}{cccc|cccc|cccc|cccc}
1&0&0&0&0&0&0&0&0&0&0&0&0&0&0&0\\
0&0&0&0&1&0&0&0&0&0&0&0&0&0&0&0\\
0&0&0&0&0&0&0&0&1&0&0&0&0&0&0&0\\
0&0&0&0&0&0&0&0&0&0&0&0&1&0&0&0\\
\hline
0&1&0&0&0&0&0&0&0&0&0&0&0&0&0&0\\
0&0&0&0&0&1&0&0&0&0&0&0&0&0&0&0\\
0&0&0&0&0&0&0&0&0&1&0&0&0&0&0&0\\
0&0&0&0&0&0&0&0&0&0&0&0&0&1&0&0\\
\hline
0&0&1&0&0&0&0&0&0&0&0&0&0&0&0&0\\
0&0&0&0&0&0&1&0&0&0&0&0&0&0&0&0\\
0&0&0&0&0&0&0&0&0&0&1&0&0&0&0&0\\
0&0&0&0&0&0&0&0&0&0&0&0&0&0&1&0\\
\hline
0&0&0&1&0&0&0&0&0&0&0&0&0&0&0&0\\
0&0&0&0&0&0&0&1&0&0&0&0&0&0&0&0\\
0&0&0&0&0&0&0&0&0&0&0&1&0&0&0&0\\
0&0&0&0&0&0&0&0&0&0&0&0&0&0&0&1
\end{array}\right). \tag{A.57}$$

Permutation matrices have the following characteristics [4]:

$$U_{p\times q}^{\mathsf{T}} = U_{q\times p}, \tag{A.58}$$

$$U_{p\times q}^{-1} = U_{q\times p}, \tag{A.59}$$

$$U_{p\times 1} = U_{1\times p} = I_p, \tag{A.60}$$

$$U_{n\times n} = U_{n\times n}^{\mathsf{T}} = U_{n\times n}^{-1}. \tag{A.61}$$

Permutation matrices are mainly used to change the order of the factors in a Kronecker-product because

$$\boxed{\boldsymbol{U}_{s\times p}(\boldsymbol{B}\otimes\boldsymbol{A})\boldsymbol{U}_{q\times t}=\boldsymbol{A}\otimes\boldsymbol{B} \quad \text{if } \boldsymbol{A}\in\mathbb{R}^{p\times q} \text{ and } \boldsymbol{B}\in\mathbb{R}^{s\times t}.} \tag{A.62}$$

A.3.4 More Properties of the Kronecker-Product

The following important properties are listed also without proof (see [4]):

$$(\boldsymbol{A}\otimes\boldsymbol{B})\otimes\boldsymbol{C}=\boldsymbol{A}\otimes(\boldsymbol{B}\otimes\boldsymbol{C}), \tag{A.63}$$

$$(\boldsymbol{A}\otimes\boldsymbol{B})^{\mathsf{T}}=\boldsymbol{A}^{\mathsf{T}}\otimes\boldsymbol{B}^{\mathsf{T}}, \tag{A.64}$$

$$(\boldsymbol{A}\otimes\boldsymbol{B})(\boldsymbol{C}\otimes\boldsymbol{D})=\boldsymbol{A}\boldsymbol{C}\otimes\boldsymbol{B}\boldsymbol{D}. \tag{A.65}$$

A.4 Derivatives of Vectors/Matrices with Respect to Vectors/Matrices

A.4.1 Definitions

Definition: The derivative of a matrix $\boldsymbol{A}\in\mathbb{R}^{n\times m}$ with respect to a matrix $\boldsymbol{M}\in\mathbb{R}^{r\times s}$ is defined as follows:

$$\frac{\partial \boldsymbol{A}}{\partial \boldsymbol{M}} \stackrel{\text{def}}{=} \begin{pmatrix} \frac{\partial A}{\partial M_{11}} & \frac{\partial A}{\partial M_{12}} & \cdots & \frac{\partial A}{\partial M_{1s}} \\ \frac{\partial A}{\partial M_{21}} & \frac{\partial A}{\partial M_{22}} & \cdots & \frac{\partial A}{\partial M_{2s}} \\ \vdots & \vdots & \ddots & \vdots \\ \frac{\partial A}{\partial M_{r1}} & \frac{\partial A}{\partial M_{r2}} & \cdots & \frac{\partial A}{\partial M_{rs}} \end{pmatrix} \in \mathbb{R}^{nr\times ms}. \tag{A.66}$$

With the $r\times s$-operator

$$\frac{\partial}{\partial \boldsymbol{M}} \stackrel{\text{def}}{=} \begin{pmatrix} \frac{\partial}{\partial M_{11}} & \frac{\partial}{\partial M_{12}} & \cdots & \frac{\partial}{\partial M_{1s}} \\ \frac{\partial}{\partial M_{21}} & \frac{\partial}{\partial M_{22}} & \cdots & \frac{\partial}{\partial M_{2s}} \\ \vdots & \vdots & \ddots & \vdots \\ \frac{\partial}{\partial M_{r1}} & \frac{\partial}{\partial M_{r2}} & \cdots & \frac{\partial}{\partial M_{rs}} \end{pmatrix} \tag{A.67}$$

the definition of the derivative (A.66) is also written as

$$\frac{\partial \boldsymbol{A}}{\partial \boldsymbol{M}} \stackrel{\text{def}}{=} \frac{\partial}{\partial \boldsymbol{M}} \otimes \boldsymbol{A}. \tag{A.68}$$

Thus one can show that

$$\left(\frac{\partial \boldsymbol{A}}{\partial \boldsymbol{M}}\right)^{\mathsf{T}} = \left(\frac{\partial}{\partial \boldsymbol{M}} \otimes \boldsymbol{A}\right)^{\mathsf{T}} = \left(\left(\frac{\partial}{\partial \boldsymbol{M}}\right)^{\mathsf{T}} \otimes \boldsymbol{A}^{\mathsf{T}}\right) = \frac{\partial \boldsymbol{A}^{\mathsf{T}}}{\partial \boldsymbol{M}^{\mathsf{T}}}. \tag{A.69}$$

For the derivatives of vectors with respect to vectors one has:

$$\frac{\partial \boldsymbol{f}^{\mathsf{T}}}{\partial \boldsymbol{p}} \stackrel{\text{def}}{=} \frac{\partial}{\partial \boldsymbol{p}} \otimes \boldsymbol{f}^{\mathsf{T}} = \begin{pmatrix} \frac{\partial \boldsymbol{f}^{\mathsf{T}}}{\partial p_1} \\ \frac{\partial \boldsymbol{f}^{\mathsf{T}}}{\partial p_2} \\ \vdots \\ \frac{\partial \boldsymbol{f}^{\mathsf{T}}}{\partial p_r} \end{pmatrix} = \begin{pmatrix} \frac{\partial f_1}{\partial p_1} & \frac{\partial f_2}{\partial p_1} & \cdots & \frac{\partial f_n}{\partial p_1} \\ \frac{\partial f_1}{\partial p_2} & \frac{\partial f_2}{\partial p_2} & \cdots & \frac{\partial f_n}{\partial M_2} \\ \vdots & \vdots & \ddots & \vdots \\ \frac{\partial f_1}{\partial p_r} & \frac{\partial f_2}{\partial p_r} & \cdots & \frac{\partial f_n}{\partial p_s} \end{pmatrix} \in \mathbb{R}^{r \times n} \tag{A.70}$$

and

$$\frac{\partial \boldsymbol{f}}{\partial \boldsymbol{p}^{\mathsf{T}}} \stackrel{\text{def}}{=} \frac{\partial}{\partial \boldsymbol{p}^{\mathsf{T}}} \otimes \boldsymbol{f} = \left(\frac{\partial}{\partial \boldsymbol{p}} \otimes \boldsymbol{f}^{\mathsf{T}}\right)^{\mathsf{T}} = \left(\frac{\partial \boldsymbol{f}^{\mathsf{T}}}{\partial \boldsymbol{p}}\right)^{\mathsf{T}} \in \mathbb{R}^{n \times r}. \tag{A.71}$$

A.4.2 Product Rule

Let $\boldsymbol{A} = \boldsymbol{A}(\alpha)$ and $\boldsymbol{B} = \boldsymbol{B}(\alpha)$. Then we obviously have

$$\frac{\partial(\boldsymbol{AB})}{\partial \alpha} = \frac{\partial \boldsymbol{A}}{\partial \alpha}\boldsymbol{B} + \boldsymbol{A}\frac{\partial \boldsymbol{B}}{\partial \alpha}. \tag{A.72}$$

In addition, (A.66) can be rewritten as:

$$\frac{\partial \boldsymbol{A}}{\partial \boldsymbol{M}} = \sum_{i,k} \boldsymbol{E}_{ik}^{s \times t} \otimes \frac{\partial \boldsymbol{A}}{\partial m_{ik}}, \quad \boldsymbol{M} \in \mathbb{R}^{s \times t}. \tag{A.73}$$

Using (A.72) and (A.73), this product rule can be derived:

$$\begin{aligned} \underline{\underline{\frac{\partial(\boldsymbol{AB})}{\partial \boldsymbol{M}}}} &= \sum_{i,k} \boldsymbol{E}_{ik}^{s \times t} \otimes \frac{\partial(\boldsymbol{AB})}{\partial m_{ik}} = \sum_{i,k} \boldsymbol{E}_{ik}^{s \times t} \otimes \left(\frac{\partial \boldsymbol{A}}{\partial m_{ik}}\boldsymbol{B} + \boldsymbol{A}\frac{\partial \boldsymbol{B}}{\partial m_{ik}}\right) \\ &= \left(\frac{\partial}{\partial \boldsymbol{M}} \otimes \boldsymbol{A}\right)(\boldsymbol{I}_t \otimes \boldsymbol{B}) + (\boldsymbol{I}_s \otimes \boldsymbol{A})\left(\frac{\partial}{\partial \boldsymbol{M}} \otimes \boldsymbol{B}\right) \\ &= \underline{\underline{\frac{\partial \boldsymbol{A}}{\partial \boldsymbol{M}}(\boldsymbol{I}_t \otimes \boldsymbol{B}) + (\boldsymbol{I}_s \otimes \boldsymbol{A})\frac{\partial \boldsymbol{B}}{\partial \boldsymbol{M}}}}. \end{aligned} \tag{A.74}$$

A.4.3 Chain Rule

When a matrix $\boldsymbol{A} \in \mathbb{R}^{n\times m}$ is a function of a matrix $\boldsymbol{B} \in \mathbb{R}^{k\times \ell}$ which is again a function of a matrix $\boldsymbol{M} \in \mathbb{R}^{r\times s}$, then the chain rule is valid [4]:

$$\begin{aligned}\frac{\partial}{\partial \boldsymbol{M}}\boldsymbol{A}\big(\boldsymbol{B}(\boldsymbol{M})\big) &= \left(\boldsymbol{I}_r \otimes \frac{\partial \boldsymbol{A}}{\partial (\boldsymbol{vec}(\boldsymbol{B}^\mathsf{T}))^\mathsf{T}}\right)\left(\frac{\partial \boldsymbol{vec}(\boldsymbol{B}^\mathsf{T})}{\partial \boldsymbol{M}} \otimes \boldsymbol{I}_m\right)\\ &= \left(\frac{\partial (\boldsymbol{vec}(\boldsymbol{B}))^\mathsf{T}}{\partial \boldsymbol{M}} \otimes \boldsymbol{I}_n\right)\left(\boldsymbol{I}_s \otimes \frac{\partial \boldsymbol{A}}{\partial \boldsymbol{vec}(\boldsymbol{B})}\right). \end{aligned} \tag{A.75}$$

A special case of this is

$$\frac{\mathrm{d}\boldsymbol{A}(\boldsymbol{x}(t))}{\mathrm{d}t} = \frac{\partial \boldsymbol{A}}{\partial \boldsymbol{x}^\mathsf{T}}\left(\frac{\mathrm{d}\boldsymbol{x}}{\mathrm{d}t} \otimes \boldsymbol{I}_m\right) = \left(\frac{\mathrm{d}\boldsymbol{x}^\mathsf{T}}{\mathrm{d}t} \otimes \boldsymbol{I}_n\right)\frac{\partial \boldsymbol{A}}{\partial \boldsymbol{x}} \in \mathbb{R}^{n\times m}. \tag{A.76}$$

A.5 Differentiation with Respect to Time

A.5.1 Differentiation of a Function with Respect to Time

Suppose a function a that depends on the three space variables x_1, x_2 and x_3 is given. The local space variables themselves are, in turn, dependent on the time parameter t. So it is

$$a = a\big(\boldsymbol{x}(t)\big), \tag{A.77}$$

if the three space variables are summarized in the vector $\boldsymbol{x}$.

We want to find the velocity

$$\dot{a} = \frac{\mathrm{d}a}{\mathrm{d}t}. \tag{A.78}$$

To determine this velocity, we first define the total difference

$$\Delta a \stackrel{\text{def}}{=} \frac{\partial a}{\partial x_1}\Delta x_1 + \frac{\partial a}{\partial x_2}\Delta x_2 + \frac{\partial a}{\partial x_3}\Delta x_3. \tag{A.79}$$

After division by Δt, in the limit $\Delta t \to 0$, one has

$$\dot{a} = \frac{\mathrm{d}a}{\mathrm{d}t} = \lim_{\Delta t\to 0}\frac{\Delta a}{\Delta t} = \frac{\partial a}{\partial x_1}\dot{x}_1 + \frac{\partial a}{\partial x_2}\dot{x}_2 + \frac{\partial a}{\partial x_3}\dot{x}_3. \tag{A.80}$$

The right-hand side of this equation can be presented by the scalar product of the column vectors $\dot{\boldsymbol{x}}$ and $\frac{\partial a}{\partial \boldsymbol{x}}$ in two ways, namely

$$\boxed{\dot{a} = \dot{\boldsymbol{x}}^\mathsf{T}\frac{\partial a}{\partial \boldsymbol{x}} = \frac{\partial a}{\partial \boldsymbol{x}^\mathsf{T}}\dot{\boldsymbol{x}}.} \tag{A.81}$$

A.5.2 Differentiation of a Vector with Respect to Time

If two functions a_1 and a_2 are given, they have the same dependence on time t as $a(t)$ in (A.77), and are summarized in the column vector

$$\boldsymbol{a} \overset{\text{def}}{=} \begin{pmatrix} a_1(\boldsymbol{x}(t)) \\ a_2(\boldsymbol{x}(t)) \end{pmatrix}, \tag{A.82}$$

we obtain initially for the derivative with respect to time using (A.81)

$$\dot{\boldsymbol{a}} = \begin{pmatrix} \dot{a}_1 \\ \dot{a}_2 \end{pmatrix} = \begin{pmatrix} \dot{\boldsymbol{x}}^{\mathsf{T}} \frac{\partial a_1}{\partial \boldsymbol{x}} \\ \dot{\boldsymbol{x}}^{\mathsf{T}} \frac{\partial a_2}{\partial \boldsymbol{x}} \end{pmatrix} = \begin{pmatrix} \frac{\partial a_1}{\partial \boldsymbol{x}^{\mathsf{T}}} \dot{\boldsymbol{x}} \\ \frac{\partial a_2}{\partial \boldsymbol{x}^{\mathsf{T}}} \dot{\boldsymbol{x}} \end{pmatrix}. \tag{A.83}$$

The next-to-last vector in (A.83) can be decomposed as follows:

$$\dot{\boldsymbol{a}} = \begin{pmatrix} \dot{\boldsymbol{x}}^{\mathsf{T}} \frac{\partial a_1}{\partial \boldsymbol{x}} \\ \dot{\boldsymbol{x}}^{\mathsf{T}} \frac{\partial a_2}{\partial \boldsymbol{x}} \end{pmatrix} = \begin{pmatrix} \dot{\boldsymbol{x}}^{\mathsf{T}} & \boldsymbol{o}_3^{\mathsf{T}} \\ \boldsymbol{o}_3^{\mathsf{T}} & \dot{\boldsymbol{x}}^{\mathsf{T}} \end{pmatrix} \begin{pmatrix} \frac{\partial a_1}{\partial \boldsymbol{x}} \\ \frac{\partial a_2}{\partial \boldsymbol{x}} \end{pmatrix} = (\boldsymbol{I}_2 \otimes \dot{\boldsymbol{x}}^{\mathsf{T}}) \left(\boldsymbol{a} \otimes \frac{\partial}{\partial \boldsymbol{x}} \right). \tag{A.84}$$

Computing the last Kronecker-product, we would formally get $a_i \frac{\partial}{\partial \boldsymbol{x}}$ which, of course, should be understood as $\frac{\partial a_i}{\partial \boldsymbol{x}}$. With the help of the permutation matrix $\boldsymbol{U}_{\alpha\times\beta}$ and the exchange rule (A.62)

$$(\boldsymbol{A} \otimes \boldsymbol{B}) = \boldsymbol{U}_{s\times p}(\boldsymbol{B} \otimes \boldsymbol{A})\boldsymbol{U}_{q\times t} \quad \text{if } \boldsymbol{A} \in \mathbb{R}^{p\times q} \text{ and } \boldsymbol{B} \in \mathbb{R}^{s\times t},$$

in the appendix, the last product in (A.84) can be written

$$\underline{\underline{\dot{\boldsymbol{a}}}} = \big[\underbrace{\boldsymbol{U}_{1\times 2}}_{\boldsymbol{I}_2}(\dot{\boldsymbol{x}}^{\mathsf{T}} \otimes \boldsymbol{I}_2)\,\underbrace{\boldsymbol{U}_{2\times r}\big]\Big[\boldsymbol{U}_{r\times 2}}_{\boldsymbol{I}_{2r}}\underbrace{\left(\frac{\partial}{\partial \boldsymbol{x}} \otimes \boldsymbol{a}\right)}_{\frac{\partial \boldsymbol{a}}{\partial \boldsymbol{x}}}\underbrace{\boldsymbol{U}_{1\times 1}}_{1}\Big] = \underline{\underline{(\dot{\boldsymbol{x}}^{\mathsf{T}} \otimes \boldsymbol{I}_2)\frac{\partial \boldsymbol{a}}{\partial \boldsymbol{x}}}}. \tag{A.85}$$

For the second form in (A.83), one obtains

$$\underline{\underline{\dot{\boldsymbol{a}}}} = \begin{pmatrix} \frac{\partial a_1}{\partial \boldsymbol{x}^{\mathsf{T}}} \dot{\boldsymbol{x}} \\ \frac{\partial a_2}{\partial \boldsymbol{x}^{\mathsf{T}}} \dot{\boldsymbol{x}} \end{pmatrix} = \left(\boldsymbol{a} \otimes \frac{\partial}{\partial \boldsymbol{x}} \right) \dot{\boldsymbol{x}} = \Big[\underbrace{\boldsymbol{U}_{1\times 2}}_{\boldsymbol{I}_2}\left(\frac{\partial}{\partial \boldsymbol{x}^{\mathsf{T}}} \otimes \boldsymbol{a}\right)\underbrace{\boldsymbol{U}_{1\times r}}_{\boldsymbol{I}_r}\Big]\dot{\boldsymbol{x}} = \underline{\underline{\frac{\partial \boldsymbol{a}}{\partial \boldsymbol{x}^{\mathsf{T}}}}}\dot{\boldsymbol{x}}, \tag{A.86}$$

so that combined, these two possible representations are written as

$$\boxed{\dot{\boldsymbol{a}} = (\dot{\boldsymbol{x}}^{\mathsf{T}} \otimes \boldsymbol{I}_2)\frac{\partial \boldsymbol{a}}{\partial \boldsymbol{x}} = \frac{\partial \boldsymbol{a}}{\partial \boldsymbol{x}^{\mathsf{T}}}\dot{\boldsymbol{x}}.} \tag{A.87}$$

A.5.3 Differentiation of a 2 × 3-Matrix with Respect to Time

For the derivative of a 2 × 3-matrix with respect to time, with the above results, one obtains

$$\underline{\underline{\dot{A}}} = \begin{pmatrix} \dot{a}_{11} & \dot{a}_{12} & \dot{a}_{13} \\ \dot{a}_{21} & \dot{a}_{22} & \dot{a}_{23} \end{pmatrix} = \begin{pmatrix} \dot{x}^{\mathsf{T}} \frac{\partial a_{11}}{\partial x} & \dot{x}^{\mathsf{T}} \frac{\partial a_{12}}{\partial x} & \dot{x}^{\mathsf{T}} \frac{\partial a_{13}}{\partial x} \\ \dot{x}^{\mathsf{T}} \frac{\partial a_{21}}{\partial x} & \dot{x}^{\mathsf{T}} \frac{\partial a_{22}}{\partial x} & \dot{x}^{\mathsf{T}} \frac{\partial a_{23}}{\partial x} \end{pmatrix}$$

$$= \begin{pmatrix} \dot{x}^{\mathsf{T}} & o_3^{\mathsf{T}} \\ o_3^{\mathsf{T}} & \dot{x}^{\mathsf{T}} \end{pmatrix} \left(A \otimes \frac{\partial}{\partial x} \right) = (I_2 \otimes \dot{x}^{\mathsf{T}}) \left(A \otimes \frac{\partial}{\partial x} \right)$$

$$= \Big[\underbrace{U_{1\times 2}}_{I_2} (\dot{x}^{\mathsf{T}} \otimes I_2) \underbrace{U_{2\times r} \Big] \Big[U_{r\times 2}}_{I_{2r}} \underbrace{\left(\frac{\partial}{\partial x} \otimes A \right)}_{\frac{\partial A}{\partial x}} \underbrace{U_{3\times 1}}_{I_3} \Big] = \underline{\underline{(\dot{x}^{\mathsf{T}} \otimes I_2) \frac{\partial A}{\partial x}}}, \quad \text{(A.88)}$$

or, with the second representation in (A.81) for the $\dot{a}_{ij}$,

$$\underline{\underline{\dot{A}}} = \begin{pmatrix} \frac{\partial a_{11}}{\partial x^{\mathsf{T}}} \dot{x} & \frac{\partial a_{12}}{\partial x^{\mathsf{T}}} \dot{x} & \frac{\partial a_{13}}{\partial x^{\mathsf{T}}} \dot{x} \\ \frac{\partial a_{21}}{\partial x^{\mathsf{T}}} \dot{x} & \frac{\partial a_{22}}{\partial x^{\mathsf{T}}} \dot{x} & \frac{\partial a_{23}}{\partial x^{\mathsf{T}}} \dot{x} \end{pmatrix} = \left(A \otimes \frac{\partial}{\partial x^{\mathsf{T}}} \right) \begin{pmatrix} \dot{x} & o & o \\ o & \dot{x} & o \\ o & o & \dot{x} \end{pmatrix}$$

$$= \Big[\underbrace{U_{1\times 2}}_{I_2} \underbrace{\left(\frac{\partial}{\partial x^{\mathsf{T}}} \otimes A \right)}_{\frac{\partial A}{\partial x^{\mathsf{T}}}} \underbrace{U_{3\times r} \Big] \Big[U_{r\times 3}}_{I_{3r}} (\dot{x} \otimes I_3) \underbrace{U_{3\times 1}}_{I_3} \Big] = \underline{\underline{\frac{\partial A}{\partial x^{\mathsf{T}}} (\dot{x} \otimes I_3)}}. \quad \text{(A.89)}$$

Here the Kronecker-product is also present.

A.5.4 Differentiation of an n × m-Matrix with Respect to Time

In general, one gets for a matrix $A \in \mathbb{R}^{n\times m}$ and a vector $x \in \mathbb{R}^r$

$$\boxed{\dot{A} = (\dot{x}^{\mathsf{T}} \otimes I_n) \frac{\partial A}{\partial x} \stackrel{\text{and}}{=} \frac{\partial A}{\partial x^{\mathsf{T}}} (\dot{x} \otimes I_m) \in \mathbb{R}^{n\times m}.} \quad \text{(A.90)}$$

The derivation is given below without any comment.

$$\dot{A} = \begin{pmatrix} \dot{x}^{\mathsf{T}} & \dots & O \\ \vdots & \ddots & \vdots \\ O & \dots & \dot{x}^{\mathsf{T}} \end{pmatrix} \left(A \otimes \frac{\partial}{\partial x} \right) = (I_n \otimes \dot{x}^{\mathsf{T}}) \left(A \otimes \frac{\partial}{\partial x} \right)$$

$$= \Big[\underbrace{U_{1\times n}}_{I_n} (\dot{x}^{\mathsf{T}} \otimes I_n) \underbrace{U_{n\times r} \Big] \Big[U_{r\times n}}_{I_{nr}} \underbrace{\left(\frac{\partial}{\partial x} \otimes A \right)}_{\frac{\partial A}{\partial x}} \underbrace{U_{m\times 1}}_{I_m} \Big] = \underline{\underline{(\dot{x}^{\mathsf{T}} \otimes I_n) \frac{\partial A}{\partial x}}}.$$

$$
\begin{aligned}
\dot{\boldsymbol{A}} &= \left(\boldsymbol{A} \otimes \frac{\partial}{\partial \boldsymbol{x}^{\mathsf{T}}}\right) \begin{pmatrix} \dot{\boldsymbol{x}} & \dots & O \\ \vdots & \ddots & \vdots \\ O & \dots & \dot{\boldsymbol{x}} \end{pmatrix} = \left(\boldsymbol{A} \otimes \frac{\partial}{\partial \boldsymbol{x}^{\mathsf{T}}}\right)(\boldsymbol{I}_n \otimes \dot{\boldsymbol{x}}) \\
&= \Bigg[\underbrace{\underbrace{\boldsymbol{U}_{1\times n}}_{\boldsymbol{I}_n} \left(\frac{\partial}{\partial \boldsymbol{x}^{\mathsf{T}}} \otimes \boldsymbol{A}\right)}_{\frac{\partial \boldsymbol{A}}{\partial \boldsymbol{x}^{\mathsf{T}}}} \underbrace{\boldsymbol{U}_{m\times r}\Bigg]\Big[\boldsymbol{U}_{r\times m}}_{\boldsymbol{I}_{mr}}(\dot{\boldsymbol{x}} \otimes \boldsymbol{I}_m)\boldsymbol{U}_{m\times 1}\Big] = \underline{\underline{\frac{\partial \boldsymbol{A}}{\partial \boldsymbol{x}^{\mathsf{T}}}(\dot{\boldsymbol{x}} \otimes \boldsymbol{I}_n)}}.
\end{aligned}
$$

A.6 Supplements to Differentiation with Respect to a Matrix

For the derivative of a 4×4 matrix with respect to itself, one has

$$
\frac{\partial \boldsymbol{M}}{\partial \boldsymbol{M}} = \bar{\boldsymbol{U}}_{4\times 4}, \tag{A.91}
$$

where $\bar{\boldsymbol{U}}_{4\times 4}$ is defined by

$$
\bar{\boldsymbol{U}}_{4\times 4} \overset{\text{def}}{=} \sum_{i}^{4}\sum_{k}^{4} \boldsymbol{E}_{ik} \otimes \boldsymbol{E}_{ik}
$$

$$
= \left(\begin{array}{cccc|cccc|cccc|cccc}
1&0&0&0&0&1&0&0&0&0&1&0&0&0&0&1\\
0&0&0&0&0&0&0&0&0&0&0&0&0&0&0&0\\
0&0&0&0&0&0&0&0&0&0&0&0&0&0&0&0\\
0&0&0&0&0&0&0&0&0&0&0&0&0&0&0&0\\
\hline
0&0&0&0&0&0&0&0&0&0&0&0&0&0&0&0\\
1&0&0&0&0&1&0&0&0&0&1&0&0&0&0&1\\
0&0&0&0&0&0&0&0&0&0&0&0&0&0&0&0\\
0&0&0&0&0&0&0&0&0&0&0&0&0&0&0&0\\
\hline
0&0&0&0&0&0&0&0&0&0&0&0&0&0&0&0\\
0&0&0&0&0&0&0&0&0&0&0&0&0&0&0&0\\
1&0&0&0&0&1&0&0&0&0&1&0&0&0&0&1\\
0&0&0&0&0&0&0&0&0&0&0&0&0&0&0&0\\
\hline
0&0&0&0&0&0&0&0&0&0&0&0&0&0&0&0\\
0&0&0&0&0&0&0&0&0&0&0&0&0&0&0&0\\
0&0&0&0&0&0&0&0&0&0&0&0&0&0&0&0\\
1&0&0&0&0&1&0&0&0&0&1&0&0&0&0&1
\end{array}\right). \tag{A.92}
$$

This fact can be easily made clear just by the definition of the differentiation of a matrix with respect to a matrix. The result is more complex when the matrix $\boldsymbol{M} = \boldsymbol{M}^{\mathsf{T}}$

is symmetric because then

$$
\begin{aligned}
\frac{\partial \boldsymbol{M}}{\partial \boldsymbol{M}} &= \bar{\boldsymbol{U}}_{4\times 4} + \boldsymbol{U}_{4\times 4} - \sum_{i}^{4} \boldsymbol{E}_{ii} \otimes \boldsymbol{E}_{ii} \\
&= \left(\begin{array}{cccc|cccc|cccc|cccc}
1 & 0 & 0 & 0 & 0 & 1 & 0 & 0 & 0 & 0 & 1 & 0 & 0 & 0 & 0 & 1 \\
0 & 0 & 0 & 0 & 1 & 0 & 0 & 0 & 0 & 0 & 0 & 0 & 0 & 0 & 0 & 0 \\
0 & 0 & 0 & 0 & 0 & 0 & 0 & 0 & 1 & 0 & 0 & 0 & 0 & 0 & 0 & 0 \\
0 & 0 & 0 & 0 & 0 & 0 & 0 & 0 & 0 & 0 & 0 & 0 & 1 & 0 & 0 & 0 \\
\hline
0 & 1 & 0 & 0 & 0 & 0 & 0 & 0 & 0 & 0 & 0 & 0 & 0 & 0 & 0 & 0 \\
1 & 0 & 0 & 0 & 0 & 1 & 0 & 0 & 0 & 0 & 1 & 0 & 0 & 0 & 0 & 1 \\
0 & 0 & 0 & 0 & 0 & 0 & 0 & 0 & 0 & 1 & 0 & 0 & 0 & 0 & 0 & 0 \\
0 & 0 & 0 & 0 & 0 & 0 & 0 & 0 & 0 & 0 & 0 & 0 & 0 & 1 & 0 & 0 \\
\hline
0 & 0 & 1 & 0 & 0 & 0 & 0 & 0 & 0 & 0 & 0 & 0 & 0 & 0 & 0 & 0 \\
0 & 0 & 0 & 0 & 0 & 0 & 1 & 0 & 0 & 0 & 0 & 0 & 0 & 0 & 0 & 0 \\
1 & 0 & 0 & 0 & 0 & 1 & 0 & 0 & 0 & 0 & 1 & 0 & 0 & 0 & 0 & 1 \\
0 & 0 & 0 & 0 & 0 & 0 & 0 & 0 & 0 & 0 & 0 & 0 & 0 & 0 & 1 & 0 \\
\hline
0 & 0 & 0 & 1 & 0 & 0 & 0 & 0 & 0 & 0 & 0 & 0 & 0 & 0 & 0 & 0 \\
0 & 0 & 0 & 0 & 0 & 0 & 0 & 1 & 0 & 0 & 0 & 0 & 0 & 0 & 0 & 0 \\
0 & 0 & 0 & 0 & 0 & 0 & 0 & 0 & 0 & 0 & 0 & 1 & 0 & 0 & 0 & 0 \\
1 & 0 & 0 & 0 & 0 & 1 & 0 & 0 & 0 & 0 & 1 & 0 & 0 & 0 & 0 & 1
\end{array}\right).
\end{aligned}
\tag{A.93}
$$

Appendix B
Some Differential Geometry

From a sheet of letter paper, one can form a cylinder or a cone, but it is impossible to obtain a surface element of a sphere without folding, stretching or cutting. The reason lies in the geometry of the spherical surface: No part of such a surface can be isometrically mapped onto the plane.

B.1 Curvature of a Curved Line in Three Dimensions

In a plane, the tangent vector remains constant when one moves on it: the plane has no curvature. The same is true for a straight line when the tangent vector coincides with the line. If a line in a neighbourhood of one of its points is not a straight line, it is called a *curved* line. The same is valid for a curved surface. We consider curves in the three-dimensional space with the position vector $\boldsymbol{x}(q)$ of points parametrized by the length q. The direction of a curve C at the point $\boldsymbol{x}(q)$ is given by the normalized tangent vector

$$\boxed{\boldsymbol{t}(q) \stackrel{\text{def}}{=} \frac{\boldsymbol{x}'(q)}{\|\boldsymbol{x}'(q)\|},}$$

with $\boldsymbol{x}'(q) \stackrel{\text{def}}{=} \frac{\partial \boldsymbol{x}(q)}{\partial q}$. Passing through a curve from a starting point $\boldsymbol{x}(q_0)$ to an end-point $\boldsymbol{x}(q)$ does not change the tangent vector $\boldsymbol{t}$ in a straight line, the tip of the tangent vector does not move, thus describing a curve of length 0. If the curve is curved, then the tip of the tangent vector describes an arc of length not equal to zero. As *arc length* we call the integral over a curved arc with endpoints q_0 and q $(q > q_0)$:

$$\int_{q_0}^{q} \sqrt[+]{x_1'^2 + x_2'^2 + x_3'^2}\, \mathrm{d}q = \int_{q_0}^{q} \sqrt[+]{\boldsymbol{x}'^{\mathsf{T}}\boldsymbol{x}'}\, \mathrm{d}q = \int_{q_0}^{q} \|\boldsymbol{x}'\|\, \mathrm{d}q. \tag{B.1}$$

This formula arises as follows: Suppose the interval $[q_0, q]$ is divided by the points $q_0 < q_1 < \cdots < q_n = q$, then the length σ_n of the inscribed polygon in the curve

G. Ludyk, *Einstein in Matrix Form*, Graduate Texts in Physics,
DOI 10.1007/978-3-642-35798-5, © Springer-Verlag Berlin Heidelberg 2013

arch is

$$\sigma_n = \sum_{k=0}^{n} \sqrt{\left[x_1(q_k) - x_1(q_{k+1})\right]^2 + \left[x_2(q_k) - x_2(q_{k+1})\right]^2 + \left[x_3(q_k) - x_3(q_{k+1})\right]^2}. \tag{B.2}$$

By the mean value theorem of differential calculus, for every smooth curve between q_k and q_{k+1} there exists a point $q_k^{(i)}$ such that

$$x_i(q_k) - x_i(q_{k+1}) = x_i'\big(q_k^{(i)}\big)(q_{k+1} - q_k) \tag{B.3}$$

for $i = 1, 2$ and 3. Inserting (B.3) into (B.2) yields

$$\sigma_n = \sum_{k=0}^{n} [q_{k+1} - q_k] \sqrt{\left[x_1'\big(q_k^{(1)}\big)\right]^2 + \left[x_2'\big(q_k^{(2)}\big)\right]^2 + \left[x_3'\big(q_k^{(3)}\big)\right]^2}. \tag{B.4}$$

From (B.4) one gets, as $q_{k+1} - q_k \to 0$, (B.1).

A measure of the curvature of a curve is the rate of change of the direction. The curvature is larger when the change of direction of the tangent vector $\boldsymbol{t}$ is greater. Generally, we therefore define as a *curvature of a curve* C at a point $\boldsymbol{x}(q_0)$

$$\boxed{\kappa(q_0) \stackrel{\text{def}}{=} \lim_{q \to q_0} \frac{\text{length of } \boldsymbol{t}}{\text{length of } \boldsymbol{x}} = \frac{\|\boldsymbol{t}'(q_0)\|}{\|\boldsymbol{x}'(q_0)\|},} \tag{B.5}$$

where

$$\text{length of } \boldsymbol{t} \stackrel{\text{def}}{=} \int_{q_0}^{q} \|\boldsymbol{t}'\|\, dq$$

and

$$\text{length of } \boldsymbol{x} \stackrel{\text{def}}{=} \int_{q_0}^{q} \|\boldsymbol{x}'\|\, dq.$$

A straight line has zero curvature. In the case of a circle, the curvature is constant; the curvature is greater, the smaller the radius. The reciprocal value $1/\kappa$ of the curvature is called the *curvature radius*.

B.2 Curvature of a Surface in Three Dimensions

B.2.1 Vectors in the Tangent Plane

Already in the nineteenth century, Gauss investigated how from the measurements on a surface one can make conclusions about its spatial form. He then came to his main result, the *Theorema Egregium*, which states that the Gaussian curvature of a surface depends only on the internal variables g_{ij} and their derivatives. This result is deduced in the following.

Suppose an area is defined as a function $\boldsymbol{x}(q_1, q_2) \in \mathbb{R}^3$ of the two coordinates q_1 and q_2. At a point P of the surface, the tangent plane is, for example, spanned by the two tangent vectors $\boldsymbol{x}_1 \stackrel{\text{def}}{=} \frac{\partial \boldsymbol{x}}{\partial q_1}$ and $\boldsymbol{x}_2 \stackrel{\text{def}}{=} \frac{\partial \boldsymbol{x}}{\partial q_2}$. If the two tangent vectors $\boldsymbol{x}_1$ and $\boldsymbol{x}_2$ are linearly independent, any vector in the tangent plane can be decomposed in a linear combination of the two vectors, e.g. as

$$v^1 \boldsymbol{x}_1 + v^2 \boldsymbol{x}_2.$$

The scalar product of two vectors from the tangent plane is then defined as

$$\underline{\underline{(\boldsymbol{v} \cdot \boldsymbol{w})}} \stackrel{\text{def}}{=} (v^1 \boldsymbol{x}_1^\mathsf{T} + v^2 \boldsymbol{x}_2^\mathsf{T})(w^1 \boldsymbol{x}_1 + w^2 \boldsymbol{x}_2)$$

$$= \boldsymbol{v}^\mathsf{T} \begin{pmatrix} \boldsymbol{x}_1^\mathsf{T} \\ \\ \boldsymbol{x}_2^\mathsf{T} \end{pmatrix} [\boldsymbol{x}_1, \boldsymbol{x}_2] \boldsymbol{w} = \boldsymbol{v}^\mathsf{T} \begin{pmatrix} \boldsymbol{x}_1^\mathsf{T}\boldsymbol{x}_1 & \boldsymbol{x}_1^\mathsf{T}\boldsymbol{x}_2 \\ \\ \boldsymbol{x}_2^\mathsf{T}\boldsymbol{x}_1 & \boldsymbol{x}_2^\mathsf{T}\boldsymbol{x}_2 \end{pmatrix} \boldsymbol{w} = \underline{\underline{\boldsymbol{v}^\mathsf{T} \boldsymbol{G} \boldsymbol{w}}}.$$

As $\boldsymbol{x}_1^\mathsf{T}\boldsymbol{x}_2 = \boldsymbol{x}_2^\mathsf{T}\boldsymbol{x}_1$, the matrix $\boldsymbol{G}^\mathsf{T} = \boldsymbol{G}$ is symmetric. In addition, one has

$$\|\boldsymbol{v}\| = \sqrt{(\boldsymbol{v} \cdot \boldsymbol{v})} = \sqrt{\boldsymbol{v}^\mathsf{T} \boldsymbol{G} \boldsymbol{v}}.$$

Now we want to define the curvature by oriented parallelograms. Let $\boldsymbol{v} \wedge \boldsymbol{w}$ be the oriented parallelogram defined by the vectors $\boldsymbol{v}$ and $\boldsymbol{w}$ in *that order*. $\boldsymbol{w} \wedge \boldsymbol{v} = -\boldsymbol{v} \wedge \boldsymbol{w}$ is then the parallelogram with the opposite orientation, $\text{area}(\boldsymbol{w} \wedge \boldsymbol{v}) = -\text{area}(\boldsymbol{v} \wedge \boldsymbol{w})$. The determinant of the matrix $\boldsymbol{G}$ is then obtained as

$$\begin{aligned} \underline{\underline{g}} &\stackrel{\text{def}}{=} \det \boldsymbol{G} = \|\boldsymbol{x}_1\|^2 \cdot \|\boldsymbol{x}_2\|^2 - (\boldsymbol{x}_1 \cdot \boldsymbol{x}_2)^2 \\ &= \|\boldsymbol{x}_1\|^2 \cdot \|\boldsymbol{x}_2\|^2 - \|\boldsymbol{x}_1\|^2 \cdot \|\boldsymbol{x}_2\|^2 \cos^2 \Theta \\ &= \|\boldsymbol{x}_1\|^2 \cdot \|\boldsymbol{x}_2\|^2 (1 - \cos^2 \Theta) = \underline{\underline{\|\boldsymbol{x}_1\|^2 \cdot \|\boldsymbol{x}_2\|^2 \sin^2 \Theta}}. \end{aligned}$$

On the other hand, one has $\|\boldsymbol{x}_1 \times \boldsymbol{x}_2\| = \|\boldsymbol{x}_1\| \cdot \|\boldsymbol{x}_2\| \cdot \sin \Theta$, and so

$$\underline{\underline{\|\boldsymbol{x}_1 \times \boldsymbol{x}_2\| = \sqrt{g} = \text{area}(\boldsymbol{x}_1 \wedge \boldsymbol{x}_2)}}. \tag{B.6}$$

For the vector product of two vectors $\boldsymbol{v}$ and $\boldsymbol{w}$ from the tangent plane, one gets, on the other hand,

$$\begin{aligned} &(v^1 \boldsymbol{x}_1 + v^2 \boldsymbol{x}_2) \times (w^1 \boldsymbol{x}_1 + w^2 \boldsymbol{x}_2) \\ &= v^1 w^1 \underbrace{(\boldsymbol{x}_1 \times \boldsymbol{x}_1)}_{0} + v^1 w^2 (\boldsymbol{x}_1 \times \boldsymbol{x}_2) + v^2 w^1 \underbrace{(\boldsymbol{x}_2 \times \boldsymbol{x}_1)}_{-(\boldsymbol{x}_1 \times \boldsymbol{x}_2)} + v^2 w^2 \underbrace{(\boldsymbol{x}_2 \times \boldsymbol{x}_2)}_{0} \\ &= \underbrace{(v^1 w^2 - v^2 w^1)}_{\stackrel{\text{def}}{=} \det \boldsymbol{R}} (\boldsymbol{x}_1 \times \boldsymbol{x}_2), \end{aligned}$$

or

$$\text{area}(\boldsymbol{v} \wedge \boldsymbol{w}) = \det \boldsymbol{R} \cdot \text{area}(\boldsymbol{x}_1 \wedge \boldsymbol{x}_2),$$

so

$$\underline{\underline{\text{area}(\boldsymbol{v} \wedge \boldsymbol{w}) = \det \boldsymbol{R}\, \sqrt{g}.}} \tag{B.7}$$

The area of the parallelogram spanned by two vectors in the tangential surface is thus determined by the vector components and the surface defining the matrix $\boldsymbol{G}$.

B.2.2 Curvature and Normal Vectors

For two-dimensional surfaces, one should define the curvature at a point without using the tangent vectors directly because there are infinitely many of them in the tangent plane. However, any smooth surface in $\mathbb{R}^3$ has at each point a unique normal direction, which is one-dimensional, so it can be described by the unit normal vector. A *normal vector* at $\boldsymbol{x}(q)$ is defined as the normalized vector perpendicular to the tangent plane:

$$\boldsymbol{n}(q) \stackrel{\text{def}}{=} \frac{\boldsymbol{x}_1 \times \boldsymbol{x}_2}{\|\boldsymbol{x}_1 \times \boldsymbol{x}_2\|}.$$

If the surface is curved, the normal vector changes with displacement according to

$$\boldsymbol{n}_i(\boldsymbol{q}) \stackrel{\text{def}}{=} \frac{\partial \boldsymbol{n}(\boldsymbol{q})}{\partial q^i}.$$

These change vectors $\boldsymbol{n}_i$ lie in the tangent plane because

$$\frac{\partial(\boldsymbol{n} \cdot \boldsymbol{n})}{\partial q^i} = 0 = \left(\frac{\partial \boldsymbol{n}}{\partial q^i} \cdot \boldsymbol{n}\right) + \left(\boldsymbol{n} \cdot \frac{\partial \boldsymbol{n}}{\partial q^i}\right) = 2(\boldsymbol{n}_i \cdot \boldsymbol{n}).$$

The bigger the area spanned by the two change vectors $\boldsymbol{n}_1$ and $\boldsymbol{n}_2$ lying in the tangent plane, the bigger the curvature at the considered point. If Ω is an area of the tangent surface which contains the point under consideration, then the following curvature definition is obvious:

$$\begin{aligned} \kappa(\boldsymbol{q}) &\stackrel{\text{def}}{=} \lim_{\Omega \to q} \frac{\text{area of } \boldsymbol{n}(\Omega)}{\text{area of } \Omega} = \lim_{\Omega \to q} \frac{\iint_\Omega \|\boldsymbol{n}_1(\tilde{\boldsymbol{q}}) \times \boldsymbol{n}_2(\tilde{\boldsymbol{q}})\| \, \mathrm{d}\tilde{q}^1 \, \mathrm{d}\tilde{q}^2}{\iint_\Omega \|\boldsymbol{x}_1(\tilde{\boldsymbol{q}}) \times \boldsymbol{x}_2(\tilde{\boldsymbol{q}})\| \, \mathrm{d}\tilde{q}^1 \, \mathrm{d}\tilde{q}^2} \\ &= \frac{\|\boldsymbol{n}_1(\boldsymbol{q}) \times \boldsymbol{n}_2(\boldsymbol{q})\|}{\|\boldsymbol{x}_1(\boldsymbol{q}) \times \boldsymbol{x}_2(\boldsymbol{q})\|} = \frac{\text{area of } \boldsymbol{n}_1(\boldsymbol{q}) \wedge \boldsymbol{n}_2(\boldsymbol{q})}{\text{area of } \boldsymbol{x}_1(\boldsymbol{q}) \wedge \boldsymbol{x}_2(\boldsymbol{q})}. \end{aligned} \tag{B.8}$$

Since both $\boldsymbol{n}_1$ and $\boldsymbol{n}_2$ lie in the tangent plane, they can be displayed as linear combinations of the vectors $\boldsymbol{x}_1$ and $\boldsymbol{x}_2$:

$$\boldsymbol{n}_1 = -b_1^1 \boldsymbol{x}_1 - b_1^2 \boldsymbol{x}_2 \quad \text{and} \quad \boldsymbol{n}_2 = -b_2^1 \boldsymbol{x}_1 - b_2^2 \boldsymbol{x}_2. \tag{B.9}$$

Combining together the coefficients $-b_i^j$ in the matrix $\overline{\boldsymbol{B}}$, this corresponds to the matrix $\boldsymbol{R}$ in (B.7), and (B.6) then yields the Gauss-*curvature*

$$\underline{\underline{\kappa(\boldsymbol{q}) = \det \overline{\boldsymbol{B}}.}} \tag{B.10}$$

To confirm the Theorema Egregium of Gauss, one must now shown that $\overline{\boldsymbol{B}}$ depends only on the inner values g_{ij} and their derivatives!

B.2.3 Theorema Egregium and the Inner Values g_{ij}

First, we examine the changes of the tangent vectors $\boldsymbol{x}_k$ by looking at their derivatives

$$\boldsymbol{x}_{jk} \stackrel{\text{def}}{=} \frac{\partial \boldsymbol{x}_k}{\partial q^j} = \frac{\partial^2 \boldsymbol{x}}{\partial q^j \partial q^k}, \tag{B.11}$$

which implies that

$$\boldsymbol{x}_{jk} = \boldsymbol{x}_{kj}. \tag{B.12}$$

Since the two vectors $\boldsymbol{x}_1$ and $\boldsymbol{x}_2$ are a basis for the tangent plane and the vector $\boldsymbol{n}$ is orthonormal to this plane, any vector in $\mathbb{R}^3$, including the vector $\boldsymbol{x}_{jk}$, can assembled as a linear combination of these three vectors:

$$\underline{\underline{\boldsymbol{x}_{jk} = \Gamma_{jk}^1 \boldsymbol{x}_1 + \Gamma_{jk}^2 \boldsymbol{x}_2 + b_{jk} \boldsymbol{n}.}} \tag{B.13}$$

The vectors $\boldsymbol{x}_{jk}$ can be summarized in a 4×2-matrix as follows:

$$\frac{\partial^2 \boldsymbol{x}}{\partial \boldsymbol{q} \partial \boldsymbol{q}^\top} = \begin{pmatrix} \boldsymbol{x}_{11} & \boldsymbol{x}_{12} \\ \boldsymbol{x}_{21} & \boldsymbol{x}_{22} \end{pmatrix} = \boldsymbol{\Gamma}_1 \otimes \boldsymbol{x}_1 + \boldsymbol{\Gamma}_2 \otimes \boldsymbol{x}_2 + \boldsymbol{B} \otimes \boldsymbol{n}, \tag{B.14}$$

where

$$\boldsymbol{\Gamma}_i \stackrel{\text{def}}{=} \begin{pmatrix} \Gamma_{11}^i & \Gamma_{12}^i \\ \Gamma_{21}^i & \Gamma_{22}^i \end{pmatrix}$$

and

$$\boldsymbol{B} \stackrel{\text{def}}{=} \begin{pmatrix} b_{11} & b_{12} \\ b_{21} & b_{22} \end{pmatrix}.$$

Multiplying (B.14) from the left by the matrix $(\boldsymbol{I}_2 \otimes \boldsymbol{n}^\top)$, due to $\boldsymbol{n}^\top \boldsymbol{x}_i = 0$ and $(\boldsymbol{I}_2 \otimes \boldsymbol{n}^\top)(\boldsymbol{B} \otimes \boldsymbol{n}) = \boldsymbol{B} \otimes (\boldsymbol{n}^\top \boldsymbol{n}) = \boldsymbol{B} \otimes 1 = \boldsymbol{B}$, one finally obtains

$$\underline{\underline{\boldsymbol{B} = (\boldsymbol{I}_2 \otimes \boldsymbol{n}^\top) \frac{\partial^2 \boldsymbol{x}}{\partial \boldsymbol{q} \partial \boldsymbol{q}^\top} = \begin{pmatrix} \boldsymbol{n}^\top \boldsymbol{x}_{11} & \boldsymbol{n}^\top \boldsymbol{x}_{12} \\ \boldsymbol{n}^\top \boldsymbol{x}_{21} & \boldsymbol{n}^\top \boldsymbol{x}_{22} \end{pmatrix},}} \tag{B.15}$$

i.e. the 2×2-matrix $\boldsymbol{B}$ is symmetric. This matrix is called the *Second Fundamental Form* of the surface. The *First Quadratic Fundamental Form* is given by the connection

$$\underline{\underline{\boldsymbol{G} = \frac{\partial \boldsymbol{x}^{\mathsf{T}}}{\partial \boldsymbol{q}} \cdot \frac{\partial \boldsymbol{x}}{\partial \boldsymbol{q}^{\mathsf{T}}}}}, \tag{B.16}$$

or more precisely, by the right-hand side of the equation for the squared line element ds of the surface:

$$\mathrm{d}s^2 = \mathrm{d}\boldsymbol{q}^{\mathsf{T}} \boldsymbol{G}\, \mathrm{d}\boldsymbol{q}.$$

Gauss designates, as is still in elementary geometry today, the elements of the matrix $\boldsymbol{G}$ in this way:

$$\boldsymbol{G} = \begin{pmatrix} E & F \\ F & G \end{pmatrix}.$$

While $\boldsymbol{G}$ therefore plays a critical role in determining the length of a curve in an area, $\boldsymbol{B}$ is, as we will see later, decisively involved in the determination of the curvature of a surface. A further representation of the matrix $\boldsymbol{B}$ is obtained from the derivative of the scalar product of the mutually orthogonal vectors $\boldsymbol{n}$ and $\boldsymbol{x}_i$:

$$\boldsymbol{n}^{\mathsf{T}} \frac{\partial \boldsymbol{x}}{\partial \boldsymbol{q}^{\mathsf{T}}} = \boldsymbol{0}^{\mathsf{T}};$$

because this derivative is, according to (A.73),

$$\frac{\partial}{\partial \boldsymbol{q}} \left(\boldsymbol{n}^{\mathsf{T}} \frac{\partial \boldsymbol{x}}{\partial \boldsymbol{q}^{\mathsf{T}}} \right) = \frac{\partial \boldsymbol{n}^{\mathsf{T}}}{\partial \boldsymbol{q}} \cdot \frac{\partial \boldsymbol{x}}{\partial \boldsymbol{q}^{\mathsf{T}}} + (\boldsymbol{I}_2 \otimes \boldsymbol{n}^{\mathsf{T}}) \frac{\partial^2 \boldsymbol{x}}{\partial \boldsymbol{q} \partial \boldsymbol{q}^{\mathsf{T}}} = \boldsymbol{0}_{2\times 2}.$$

With (B.15) we obtain

$$\underline{\underline{\boldsymbol{B} = -\frac{\partial \boldsymbol{n}^{\mathsf{T}}}{\partial \boldsymbol{q}} \cdot \frac{\partial \boldsymbol{x}}{\partial \boldsymbol{q}^{\mathsf{T}}}}}, \tag{B.17}$$

i.e. a further interesting form for the matrix $\boldsymbol{B}$

$$\underline{\underline{\boldsymbol{B} = -\begin{pmatrix} \boldsymbol{n}_1^{\mathsf{T}} \boldsymbol{x}_1 & \boldsymbol{n}_1^{\mathsf{T}} \boldsymbol{x}_2 \\ \boldsymbol{n}_2^{\mathsf{T}} \boldsymbol{x}_1 & \boldsymbol{n}_2^{\mathsf{T}} \boldsymbol{x}_2 \end{pmatrix}}}. \tag{B.18}$$

The two equations (B.9) can be summarized to a matrix equation as follows:

$$[\boldsymbol{n}_1 | \boldsymbol{n}_2] = \underline{\underline{\frac{\partial \boldsymbol{n}}{\partial \boldsymbol{q}^{\mathsf{T}}}}} = \frac{\partial \boldsymbol{x}}{\partial \boldsymbol{q}^{\mathsf{T}}} \begin{pmatrix} -b_1^1 & -b_2^1 \\ -b_1^2 & -b_2^2 \end{pmatrix} = \underline{\underline{\frac{\partial \boldsymbol{x}}{\partial \boldsymbol{q}^{\mathsf{T}}} \cdot \overline{\boldsymbol{B}}}}. \tag{B.19}$$

This used in (B.18) together with (B.16) yields

$$\boldsymbol{B} = -\overline{\boldsymbol{B}}^{\mathsf{T}} \cdot \frac{\partial \boldsymbol{x}^{\mathsf{T}}}{\partial \boldsymbol{q}} \cdot \frac{\partial \boldsymbol{x}}{\partial \boldsymbol{q}^{\mathsf{T}}} = -\overline{\boldsymbol{B}}^{\mathsf{T}} \boldsymbol{G},$$

or transposed

$$\underline{\underline{\boldsymbol{B} = -\boldsymbol{G}\overline{\boldsymbol{B}}}}, \tag{B.20}$$

since $\boldsymbol{B}$ and $\boldsymbol{G}$ are both symmetric matrices. Our goal remains to show that the Gaussian curvature depends only on the g_{ij}'s and their derivatives with respect to q_k, i.e., according to (B.10), one must show that for the matrix $\overline{\boldsymbol{B}}$

$$\kappa(\boldsymbol{q}) = \det \overline{\boldsymbol{B}}$$

is valid. We first examine the matrix $\boldsymbol{G}$. For this purpose, its elements are differentiated:

$$\frac{\partial g_{ij}}{\partial q_k} = \boldsymbol{x}_{ik}^{\mathsf{T}} \boldsymbol{x}_j + \boldsymbol{x}_{jk}^{\mathsf{T}} \boldsymbol{x}_i.$$

With (B.13) we obtain

$$\begin{aligned} \boldsymbol{x}_j^{\mathsf{T}} \boldsymbol{x}_{ik} &= \Gamma_{ik}^1 \boldsymbol{x}_j^{\mathsf{T}} \boldsymbol{x}_1 + \Gamma_{ik}^2 \boldsymbol{x}_j^{\mathsf{T}} \boldsymbol{x}_2 + b_{ik} \boldsymbol{x}_j^{\mathsf{T}} \boldsymbol{n} \\ &= \Gamma_{ik}^1 g_{j1} + \Gamma_{ik}^2 g_{j2}. \end{aligned}$$

If we define

$$\check{\Gamma}_{ik}^{j} \overset{\text{def}}{=} \Gamma_{ik}^1 g_{j1} + \Gamma_{ik}^2 g_{j2}$$

and assemble all four components into a matrix $\check{\boldsymbol{\Gamma}}_j$, we obtain

$$\begin{pmatrix} \check{\Gamma}_{j1}^1 & \check{\Gamma}_{j1}^2 \\ \check{\Gamma}_{j2}^1 & \check{\Gamma}_{j2}^2 \end{pmatrix} = \begin{pmatrix} \Gamma_{j1}^1 & \Gamma_{j1}^2 \\ \Gamma_{j2}^1 & \Gamma_{j2}^2 \end{pmatrix} \begin{pmatrix} g_{11} & g_{12} \\ g_{21} & g_{22} \end{pmatrix} = \boldsymbol{\Gamma}_j \boldsymbol{G}^{\mathsf{T}} \tag{B.21}$$

or, because of $\boldsymbol{G}^{\mathsf{T}} = \boldsymbol{G}$,

$$\check{\boldsymbol{\Gamma}}_j = \boldsymbol{\Gamma}_j \boldsymbol{G}. \tag{B.22}$$

It is therefore true that

$$\frac{\partial g_{ij}}{\partial q_k} = \check{\Gamma}_{ik}^{j} + \check{\Gamma}_{jk}^{i}. \tag{B.23}$$

With the following expression of three different derivatives, one obtains

$$\frac{\partial g_{ij}}{\partial q_k} + \frac{\partial g_{ik}}{\partial q_j} - \frac{\partial g_{jk}}{\partial q_i} = \check{\Gamma}_{ik}^{j} + \check{\Gamma}_{jk}^{i} + \check{\Gamma}_{ij}^{k} + \check{\Gamma}_{jk}^{i} - \check{\Gamma}_{ik}^{j} - \check{\Gamma}_{ji}^{k}, \tag{B.24}$$

so

$$\underline{\underline{\check{\Gamma}_{jk}^{i} = \frac{1}{2}\left(\frac{\partial g_{ij}}{\partial q_k} + \frac{\partial g_{ik}}{\partial q_j} - \frac{\partial g_{jk}}{\partial q_i} \right)}}. \tag{B.25}$$

Multiplying (B.22) from the right with

$$G^{-1} \stackrel{\text{def}}{=} \begin{pmatrix} g_{11}^{[-1]} & g_{12}^{[-1]} \\ g_{21}^{[-1]} & g_{22}^{[-1]} \end{pmatrix},$$

one gets the relation

$$\boldsymbol{\Gamma}_j = \check{\boldsymbol{\Gamma}}_j \boldsymbol{G}^{-1}, \tag{B.26}$$

i.e. element by element

$$\Gamma_{jk}^{\ell} = \frac{1}{2} \sum_i g_{i\ell}^{[-1]} \check{\Gamma}_{jk}^i, \tag{B.27}$$

so with (B.25)

$$\boxed{\Gamma_{jk}^{\ell} = \frac{1}{2} \sum_i g_{i\ell}^{[-1]} \left(\frac{\partial g_{ij}}{\partial q_k} + \frac{\partial g_{ik}}{\partial q_j} - \frac{\partial g_{jk}}{\partial q_i} \right).} \tag{B.28}$$

This clarifies the relationship of the Christoffel-symbols Γ_{jk}^{ℓ} with the g_{ij} and their derivatives. Now the direct relationship of these variables with the Gaussian curvature κ has to be made. One gets this finally by repeated differentiation of $\boldsymbol{x}_{jk}$ with respect to q^{ℓ}:

$$\begin{aligned} \boldsymbol{x}_{jk\ell} &\stackrel{\text{def}}{=} \frac{\partial \boldsymbol{x}_{jk}}{\partial q^{\ell}} = \sum_i \frac{\partial \Gamma_{jk}^i}{\partial q^{\ell}} \boldsymbol{x}_i + \sum_i \Gamma_{j\ell}^i \boldsymbol{x}_{i\ell} + \frac{\partial b_{jk}}{\partial q^{\ell}} \boldsymbol{n} + b_{jk} \boldsymbol{n}_{\ell} \\ &= \sum_i \left(\frac{\partial \Gamma_{jk}^i}{\partial q^{\ell}} + \Gamma_{j\ell}^i \Gamma_{p\ell}^i - b_{jk} b_{\ell}^i \right) \boldsymbol{x}_i + \left(\frac{\partial b_{jk}}{\partial q^{\ell}} + \sum_p \Gamma_{jk}^p b_{p\ell} \right) \boldsymbol{n}. \end{aligned} \tag{B.29}$$

Interchanging in (B.29) k and ℓ, we obtain

$$\boldsymbol{x}_{j\ell k} = \sum_i \left(\frac{\partial \Gamma_{j\ell}^i}{\partial q^k} + \Gamma_{jk}^i \Gamma_{pk}^i - b_{j\ell} b_k^i \right) \boldsymbol{x}_i + \left(\frac{\partial b_{j\ell}}{\partial q^k} + \sum_p \Gamma_{j\ell}^p b_{pk} \right) \boldsymbol{n}. \tag{B.30}$$

Subtracting the two third-order derivatives, we obtain

$$\boldsymbol{0} = \boldsymbol{x}_{j\ell k} - \boldsymbol{x}_{jk\ell} = \sum_i \left[R_{jk}^{i\ell} - \left(b_{j\ell} b_k^i - b_{jk} b_{\ell}^i \right) \right] \boldsymbol{x}_i + (\cdots) \boldsymbol{n}, \tag{B.31}$$

with

$$R_{jk}^{i\ell} \stackrel{\text{def}}{=} \frac{\partial \Gamma_{j\ell}^i}{\partial q^k} - \frac{\partial \Gamma_{jk}^i}{\partial q^{\ell}} + \sum_p \Gamma_{j\ell}^p \Gamma_{p\ell}^i - \sum_p \Gamma_{jk}^p \Gamma_{pk}^i. \tag{B.32}$$

Since the vectors $\boldsymbol{x}_1, \boldsymbol{x}_2$ and $\boldsymbol{n}$ are linearly independent, the square bracket in (B.31) must be zero, which implies

$$R^{i\ell}_{jk} = b_{j\ell} b^i_k - b_{jk} b^i_\ell. \tag{B.33}$$

Defining

$$\check{R}^{i\ell}_{jk} = \sum_i g_{ih} R^{i\ell}_{jk}, \tag{B.34}$$

one gets

$$\check{R}^{i\ell}_{jk} = g_{1h} b_{j\ell} b^1_k - g_{1h} b_{jk} b^1_\ell + g_{2h} b_{j\ell} b^2_k - g_{2h} b_{jk} b^2_\ell = b_{j\ell} b_{kh} - b_{jk} b_{\ell h}. \tag{B.35}$$

In particular,

$$\underline{\underline{\check{R}^{12}_{12} = b_{22} b_{11} - b_{21} b_{21} = \det \boldsymbol{B}.}} \tag{B.36}$$

It is therefore true that

$$\kappa(\boldsymbol{q}) = \det \bar{\boldsymbol{B}} = \frac{\det \boldsymbol{B}}{\det \boldsymbol{G}} = \frac{\check{R}^{12}_{12}}{g},$$

$$\boxed{\kappa(\boldsymbol{q}) = \frac{\check{R}^{12}_{12}}{g},} \tag{B.37}$$

which has finally proved the Theorema Egregium because, due to (B.34), $\check{R}^{i\ell}_{jk}$ depends on $R^{i\ell}_{jk}$; according to (B.32), $R^{i\ell}_{jk}$ only depends on Γ^k_{ij} and their derivatives and, in accordance with (B.25), the Γ^k_{ij}'s depend only on the g_{ik}'s and their derivatives. In the form

$$\boxed{\kappa(\boldsymbol{q}) = \frac{\det \boldsymbol{B}}{\det \boldsymbol{G}},} \tag{B.38}$$

the paramount importance of the two fundamental forms is expressed.

Remarks

1. Euclidean geometry of is based on a number of Axioms that require no proofs. One is the parallel postulate stating that to every line one can draw through a point not belonging to it one and only one other line which lies in the same plane and does not intersect the former line. This axiom is replaced in the *hyperbolic* geometry in that it admits infinitely many parallels. An example is the surface of a hyperboloid. In the *elliptical* geometry, for example, on the surface of an ellipsoid and, as a special case, on a spherical surface, there are absolutely no parallels because all great circles, which are here the "straight lines", meet in

two points. In Euclidean geometry, the distance between two points with the Cartesian coordinates x_1, x_2, x_3 and $x_1 + \mathrm{d}x_1, x_2 + \mathrm{d}x_2, x_3 + \mathrm{d}x_3$ is simply

$$\mathrm{d}s = \sqrt{\mathrm{d}x_1^2 + \mathrm{d}x_2^2 + \mathrm{d}x_3^2},$$

and in the other two geometries this formula is replaced by

$$\mathrm{d}s^2 = a_1\,\mathrm{d}x_1^2 + a_2\,\mathrm{d}x_2^2 + a_3\,\mathrm{d}x_3^2,$$

where the coefficients a_i are certain simple functions of x_i, in the hyperbolic case, of course, different than in the elliptic case. A convenient analytical representation of curved surfaces is the above used Gaussian parameter representation $\boldsymbol{x} = \boldsymbol{x}(q_1, q_2)$, where Gauss attaches as curved element:

$$\mathrm{d}s^2 = E\,\mathrm{d}q_1^2 + 2F\,\mathrm{d}q_1\,\mathrm{d}q_2 + G\,\mathrm{d}q_2^2.$$

As an example, we introduce the Gauss-specific parameter representation of the unit sphere, with $\theta = q_1$ and $\varphi = q_2$:

$$x_1 = \sin\theta\,\cos\varphi, \qquad x_2 = \sin\theta\,\sin\varphi, \qquad x_3 = \cos\varphi.$$

For the arc element of the unit sphere, we obtain

$$\mathrm{d}s^2 = \mathrm{d}\theta^2 + \sin^2\theta\,(\mathrm{d}\varphi)^2.$$

Riemann generalized the Gaussian theory of surfaces, which is valid for two-dimensional surfaces in three-dimensional spaces, to p-dimensional hypersurfaces in n-dimensional spaces, i.e. where

$$\boldsymbol{x} = \boldsymbol{x}(q_1, \ldots, q_p) \in \mathbb{R}^n$$

is a point on the hypersurface. He made in addition the fundamentally important step, to set up a homogeneous quadratic function of $\mathrm{d}q_i$ with arbitrary functions of the q_i as coefficients, as the square of the line elements (quadratic form)

$$\mathrm{d}s^2 = \sum_{ik} g_{ik}\,\mathrm{d}q_i\,\mathrm{d}q_k = \mathrm{d}\boldsymbol{q}^{\mathsf{T}}\boldsymbol{G}\,\mathrm{d}\boldsymbol{q}.$$

2. The above-occurring $R^{i\ell}_{jk}$ can be used as matrix elements of the 4×4-matrix $\boldsymbol{R}$, the Riemannian *Curvature Matrix*, to be constructed as a block matrix as follows:

$$\boldsymbol{R} = \begin{pmatrix} \boldsymbol{R}^{11} & \boldsymbol{R}^{12} \\ \boldsymbol{R}^{21} & \boldsymbol{R}^{22} \end{pmatrix},$$

where the 2×2 sub-matrices have the form:

$$\boldsymbol{R}^{i\ell} = \begin{pmatrix} R^{i\ell}_{11} & R^{i\ell}_{12} \\ R^{i\ell}_{21} & R^{i\ell}_{22} \end{pmatrix}.$$

In particular, $\check{R}^{12}_{12}$ is the element in the top right corner of the matrix $\check{\boldsymbol{R}} = \boldsymbol{G}\boldsymbol{R}$.

3. Expanding the representation of $\boldsymbol{x}(\boldsymbol{q} + \Delta\boldsymbol{q})$ in a Taylor series, one obtains

$$\boldsymbol{x}(\boldsymbol{q} + \Delta\boldsymbol{q}) = \boldsymbol{x}(\boldsymbol{q}) + \sum_i \boldsymbol{x}_i \Delta q_i + \frac{1}{2}\sum_{i,k} \boldsymbol{x}_{ik} \Delta q_i \Delta q_k + \sigma(3).$$

Subtracting $\boldsymbol{x}(\boldsymbol{q})$ on both sides of this equation and multiplying the result from the left with the transposed normal vector $\boldsymbol{n}^{\mathsf{T}}$, we obtain

$$\begin{aligned}\boldsymbol{n}^{\mathsf{T}}\big[\boldsymbol{x}(\boldsymbol{q} + \Delta\boldsymbol{q}) - \boldsymbol{x}(\boldsymbol{q})\big] &= \sum_i \underbrace{\boldsymbol{n}^{\mathsf{T}}\boldsymbol{x}_i}_{0} \Delta q_i + \frac{1}{2}\sum_{i,k} \underbrace{\boldsymbol{n}^{\mathsf{T}}\boldsymbol{x}_{ik}}_{b_{ik}} \Delta q_i \Delta q_k + \sigma(3) \\ &= \boldsymbol{n}^{\mathsf{T}}\Delta\boldsymbol{x}(\boldsymbol{q}) \stackrel{\text{def}}{=} \Delta\ell.\end{aligned}$$

Thus

$$\mathrm{d}\ell \approx \frac{1}{2}\sum_{i,k} b_{ik}\, \mathrm{d}q_i\, \mathrm{d}q_k.$$

The coefficients of the second fundamental form, i.e. the elements of the matrix $\boldsymbol{B}$, Gauss denotes by L, M and N. Then the distance $\mathrm{d}\ell$ of the point $\boldsymbol{x}(q_1 + \mathrm{d}q_1, q_2 + \mathrm{d}q_2)$ to the tangent surface at the point $\boldsymbol{x}(q_1, q_2)$ is

$$\mathrm{d}\ell \approx \frac{1}{2}\big(L\,\mathrm{d}q_1^2 + 2M\,\mathrm{d}q_1\,\mathrm{d}q_2 + N\,\mathrm{d}q_2^2\big).$$

The *normal curvature* κ of a surface at a given point P and in a given direction $\boldsymbol{q}$ is defined as

$$\kappa \stackrel{\text{def}}{=} \frac{L\,\mathrm{d}q_1^2 + 2M\,\mathrm{d}q_1\,\mathrm{d}q_2 + N\,\mathrm{d}q_2^2}{E\,\mathrm{d}q_1^2 + 2F\,\mathrm{d}q_1\,\mathrm{d}q_2 + G\,\mathrm{d}q_2^2}. \tag{B.39}$$

The so-defined normal curvature depends, in general, on the chosen direction $\mathrm{d}\boldsymbol{q}$. Those directions, in which the normal curvatures at a given point assume an extreme value, are named the *main* directions of the surface at this point. As long as we examine real surfaces, the quadratic differential form $E\,\mathrm{d}q_1^2 + 2F\,\mathrm{d}q_1\,\mathrm{d}q_2 + G\,\mathrm{d}q_2^2$ is positive definite, i.e. it is always positive for $\mathrm{d}\boldsymbol{q} \neq \boldsymbol{0}$. Thus the sign of the curvature depends only on the quadratic differential form $L\,\mathrm{d}q_1^2 + 2M\,\mathrm{d}q_1\,\mathrm{d}q_2 + N\,\mathrm{d}q_2^2$ in the numerator of (B.39). There are three cases:

(a) $LN - M^2 > 0$, i.e. $\boldsymbol{B}$ is positive definite, and the numerator retains the same sign, in each direction one is looking. Such a point is called an *elliptical* point. An example is any point on an ellipsoid, in particular, of course, on a sphere.

(b) $LN - M^2 = 0$, i.e. $\boldsymbol{B}$ is semi-definite. The surface behaves at this point as at an elliptical point except in one direction where is $\kappa = 0$. This point is called *parabolic*. An example is any point on a cylinder.

(c) $LN - M^2 < 0$, i.e. $\boldsymbol{B}$ is indefinite. The numerator does not keep the same sign for all directions. Such a point is called *hyperbolic*, or a *saddle point*. An example is a point on a hyperbolic paraboloid.

Dividing the numerator and the denominator in (B.39) by $\mathrm{d}q_2$ and introducing $\mathrm{d}q_1/\mathrm{d}q_2 \overset{\text{def}}{=} \lambda$, we obtain

$$\kappa(\lambda) = \frac{L + 2M\lambda + N\lambda^2}{E + 2F\lambda + G\lambda^2} \tag{B.40}$$

and from this the extreme values from

$$\frac{\mathrm{d}\kappa}{\mathrm{d}\lambda} = 0$$

as those satisfying

$$\left(E + 2F\lambda + G\lambda^2\right)(M + N\lambda) - \left(L + 2M\lambda + N\lambda^2\right)(F + G\lambda) = 0. \tag{B.41}$$

In this case, the resulting expression for κ is

$$\kappa = \frac{L + 2M\lambda + N\lambda^2}{E + 2F\lambda + G\lambda^2} = \frac{M + N\lambda}{F + G\lambda}. \tag{B.42}$$

Since furthermore

$$E + 2F\lambda + G\lambda^2 = (E + F\lambda) + \lambda(F + G\lambda)$$

and

$$L + 2M\lambda + N\lambda^2 = (L + M\lambda) + \lambda(M + N\lambda),$$

(B.40) can be transformed into the simpler form

$$\kappa = \frac{L + M\lambda}{E + F\lambda}. \tag{B.43}$$

From this the two equations for κ follow:

$$(\kappa E - L) + (\kappa F - M)\lambda = 0,$$

$$(\kappa F - M) + (\kappa G - N)\lambda = 0.$$

These equations are simultaneously satisfied if and only if

$$\det\begin{pmatrix} \kappa E - L & \kappa F - M \\ \kappa F - M & \kappa G - N \end{pmatrix} = 0. \tag{B.44}$$

This can also be written as

$$\det(\kappa \boldsymbol{G} - \boldsymbol{B}) = 0. \tag{B.45}$$

This is the solvability condition for the eigenvalue equation

$$\kappa \boldsymbol{G} - \boldsymbol{B} = \boldsymbol{0},$$

which can be transformed into

$$\kappa \boldsymbol{I} - \boldsymbol{G}^{-1}\boldsymbol{B} = \boldsymbol{0}. \tag{B.46}$$

This results in a quadratic equation for κ. The two solutions are called the *principal curvatures* and are denoted as κ_1 and κ_2. The Gaussian *curvature* κ of a surface at a given point is the product of the principal curvatures κ_1 and κ_2 of the surface in this point. According to Vieta's root theorem, the product of the solutions is equal to the determinant of the matrix $\boldsymbol{G}^{-1}\boldsymbol{B}$, so finally,

$$\kappa = \kappa_1 \kappa_2 = \det\left(\boldsymbol{G}^{-1}\boldsymbol{B}\right) = \frac{\det \boldsymbol{B}}{\det \boldsymbol{G}} = \frac{LN - M^2}{EG - F^2}.$$

Appendix C
Geodesic Deviation

Geodesics are the lines of general manifolds along which, for example, free particles move. In a flat space the relative velocity of each pair of particles is constant, so that their relative acceleration is always equal to zero. Generally, due to the curvature of space, the relative acceleration is not equal to zero.

The curvature of a surface can be illustrated as follows [21]. Suppose there are two ants on an apple which leave a starting line at the same time and follow with the same speed geodesics which are initially perpendicular to the start line. Initially, their paths are parallel, but, due to the curvature of the apple, they are approaching each other from the beginning. Their distance $\boldsymbol{\xi}$ from one another is not constant, i.e., in general, the relative acceleration of the ants moving on geodesics with constant velocity is not equal zero if the area over which they move is curved. So the curvature can be indirectly perceived through the so-called *geodesic deviation* $\boldsymbol{\xi}$.

The two neighboring geodesics $\boldsymbol{x}(u)$ and $\check{\boldsymbol{x}}(u)$ have the distance

$$\boldsymbol{\xi}(u) \stackrel{\text{def}}{=} \check{\boldsymbol{x}}(u) - \boldsymbol{x}(u), \tag{C.1}$$

where u is the proper time or distance.

The mathematical descriptions of these geodesics are

$$\ddot{\check{\boldsymbol{x}}} + \left(\boldsymbol{I}_4 \otimes \dot{\check{\boldsymbol{x}}}^{\mathsf{T}}\right)\check{\boldsymbol{\Gamma}}\dot{\check{\boldsymbol{x}}} = \boldsymbol{0}, \tag{C.2}$$

$$\ddot{\boldsymbol{x}} + \left(\boldsymbol{I}_4 \otimes \dot{\boldsymbol{x}}^{\mathsf{T}}\right)\boldsymbol{\Gamma}\dot{\boldsymbol{x}} = \boldsymbol{0}. \tag{C.3}$$

The Christoffel-matrix $\check{\boldsymbol{\Gamma}}$ is approximated by

$$\check{\boldsymbol{\Gamma}} \approx \boldsymbol{\Gamma} + \frac{\partial \boldsymbol{\Gamma}}{\partial \boldsymbol{x}^{\mathsf{T}}}(\boldsymbol{\xi} \otimes \boldsymbol{I}_4). \tag{C.4}$$

Subtracting (C.3) from (C.2) and considering (C.1) and (C.4), one obtains

$$\ddot{\boldsymbol{\xi}} + \left(\boldsymbol{I}_4 \otimes \dot{\check{\boldsymbol{x}}}^{\mathsf{T}}\right)\boldsymbol{\Gamma}\dot{\check{\boldsymbol{x}}} - \left(\boldsymbol{I}_4 \otimes \dot{\boldsymbol{x}}^{\mathsf{T}}\right)\boldsymbol{\Gamma}\dot{\boldsymbol{x}} + \left(\boldsymbol{I}_4 \otimes \dot{\check{\boldsymbol{x}}}^{\mathsf{T}}\right)\frac{\partial \boldsymbol{\Gamma}}{\partial \boldsymbol{x}^{\mathsf{T}}}(\boldsymbol{\xi} \otimes \boldsymbol{I}_4)\dot{\check{\boldsymbol{x}}} = \boldsymbol{0}. \tag{C.5}$$

G. Ludyk, *Einstein in Matrix Form*, Graduate Texts in Physics,
DOI 10.1007/978-3-642-35798-5, © Springer-Verlag Berlin Heidelberg 2013

With $\dot{\boldsymbol{x}} = \dot{\boldsymbol{\xi}} + \dot{\boldsymbol{x}}$ and neglecting quadratic and higher powers of $\boldsymbol{\xi}$ and $\dot{\boldsymbol{\xi}}$, one obtains from (C.5)

$$\ddot{\boldsymbol{\xi}} + (\boldsymbol{I}_4 \otimes \dot{\boldsymbol{\xi}}^\top)\boldsymbol{\Gamma}\dot{\boldsymbol{x}} + (\boldsymbol{I}_4 \otimes \dot{\boldsymbol{x}}^\top)\boldsymbol{\Gamma}\dot{\boldsymbol{\xi}} + (\boldsymbol{I}_4 \otimes \dot{\boldsymbol{x}}^\top)\frac{\partial \boldsymbol{\Gamma}}{\partial \boldsymbol{x}^\top}(\boldsymbol{\xi} \otimes \boldsymbol{I}_4)\dot{\boldsymbol{x}} = \boldsymbol{0}. \tag{C.6}$$

Hence

$$\frac{\mathrm{D}\boldsymbol{\xi}}{\mathrm{d}u} = \dot{\boldsymbol{\xi}} + (\boldsymbol{I}_4 \otimes \boldsymbol{\xi}^\top)\boldsymbol{\Gamma}\dot{\boldsymbol{x}} \tag{C.7}$$

and

$$\begin{aligned}\frac{\mathrm{D}^2\boldsymbol{\xi}}{\mathrm{d}u^2} &= \frac{\mathrm{D}}{\mathrm{d}u}\left(\dot{\boldsymbol{\xi}} + (\boldsymbol{I}_4 \otimes \boldsymbol{\xi}^\top)\boldsymbol{\Gamma}\dot{\boldsymbol{x}}\right)\\ &= \ddot{\boldsymbol{\xi}} + \frac{\mathrm{d}}{\mathrm{d}u}\left\{(\boldsymbol{I}_4 \otimes \boldsymbol{\xi}^\top)\boldsymbol{\Gamma}\dot{\boldsymbol{x}}\right\} + \left(\boldsymbol{I}_4 \otimes \left[\dot{\boldsymbol{\xi}} + (\boldsymbol{I}_4 \otimes \boldsymbol{\xi}^\top)\boldsymbol{\Gamma}\dot{\boldsymbol{x}}\right]^\top\right)\boldsymbol{\Gamma}\dot{\boldsymbol{x}}\\ &= \ddot{\boldsymbol{\xi}} + \frac{\mathrm{d}}{\mathrm{d}u}\left\{(\boldsymbol{I}_4 \otimes \boldsymbol{\xi}^\top)\boldsymbol{\Gamma}\dot{\boldsymbol{x}}\right\} + (\boldsymbol{I}_4 \otimes \dot{\boldsymbol{\xi}}^\top)\boldsymbol{\Gamma}\dot{\boldsymbol{x}} + \left(\boldsymbol{I}_4 \otimes \left[(\boldsymbol{I}_4 \otimes \boldsymbol{\xi}^\top)\boldsymbol{\Gamma}\dot{\boldsymbol{x}}\right]^\top\right)\boldsymbol{\Gamma}\dot{\boldsymbol{x}}.\end{aligned} \tag{C.8}$$

For the second term, by (C.3), one gets

$$\begin{aligned}\frac{\mathrm{d}}{\mathrm{d}u}\left\{(\boldsymbol{I}_4 \otimes \boldsymbol{\xi}^\top)\boldsymbol{\Gamma}\dot{\boldsymbol{x}}\right\} &= (\boldsymbol{I}_4 \otimes \dot{\boldsymbol{\xi}}^\top)\boldsymbol{\Gamma}\dot{\boldsymbol{x}} + (\boldsymbol{I}_4 \otimes \boldsymbol{\xi}^\top)\frac{\partial \boldsymbol{\Gamma}}{\partial \boldsymbol{x}^\top}(\dot{\boldsymbol{x}} \otimes \boldsymbol{I}_4)\dot{\boldsymbol{x}} + (\boldsymbol{I}_4 \otimes \boldsymbol{\xi}^\top)\boldsymbol{\Gamma}\ddot{\boldsymbol{x}}\\ &= (\boldsymbol{I}_4 \otimes \dot{\boldsymbol{\xi}}^\top)\boldsymbol{\Gamma}\dot{\boldsymbol{x}} + (\boldsymbol{I}_4 \otimes \boldsymbol{\xi}^\top)\frac{\partial \boldsymbol{\Gamma}}{\partial \boldsymbol{x}^\top}(\dot{\boldsymbol{x}} \otimes \boldsymbol{I}_4)\dot{\boldsymbol{x}}\\ &\quad - (\boldsymbol{I}_4 \otimes \boldsymbol{\xi}^\top)\boldsymbol{\Gamma}(\boldsymbol{I}_4 \otimes \dot{\boldsymbol{x}}^\top)\boldsymbol{\Gamma}\dot{\boldsymbol{x}}.\end{aligned} \tag{C.9}$$

Equation (C.9) used in (C.8) yields

$$\begin{aligned}\frac{\mathrm{D}^2\boldsymbol{\xi}}{\mathrm{d}u^2} &= \ddot{\boldsymbol{\xi}} + (\boldsymbol{I}_4 \otimes \dot{\boldsymbol{\xi}}^\top)\boldsymbol{\Gamma}\dot{\boldsymbol{x}} + (\boldsymbol{I}_4 \otimes \boldsymbol{\xi}^\top)\frac{\partial \boldsymbol{\Gamma}}{\partial \boldsymbol{x}^\top}(\dot{\boldsymbol{x}} \otimes \boldsymbol{I}_4)\dot{\boldsymbol{x}} - (\boldsymbol{I}_4 \otimes \boldsymbol{\xi}^\top)\boldsymbol{\Gamma}(\boldsymbol{I}_4 \otimes \dot{\boldsymbol{x}}^\top)\boldsymbol{\Gamma}\dot{\boldsymbol{x}}\\ &\quad + (\boldsymbol{I}_4 \otimes \dot{\boldsymbol{\xi}}^\top)\boldsymbol{\Gamma}\dot{\boldsymbol{x}} + \left(\boldsymbol{I}_4 \otimes \left[(\boldsymbol{I}_4 \otimes \boldsymbol{\xi}^\top)\boldsymbol{\Gamma}\dot{\boldsymbol{x}}\right]^\top\right)\boldsymbol{\Gamma}\dot{\boldsymbol{x}}.\end{aligned} \tag{C.10}$$

Remark Since the sub-matrices $\boldsymbol{\Gamma}_i$ are symmetric, it is generally true that

$$(\boldsymbol{I}_4 \otimes \boldsymbol{a}^\top)\boldsymbol{\Gamma}\boldsymbol{b} = (\boldsymbol{I}_4 \otimes \boldsymbol{b}^\top)\boldsymbol{\Gamma}\boldsymbol{a}. \tag{C.11}$$

In addition, one has $(\boldsymbol{I}_4 \otimes \boldsymbol{a}^\top)\boldsymbol{\Gamma}\boldsymbol{b} = \overline{\boldsymbol{\Gamma}}(\boldsymbol{I}_4 \otimes \boldsymbol{a})\boldsymbol{b} = \overline{\boldsymbol{\Gamma}}(\boldsymbol{b} \otimes \boldsymbol{a})$ and $(\boldsymbol{I}_4 \otimes \boldsymbol{b}^\top)\boldsymbol{\Gamma}\boldsymbol{a} = \overline{\boldsymbol{\Gamma}}(\boldsymbol{I}_4 \otimes \boldsymbol{b})\boldsymbol{a} = \overline{\boldsymbol{\Gamma}}(\boldsymbol{a} \otimes \boldsymbol{b})$, thus, due to (C.11),

$$\overline{\boldsymbol{\Gamma}}(\boldsymbol{b} \otimes \boldsymbol{a}) = \overline{\boldsymbol{\Gamma}}(\boldsymbol{a} \otimes \boldsymbol{b}). \tag{C.12}$$

With (C.11), one has from (C.10)

$$\ddot{\boldsymbol{\xi}} + (\boldsymbol{I}_4 \otimes \dot{\boldsymbol{\xi}}^\top)\boldsymbol{\Gamma}\dot{\boldsymbol{x}} + (\boldsymbol{I}_4 \otimes \dot{\boldsymbol{x}}^\top)\boldsymbol{\Gamma}\dot{\boldsymbol{\xi}}$$

$$= \frac{\mathrm{D}^2\boldsymbol{\xi}}{\mathrm{d}u^2} - (\boldsymbol{I}_4 \otimes \boldsymbol{\xi}^\mathsf{T})\frac{\partial \boldsymbol{\Gamma}}{\partial \boldsymbol{x}^\mathsf{T}}(\dot{\boldsymbol{x}} \otimes \boldsymbol{I}_4)\dot{\boldsymbol{x}}$$
$$+ (\boldsymbol{I}_4 \otimes \boldsymbol{\xi}^\mathsf{T})\boldsymbol{\Gamma}(\boldsymbol{I}_4 \otimes \dot{\boldsymbol{x}}^\mathsf{T})\boldsymbol{\Gamma}\dot{\boldsymbol{x}} - (\boldsymbol{I}_4 \otimes [(\boldsymbol{I}_4 \otimes \boldsymbol{\xi}^\mathsf{T})\boldsymbol{\Gamma}\dot{\boldsymbol{x}}]^\mathsf{T})\boldsymbol{\Gamma}\dot{\boldsymbol{x}}. \tag{C.13}$$

For $(\boldsymbol{I}_4 \otimes \boldsymbol{\xi}^\mathsf{T})\boldsymbol{\Gamma}(\boldsymbol{I}_4 \otimes \dot{\boldsymbol{x}}^\mathsf{T})\boldsymbol{\Gamma}\dot{\boldsymbol{x}}$ one can write

$$(\boldsymbol{I}_4 \otimes \boldsymbol{\xi}^\mathsf{T})\boldsymbol{\Gamma}(\boldsymbol{I}_4 \otimes \dot{\boldsymbol{x}}^\mathsf{T})\boldsymbol{\Gamma}\dot{\boldsymbol{x}} = \underline{\underline{(\boldsymbol{I}_4 \otimes \boldsymbol{\xi}^\mathsf{T})\boldsymbol{\Gamma}\overline{\boldsymbol{\Gamma}}(\boldsymbol{I}_4 \otimes \dot{\boldsymbol{x}})}}\dot{\boldsymbol{x}}, \tag{C.14}$$

and the expression $(\boldsymbol{I}_4 \otimes [(\boldsymbol{I}_4 \otimes \boldsymbol{\xi}^\mathsf{T})\boldsymbol{\Gamma}\dot{\boldsymbol{x}}]^\mathsf{T})\boldsymbol{\Gamma}\dot{\boldsymbol{x}}$ can be rewritten as

$$(\boldsymbol{I}_4 \otimes [(\boldsymbol{I}_4 \otimes \boldsymbol{\xi}^\mathsf{T})\boldsymbol{\Gamma}\dot{\boldsymbol{x}}]^\mathsf{T})\boldsymbol{\Gamma}\dot{\boldsymbol{x}}$$
$$= \overline{\boldsymbol{\Gamma}}(\boldsymbol{I}_4 \otimes (\boldsymbol{I}_4 \otimes \boldsymbol{\xi}^\mathsf{T})\boldsymbol{\Gamma}\dot{\boldsymbol{x}})\dot{\boldsymbol{x}}$$
$$= \overline{\boldsymbol{\Gamma}}(\boldsymbol{I}_{16} \otimes \boldsymbol{\xi}^\mathsf{T})(\boldsymbol{I}_4 \otimes \boldsymbol{\Gamma}\dot{\boldsymbol{x}})\dot{\boldsymbol{x}} = \underline{\underline{(\boldsymbol{I}_4 \otimes \boldsymbol{\xi}^\mathsf{T})(\overline{\boldsymbol{\Gamma}} \otimes \boldsymbol{I}_4)(\boldsymbol{I}_4 \otimes \boldsymbol{\Gamma})(\boldsymbol{I}_4 \otimes \dot{\boldsymbol{x}})}}\dot{\boldsymbol{x}}. \tag{C.15}$$

With (C.14) (in somewhat modified form) and (C.15) one obtains for (C.13)

$$\ddot{\boldsymbol{\xi}} + (\boldsymbol{I}_4 \otimes \dot{\boldsymbol{\xi}}^\mathsf{T})\boldsymbol{\Gamma}\dot{\boldsymbol{x}} + (\boldsymbol{I}_4 \otimes \dot{\boldsymbol{x}}^\mathsf{T})\boldsymbol{\Gamma}\dot{\boldsymbol{\xi}}$$
$$= \frac{\mathrm{D}^2\boldsymbol{\xi}}{\mathrm{d}u^2} - (\boldsymbol{I}_4 \otimes \boldsymbol{\xi}^\mathsf{T})\frac{\partial \boldsymbol{\Gamma}}{\partial \boldsymbol{x}^\mathsf{T}}(\dot{\boldsymbol{x}} \otimes \boldsymbol{I}_4)\dot{\boldsymbol{x}}$$
$$+ (\boldsymbol{I}_4 \otimes \boldsymbol{\xi}^\mathsf{T})[\boldsymbol{\Gamma}\overline{\boldsymbol{\Gamma}} - (\overline{\boldsymbol{\Gamma}} \otimes \boldsymbol{I}_4)(\boldsymbol{I}_4 \otimes \boldsymbol{\Gamma})](\dot{\boldsymbol{x}} \otimes \dot{\boldsymbol{x}}). \tag{C.16}$$

Equation (C.16) used in (C.6) provides

$$\frac{\mathrm{D}^2\boldsymbol{\xi}}{\mathrm{d}u^2} = -(\boldsymbol{I}_4 \otimes \dot{\boldsymbol{x}}^\mathsf{T})\frac{\partial \boldsymbol{\Gamma}}{\partial \boldsymbol{x}^\mathsf{T}}(\boldsymbol{\xi} \otimes \boldsymbol{I}_4)\dot{\boldsymbol{x}} + (\boldsymbol{I}_4 \otimes \boldsymbol{\xi}^\mathsf{T})\frac{\partial \boldsymbol{\Gamma}}{\partial \boldsymbol{x}^\mathsf{T}}(\dot{\boldsymbol{x}} \otimes \boldsymbol{I}_4)\dot{\boldsymbol{x}}$$
$$+ (\boldsymbol{I}_4 \otimes \boldsymbol{\xi}^\mathsf{T})[\boldsymbol{\Gamma}\overline{\boldsymbol{\Gamma}} - (\overline{\boldsymbol{\Gamma}} \otimes \boldsymbol{I}_4)(\boldsymbol{I}_4 \otimes \boldsymbol{\Gamma})](\dot{\boldsymbol{x}} \otimes \dot{\boldsymbol{x}}). \tag{C.17}$$

As the 16×16-matrix $\frac{\partial \boldsymbol{\Gamma}}{\partial \boldsymbol{x}^\mathsf{T}}$ is symmetric, the first term of the right-hand side can transformed as follows:

$$(\boldsymbol{I}_4 \otimes \dot{\boldsymbol{x}}^\mathsf{T})\frac{\partial \boldsymbol{\Gamma}}{\partial \boldsymbol{x}^\mathsf{T}}(\boldsymbol{\xi} \otimes \boldsymbol{I}_4)\dot{\boldsymbol{x}} = (\boldsymbol{I}_4 \otimes \dot{\boldsymbol{x}}^\mathsf{T})\frac{\partial \boldsymbol{\Gamma}}{\partial \boldsymbol{x}^\mathsf{T}}\boldsymbol{U}_{4\times 4}(\boldsymbol{I}_4 \otimes \boldsymbol{\xi})\dot{\boldsymbol{x}}$$
$$= (\boldsymbol{I}_4 \otimes \boldsymbol{\xi}^\mathsf{T})\frac{\partial \boldsymbol{\Gamma}}{\partial \boldsymbol{x}^\mathsf{T}}\boldsymbol{U}_{4\times 4}(\boldsymbol{I}_4 \otimes \dot{\boldsymbol{x}})\dot{\boldsymbol{x}}.$$

This in (C.17) provides

$$\frac{\mathrm{D}^2\boldsymbol{\xi}}{\mathrm{d}u^2} = (\boldsymbol{I}_4 \otimes \boldsymbol{\xi}^\mathsf{T})\left[\frac{\partial \boldsymbol{\Gamma}}{\partial \boldsymbol{x}^\mathsf{T}} - \frac{\partial \boldsymbol{\Gamma}}{\partial \boldsymbol{x}^\mathsf{T}}\boldsymbol{U}_{4\times 4}\right](\dot{\boldsymbol{x}} \otimes \boldsymbol{I}_4)\dot{\boldsymbol{x}}$$
$$+ (\boldsymbol{I}_4 \otimes \boldsymbol{\xi}^\mathsf{T})[\boldsymbol{\Gamma}\overline{\boldsymbol{\Gamma}} - (\overline{\boldsymbol{\Gamma}} \otimes \boldsymbol{I}_4)(\boldsymbol{I}_4 \otimes \boldsymbol{\Gamma})](\dot{\boldsymbol{x}} \otimes \dot{\boldsymbol{x}}), \tag{C.18}$$

and finally,

$$\frac{\mathrm{D}^2\boldsymbol{\xi}}{\mathrm{d}u^2} = \left(\boldsymbol{I}_4 \otimes \boldsymbol{\xi}^\mathsf{T}\right) \underbrace{\left[\frac{\partial \boldsymbol{\Gamma}}{\partial \boldsymbol{x}^\mathsf{T}}(\boldsymbol{I}_{16} - \boldsymbol{U}_{4\times4}) + \left(\boldsymbol{\Gamma}\overline{\boldsymbol{\Gamma}} - (\overline{\boldsymbol{\Gamma}} \otimes \boldsymbol{I}_4)(\boldsymbol{I}_4 \otimes \boldsymbol{\Gamma})\right)\right]}_{-\boldsymbol{R}} (\dot{\boldsymbol{x}} \otimes \dot{\boldsymbol{x}}). \tag{C.19}$$

Using a slightly modified Riemannian curvature matrix $\boldsymbol{R}$, we finally obtain for the dynamic behavior of the geodesic deviation

$$\boxed{\frac{\mathrm{D}^2\boldsymbol{\xi}}{\mathrm{d}u^2} + \left(\boldsymbol{I}_4 \otimes \boldsymbol{\xi}^\mathsf{T}\right)\boldsymbol{R}(\dot{\boldsymbol{x}} \otimes \dot{\boldsymbol{x}}) = \boldsymbol{0}.} \tag{C.20}$$

In a flat manifold, i.e. in a gravity-free space one has $\boldsymbol{R} \equiv \boldsymbol{0}$ and in Cartesian coordinates $\mathrm{D}/\mathrm{d}u = \mathrm{d}/\mathrm{d}u$ so that (C.20) reduces to the equation $\mathrm{d}^2\boldsymbol{\xi}/\mathrm{d}u^2 = \boldsymbol{0}$ whose solution is the linear relationship $\boldsymbol{\xi}(u) = \dot{\boldsymbol{\xi}}_0 \cdot u + \boldsymbol{\xi}_0$. If $\boldsymbol{R} \neq \boldsymbol{0}$, gravity exists and the solution of (C.20) is nonlinear, curved.

Appendix D
Another Ricci-Matrix

The Ricci-matrix $\boldsymbol{R}_{\text{Ric}}$ is now defined as the sum of the sub-matrices on the main diagonal of $\boldsymbol{R}$

$$\boldsymbol{R}_{\text{Ric}} \stackrel{\text{def}}{=} \sum_{\nu=0}^{3} \boldsymbol{R}^{\nu\nu}. \tag{D.1}$$

Analogously, we define

$$\check{\boldsymbol{R}}_{\text{Ric}} \stackrel{\text{def}}{=} \sum_{\nu=0}^{3} \check{\boldsymbol{R}}^{\nu\nu}. \tag{D.2}$$

From (2.194) it can immediately be read that the Ricci-matrix $\check{\boldsymbol{R}}_{\text{Ric}}$ is *symmetric* because $\check{R}^{\gamma\gamma}_{\alpha\beta} = \check{R}^{\gamma\gamma}_{\beta\alpha}$. It is also true that

$$\boldsymbol{R} = \left(\boldsymbol{G}^{-1} \otimes \boldsymbol{I}_4\right)\check{\boldsymbol{R}},$$

so

$$\boldsymbol{R}^{\gamma\delta} = \left(\boldsymbol{g}_{\gamma}^{-T} \otimes \boldsymbol{I}_4\right)\check{\boldsymbol{R}}^{\delta} = \sum_{\nu=0}^{3} g^{[-1]}_{\gamma\nu}\check{\boldsymbol{R}}^{\nu\delta}, \tag{D.3}$$

where $\boldsymbol{g}_{\gamma}^{-T}$ is the γth row of $\boldsymbol{G}^{-1}$ and $\check{\boldsymbol{R}}^{\delta}$ is the matrix consisting of the sub-matrices in the δth block column of $\check{\boldsymbol{R}}$, i.e. the matrix elements are

$$R^{\gamma\delta}_{\alpha\beta} = \sum_{\nu=0}^{3} g^{[-1]}_{\gamma\nu}\check{R}^{\nu\delta}_{\alpha\beta}. \tag{D.4}$$

With the help of (D.3), the Ricci-matrix is obtained as

$$\boldsymbol{R}_{\text{Ric}} = \sum_{\gamma} \boldsymbol{R}^{\gamma\gamma} = \sum_{\gamma}\sum_{\nu} g^{[-1]}_{\gamma\nu}\check{\boldsymbol{R}}^{\nu\nu}, \tag{D.5}$$

G. Ludyk, *Einstein in Matrix Form*, Graduate Texts in Physics,
DOI 10.1007/978-3-642-35798-5, © Springer-Verlag Berlin Heidelberg 2013

i.e. for the components one has

$$R_{\text{Ric},\alpha\beta} = \sum_{\gamma}\sum_{\nu} g_{\gamma\nu}^{[-1]} \check{R}_{\alpha\beta}^{\nu\gamma}, \tag{D.6}$$

or with (2.173)

$$R_{\text{Ric},\alpha\beta} = \sum_{\gamma}\sum_{\nu} g_{\gamma\nu}^{[-1]} \check{R}_{\nu\gamma}^{\alpha\beta}. \tag{D.7}$$

The *curvature scalar* R is obtained from the Ricci-matrix by taking the trace

$$R \stackrel{\text{def}}{=} \sum_{\alpha} R_{\text{Ric},\alpha\alpha} = \sum_{\alpha}\sum_{\gamma}\sum_{\nu} g_{\gamma\nu}^{[-1]} \check{R}_{\nu\gamma}^{\alpha\alpha} = \sum_{\gamma}\sum_{\nu} g_{\gamma\nu}^{[-1]} \check{R}_{\text{Ric},\nu\gamma}. \tag{D.8}$$

Conversely, we obtain a corresponding relationship

$$\check{R}_{\alpha\beta}^{\gamma\delta} = \sum_{\nu=0}^{3} g_{\gamma\nu} R_{\alpha\beta}^{\nu\delta}. \tag{D.9}$$

From (2.165) it directly follows that

$$\underline{\underline{R_{\text{Ric},\alpha\beta} = \sum_{\gamma=0}^{3}\left(\frac{\partial}{\partial x_\beta}\Gamma_{\alpha\gamma}^{\gamma} - \frac{\partial}{\partial x_\gamma}\Gamma_{\alpha\beta}^{\gamma} + \sum_{\nu=0}^{3}\Gamma_{\beta\nu}^{\gamma}\Gamma_{\gamma\alpha}^{\nu} - \sum_{\nu=0}^{3}\Gamma_{\gamma\nu}^{\gamma}\Gamma_{\alpha\beta}^{\nu}\right)}} \tag{D.10}$$

and from (2.168)

$$\underline{\underline{\check{R}_{\text{Ric},\alpha\beta} = \sum_{\gamma=0}^{3}\left(\frac{\partial}{\partial x_\beta}\check{\Gamma}_{\alpha\gamma}^{\gamma} - \frac{\partial}{\partial x_\gamma}\check{\Gamma}_{\alpha\beta}^{\gamma} + \sum_{\nu=0}^{3}\Gamma_{\alpha\beta}^{\nu}\check{\Gamma}_{\gamma\gamma}^{\nu} - \sum_{\nu=0}^{3}\Gamma_{\alpha\gamma}^{\nu}\check{\Gamma}_{\gamma\beta}^{\nu}\right)}}. \tag{D.11}$$

Symmetry of the Ricci-Matrix $\boldsymbol{R}_{\text{Ric}}$ Even if $\boldsymbol{R}$ itself is not symmetric, the from $\boldsymbol{R}$ derived Ricci-matrix $\boldsymbol{R}_{\text{Ric}}$ is symmetric; this will be shown in the following. The symmetry will follow from the components equation (D.10) of the Ricci-matrix. One sees immediately that the second and fourth summands are symmetric in α and β.

The symmetry of the term $\sum_{\gamma=0}^{3}\frac{\partial}{\partial x_\beta}\Gamma_{\alpha\gamma}^{\gamma}$ in α and β is not seen directly. This can be checked using the Laplace-expansion theorem for determinants.[1] Developing the determinant of $\boldsymbol{G}$ along the γth row yields

$$g \stackrel{\text{def}}{=} \det(\boldsymbol{G}) = g_{\gamma 1}A_{\gamma 1} + \cdots + g_{\gamma\beta}A_{\gamma\beta} + \cdots + g_{\gamma n}A_{\gamma n},$$

[1]The sum of the products of all elements of a row (or column) with their adjuncts is equal to the determinant's value.

where $A_{\gamma\beta}$ is the element in the γth row and βth column of the adjoint of $\boldsymbol{G}$. If $g_{\beta\gamma}^{[-1]}$ is the $(\beta\gamma)$th element of the inverse of $\boldsymbol{G}$, then $g_{\beta\gamma}^{[-1]} = \frac{1}{g} A_{\gamma\beta}$, so $A_{\gamma\beta} = g\, g_{\beta\gamma}^{[-1]}$. Thus we obtain

$$\frac{\partial g}{\partial g_{\gamma\beta}} = A_{\gamma\beta} = g\, g_{\beta\gamma}^{[-1]},$$

or

$$\delta g = g\, g_{\beta\gamma}^{[-1]} \delta g_{\gamma\beta},$$

or

$$\frac{\partial g}{\partial x_\alpha} = g\, g_{\beta\gamma}^{[-1]} \frac{\partial g_{\gamma\beta}}{\partial x_\alpha},$$

i.e.

$$\frac{1}{g}\frac{\partial g}{\partial x_\alpha} = g_{\beta\gamma}^{[-1]} \frac{\partial g_{\gamma\beta}}{\partial x_\alpha}. \tag{D.12}$$

Using (2.62), on the other hand, one has

$$\sum_{\gamma=0}^{3} \Gamma_{\alpha\gamma}^{\gamma} = \sum_{\gamma=0}^{3}\sum_{\beta=0}^{3} \frac{g_{\beta\gamma}^{[-1]}}{2} \left(\frac{\partial g_{\gamma\beta}}{\partial x_\alpha} + \frac{\partial g_{\alpha\beta}}{\partial x_\gamma} - \frac{\partial g_{\alpha\gamma}}{\partial x_\beta} \right),$$

i.e. the last two summands cancel out and it remains to deal with

$$\sum_{\gamma=0}^{3} \Gamma_{\alpha\gamma}^{\gamma} = \sum_{\gamma=0}^{3}\sum_{\beta=0}^{3} \frac{1}{2} g_{\beta\gamma}^{[-1]} \frac{\partial g_{\gamma\beta}}{\partial x_\alpha}.$$

It follows from (D.12) that

$$\sum_{\gamma=0}^{3} \frac{\partial}{\partial x_\beta} \Gamma_{\alpha\gamma}^{\gamma} = \sum_{\gamma=0}^{3}\sum_{\beta=0}^{3} \frac{1}{\sqrt{|g|}} \frac{\partial^2 \sqrt{|g|}}{\partial x_\alpha \partial x_\beta}. \tag{D.13}$$

But this form is immediately seen symmetric in α and β.

Now it remains to shown that the third term in (D.10) is symmetric. Its expression is

$$\sum_{\gamma=0}^{3}\sum_{\nu=0}^{3} \Gamma_{\beta\nu}^{\gamma} \Gamma_{\gamma\alpha}^{\nu}.$$

Now one can see that this term is symmetric because

$$\sum_{\gamma,\nu=0}^{3} \Gamma_{\beta\nu}^{\gamma} \Gamma_{\gamma\alpha}^{\nu} = \sum_{\gamma,\nu=0}^{3} \Gamma_{\nu\beta}^{\gamma} \Gamma_{\alpha\gamma}^{\nu} = \sum_{\nu,\gamma=0}^{3} \Gamma_{\gamma\beta}^{\nu} \Gamma_{\alpha\nu}^{\gamma}.$$

All this shows that the Ricci-matrix $\boldsymbol{R}_{\text{Ric}}$ is symmetric.

Divergence of the Ricci-Matrix $\boldsymbol{R}_{\mathrm{Ric}}$ Multiplying the Bianchi-identities (2.199) in the form of

$$\frac{\partial}{\partial x_\kappa} R^{\nu\delta}_{\alpha\beta} + \frac{\partial}{\partial x_\beta} R^{\nu\kappa}_{\alpha\delta} + \frac{\partial}{\partial x_\delta} R^{\nu\beta}_{\alpha\kappa} = 0$$

with $g_{\gamma\nu}$ and summing over ν, we obtain in $\mathcal{P}$, since there $\frac{\partial \boldsymbol{G}}{\partial \boldsymbol{x}} = \boldsymbol{0}$,

$$\frac{\partial}{\partial x_\kappa} \sum_{\nu=0}^{3} g_{\gamma\nu} R^{\nu\delta}_{\alpha\beta} + \frac{\partial}{\partial x_\beta} \sum_{\nu=0}^{3} g_{\gamma\nu} R^{\nu\kappa}_{\alpha\delta} + \frac{\partial}{\partial x_\delta} \sum_{\nu=0}^{3} g_{\gamma\nu} R^{\nu\beta}_{\alpha\kappa} = 0.$$

Combined with (D.9) this becomes

$$\frac{\partial}{\partial x_\kappa} \check{R}^{\gamma\delta}_{\alpha\beta} + \frac{\partial}{\partial x_\beta} \check{R}^{\gamma\kappa}_{\alpha\delta} + \frac{\partial}{\partial x_\delta} \check{R}^{\gamma\beta}_{\alpha\kappa} = 0. \tag{D.14}$$

The second term, according to (2.172), may be written as

$$-\frac{\partial}{\partial x_\beta} \check{R}^{\gamma\delta}_{\alpha\kappa}.$$

Substituting now $\gamma = \delta$ and summing over γ, one obtains

$$\frac{\partial}{\partial x_\kappa} \check{R}_{\mathrm{Ric},\alpha\beta} - \frac{\partial}{\partial x_\beta} \check{R}_{\mathrm{Ric},\alpha\kappa} + \sum_{\gamma=0}^{3} \frac{\partial}{\partial x_\gamma} \check{R}^{\gamma\beta}_{\alpha\kappa} = 0. \tag{D.15}$$

In the third summand, one can, according to (2.171), replace $\check{R}^{\gamma\beta}_{\alpha\kappa}$ by $-\check{R}^{\alpha\beta}_{\gamma\kappa}$. If we set $\alpha = \beta$ and sum over α, we obtain for (D.15) with the trace $\check{R} \stackrel{\mathrm{def}}{=} \sum_{\alpha=0}^{3} \check{R}_{\mathrm{Ric},\alpha\alpha}$ of the Riccati-matrix $\check{\boldsymbol{R}}_{\mathrm{Ric}}$

$$\frac{\partial}{\partial x_\kappa} \check{R} - \sum_{\alpha=0}^{3} \frac{\partial}{\partial x_\alpha} \check{R}_{\mathrm{Ric},\alpha\kappa} - \sum_{\gamma=0}^{3} \frac{\partial}{\partial x_\gamma} \check{R}_{\mathrm{Ric},\gamma\kappa} = 0. \tag{D.16}$$

If in the last sum the summation index γ is replaced by α, we can finally summarize:

$$\frac{\partial}{\partial x_\kappa} \check{R} - 2 \sum_{\alpha=0}^{3} \frac{\partial}{\partial x_\alpha} \check{R}_{\mathrm{Ric},\alpha\kappa} = 0. \tag{D.17}$$

One would get the same result, if one were to start with the equation:

$$\frac{\partial}{\partial x_\kappa} \check{R}^{\gamma\delta}_{\alpha\beta} - 2 \frac{\partial}{\partial x_\beta} \check{R}^{\gamma\delta}_{\alpha\kappa} = 0. \tag{D.18}$$

Indeed, if we set $\delta = \gamma$ and sum over γ, we get first

$$\frac{\partial}{\partial x_\kappa} \check{R}_{\mathrm{Ric},\alpha\beta} - 2\frac{\partial}{\partial x_\beta} \check{R}_{\mathrm{Ric},\alpha\kappa} = 0.$$

If we now set $\alpha = \beta$ and sum over α, we will again arrive at (D.17).

A different result is obtained when starting from (D.18) (with ν instead of γ) first, multiplying this equation by $g_{\gamma\nu}^{[-1]}$ to get

$$\frac{\partial}{\partial x_\kappa} g_{\gamma\nu}^{[-1]} \check{R}_{\alpha\beta}^{\nu\delta} - 2\frac{\partial}{\partial x_\beta} g_{\gamma\nu}^{[-1]} \check{R}_{\alpha\kappa}^{\nu\delta} = 0,$$

then again setting $\gamma = \delta$ and summing over γ and ν and noting (D.6):

$$\sum_\gamma \sum_\nu \frac{\partial}{\partial x_\kappa} g_{\gamma\nu}^{[-1]} \check{R}_{\alpha\beta}^{\nu\gamma} - 2\sum_\gamma \sum_\nu \frac{\partial}{\partial x_\beta} g_{\gamma\nu}^{[-1]} \check{R}_{\alpha\kappa}^{\nu\gamma}$$
$$= \frac{\partial}{\partial x_\kappa} R_{\mathrm{Ric},\alpha\beta} - 2\frac{\partial}{\partial x_\beta} R_{\mathrm{Ric},\alpha\kappa} = 0.$$

If we now take $\alpha = \eta$ and sum over α, we finally obtain the important relationship

$$\underline{\underline{\frac{\partial}{\partial x_\kappa} R - 2\sum_\alpha \frac{\partial}{\partial x_\alpha} R_{\mathrm{Ric},\alpha\kappa} = 0.}} \tag{D.19}$$

These are the four equations for the four spacetime coordinates $x_0, \ldots, x_3$.

Finally, this overall result can be represented as

$$\underline{\underline{\vec{\nabla}^{\mathsf{T}} \left(\boldsymbol{R}_{\mathrm{Ric}} - \frac{1}{2} R \boldsymbol{I}_4 \right) = \boldsymbol{0}^{\mathsf{T}}.}} \tag{D.20}$$

References

1. V.I. Arnold, *Mathematical Methods of Classical Mechanics* (Springer, Berlin, 1978)
2. G. Barton, *Introduction to the Relativity Principle* (Wiley, New York, 1999)
3. H. Bondi, *Relativity and Common Sense* (Dover Publications, Dover, 1986)
4. J.W. Brewer, Kronecker products and matrix calculus in system theory. IEEE Trans. Circuits Syst. 772–781 (1978)
5. J.J. Callahan, *The Geometry of Spacetime* (Springer, Berlin, 2000)
6. S. Carrol, *Spacetime and Geometry: An Introduction to General Relativity* (Addison Wesley, San Francisco, 2003)
7. S. Chandrasekhar, *The Mathematical Theory of Black Holes* (Clarendron Press, Oxford, 1983)
8. I. Ciufolini, J.A. Wheeler, *Gravitation and Inertia* (Princeton University Press, Princeton, 1995)
9. R. D'Inverno, *Introducing Einstein's Relativity* (Clarendon Press, New York, 1992)
10. A. Einstein, *Relativity, The Special and the General Theory*, 2nd edn. (Edit Benei Noaj, 2007)
11. A. Föppl, Über einen Kreiselversuch zur Messung der Umdrehungsgeschwindigkeit der Erde. Sitzungsber. Bayer. Akad. Wiss. **34**, 5–28 (1904)
12. J. Foster, J.D. Nightingale, *A Short Course in General Relativity* (Springer, Berlin, 1995)
13. R.G. Gass, F.P. Esposito, L.C.R. Wijewardhansa, L. Witten, Detecting event horizons and stationary surfaces (1998). arXiv:gr-qc/9808055v
14. R. Geroch, *Relativity from A to B* (Chicago University Press, Chicago, 1981)
15. O. Gron, S. Hervik, *Einstein's General Theory of Relativity* (Springer, New York, 2007)
16. O. Gron, A. Naess, *Einstein's Theory* (Springer, New York, 2011)
17. J.B. Hartle, *Gravity: An Introduction to Einstein's General Relativity* (Addison Wesley, San Francisco, 2003)
18. S.W. Hawking, G.F.R. Ellis, *The Large Scale Structure of Space-Time* (Cambridge University Press, Cambridge, 1973)
19. I.R. Kenyon, *General Relativity* (Oxford University Press, Oxford, 1990)
20. L.D. Landau, E.M. Lifschitz, *Classical Theory of Fields: 2.* Course of Theoretical Physics Series (Butterworth Heinemann, Stoneham, 1987)
21. C.W. Misner, K.S. Thorne, J.A. Wheeler, *Gravitation* (Freeman, San Francisco, 1973)
22. R.A. Mould, *Basic Relativity* (Springer, Berlin, 2002)
23. J. Natario, *General Relativity Without Calculus* (Springer, Berlin, 2011)
24. W. Rindler, *Essential Relativity* (Springer, Berlin, 1977)
25. L. Sartori, *Understanding Relativity: A Simplified Approach to Einstein's Theories* (University of California Press, Berkeley, 1996)
26. B.F. Schutz, *A First Course in General Relativity* (Cambridge University Press, Cambridge, 1985)
27. H. Stephani, *General Relativity* (Cambridge University Press, Cambridge, 1977)

G. Ludyk, *Einstein in Matrix Form*, Graduate Texts in Physics,
DOI 10.1007/978-3-642-35798-5,

28. J. Stewart, *Advanced General Relativity* (Cambridge University Press, Cambridge, 1991)
29. T. Takeuchi, *An Illustrated Guide to Relativity* (Cambridge University Press, Cambridge, 2010)
30. E.F. Taylor, J.A. Wheeler, *Spacetime Physics* (Freeman, New York, 1992)
31. E.F. Taylor, J.A. Wheeler, *Black Holes* (Addison Wesley, Reading, 2000)
32. E.F. Taylor, J.A. Wheeler, *Exploring Black Holes: Introduction to General Relativity* (Addison Wesley, San Francisco, 2000)
33. M. von Laue, *Die Relativitätstheorie. Erster Band: Die spezielle Relativitätstheorie*, 7th edn. (Vieweg, Wiesbaden, 1961)
34. R. Wald, *General Relativity* (Chicago University Press, Chicago, 1984)
35. R. Wald, *Space, Time and Gravity: Theory of the Big Bang and Black Holes* (University Press, Chicago, 1992)
36. S. Weinberg, *Gravitation and Cosmology* (Wiley, New York, 1972)
37. W.J. Wild, A matrix formulation of Einstein's vacuum field equations, gr-qc/9812095, 31 Dec. 1998

Index

G. Ludyk, *Einstein in Matrix Form*, Graduate Texts in Physics,
DOI 10.1007/978-3-642-35798-5,